DATE DUE

Demco, Inc. 38-293

D1087577

This book is devoted to our late friend and teacher,
Raphael Hoegh-Kröhn

XXIV Karpacz Winter School on Theoretical Physics

STOCHASTIC METHODS IN MATHEMATICS AND PHYSICS

Karpacz, Poland, 13 — 27 Jan 1988

Editors
R GIELERAK
W KARWOWSKI

World Scientific
Singapore • New Jersey • London • Hong Kong

Published by

World Scientific Publishing Co. Pte. Ltd.
P O Box 128, Farrer Road, Singapore 9128

USA office: World Scientific Publishing Co., Inc.
687 Hartwell Street, Teaneck, NJ 07666, USA

UK office: World Scientific Publishing Co. Pte. Ltd.
73 Lynton Mead, Totteridge, London N20 8DH, England

VANDERBILT UNIVERSITY
LIBRARY
NASHVILLE, TENNESSEE

STOCHASTIC METHODS IN MATHEMATICS AND PHYSICS

Copyright © 1989 by World Scientific Publishing Co. Pte. Ltd.

All rights reserved. This book, or parts thereof, may not be reproduced in any form or by any means, electronic or mechanical, including photocopying, recording or any information storage and retrieval system now known or to be invented, without written permission from the Publisher.

ISBN 9971-50-648-3

Printed in Singapore by JBW Printers & Binders Pte. Ltd.

FOREWORD

Ladies and Gentlemen,

Welcome to the XXIV Winter School of Theoretical Physics at Karpacz. The organizers feel very much honoured by the fact that so many outstanding scientists have accepted their invitation to come here to Karpacz.

The School is organized by the Institute of Theoretical Physics of Wroclaw University. Wroclaw is the capital of an administrative district called wojewodztwo. We say in Polish województwo wroclawskie. Here we are in another district, województwo jeleniogórskie. So formally we are not at home here. However for many years our Winter School enjoy reliable and very valuable sponsorship of the authorities of Jelenia Góra and Karpacz.

The well-being of a group as this one is of course an important factor for the productivity of our joint work, and thus, it is correct to extend at this point our particular thanks to those authorities and specially to Mr. Sylwester Samol Wojewoda Jeleniogórski for their help and support.

The primary aim of the Winter School had always been to provide young theoretical physicists with opportunities to see the wide panorama of important subjects of contemporary research and with the opportunity of contacts with actively working scientists.

Though originally we had in mind Polish physicists, we are happy to have here young people from other countries and we welcome them heartily.

We hope that the more senior participants of the School will also find it interesting as an opportunity to get together with the colleagues working in the same or close fields. This is so in particular because we have guests here from the West and East, and Karpacz plays (once again) its role as an important meeting point for scientists and students from different parts of the world.

The stochastic processes are being studied as an abstract mathematical theory as well as a powerful tool in numerous applications. However I dare say it is much more than a mere tool, but rather an ideology at least if physics is concerned.

One of the marks of recognition of importance of the stochastic processes is the world-wide interest in the activities of the research center Bielefeld-Bochum-Stochastics, which in these first few years of its existence has rapidly developed into an extremaly active international research institute.

In fact the idea of this School and its scientific program has been worked out jointly by the people from our Institute and from BiBoS. Let me mention here Sergio Albeverio, Philippe Blanchard, Roman Gielerak, Zbigniew Haba, Raphael Høegh-Krohn and Ludwig Streit.

On behalf of the organizers I wish you a good and enjoyable stay at Karpacz and much scientific profit from participating in the School.

Witold Karwowski

Dedication

This proceedings volume is dedicated to the memory of Jan Raphael Høegh-Krohn. Raphael was born February 2, 1938, in Ålesund, on the west coast of Norway. He studied mathematics and physics at the Universities of Oslo (Norway) and Aarhus (Danmark). He took this ph.d. in 1966 at the Courant Institute of Mathematical Sciences in New York with K.O. Friedrichs as thesis advisor. He then returned to Oslo and remained since associated with the Mathematics Department of Oslo University, becoming a full professor in 1982. In 1977 he was awarded the Fritjof Nansen prise of the Norwegian Academy of Sciences, of which he also became a member. He spent two years, 1969-1970, in Princeton and had many long research stays in Bielefeld and Bochum (West Germany), as well as in Marseille (France). He also held for several years the position of full professor at the Centre de Physique Theorique, CNRS, Marseille-Luminy, and at the Universite de Provence.

He was married and had five children.

By a tragic destiny he succumbed, less than fifty years old, he who was always so strong, healthy and dynamic, to a sudden heart attact on January 24, 1988, near his beloved home town Ålesund.

He was a great man and an extremaly original, intuitive and gifted mathematician, with an extraordinary broad spectrum of activities and interests in mathematics and all its applications. He deeply believed (setting forth a great tradition to which his teacher Friedrichs also belonged) in mathematics as the language in which "the great book of nature is written", and this has been a constant inspiration for him.

He has done deep outstanding work particularly in such domains as quantum field theory and quantum mechanics, functional and stochastic analysis, non standard analysis, topological groups, operator algebras, partial differential equations. Applications of this work include solid state, atomic nuclear, plasma and high energy physics, statistical mechanics, hydrodynamics, astrophysics. In recent years he also engaged in an extensive numerical and computer program in oil resorvoir analysis.

He has published well over hundred research papers and three research monographs. His influence goes however far beyond this.

He has been a permanent inspirator for very many scientists all over the world (his list of publications includes over forty coworkers, leading experts in these fields as well as students and younger scientists). He was a truly exceptional person, combining high technical skills with a wealth of imagination and ideas, which he freely shared with others. It was a great human experience to come in close contact with him. He was always enthusiastic, full of optimism, good humour, joviality, concerned with making the other feel at ease. Everything in life, culture, and nature interested him deeply and he had original ideas about everything he considered.

He was a generous, very dear friend for many of us. He always put friendship in the first place, a much superior goal than achieving this or that particular result. It is hard to accept that he is no longer here, that he has departed in his best years. His ideas and work will continue to inspire our own work, we remember his as a great and good man.

In great sorrows we deeply mourn him.

Sergio Albeverio

CONTENTS

III. MATHEMATICS

24

STOCHASTIC MECHANICS

Progress and Problems in Stochastic Mechanics

Eric A. Carlen
Department of Mathematics
Princeton University
Princeton, N.J. 08544
USA

Karpacz Winter School
1988

In orthodox quantum mechanics, the complete description of the state of a system is given by specifying a square integrable function ψ on the classical configuration space M of the system. Thought of as a function, ψ is referred to as the wave function; and when thought of as an element of $L^2(M)$, ψ is referred to as the state vector. Either way, the dynamics; that is, the time evolution of the state ψ, is given by the Schrödinger equation. The initial value problem for the Schrödinger equation is well posed in cases of physical interest, and so the state ψ evolves completely deterministically. There is no randomness whatsoever in the dynamics of orthodox quantum mechanics. Randomness enters only when a measurement is made.

It is difficult for many people to be satisfied with the manner in which the randomness we observe in nature is incorporated into orthodox quantum mechanics. For if one applys the dynamical laws of quantum mechanics to the observer as well as to the observed system, the result of the experiment is deterministic with the final state corresponding to no single "outcome". Schrödinger's familiar cat paradox is based on just this quandry.

Nevertheless, it is almost impossible to be dissatisfied with the Schrödinger equation insofar as one is interested in predicting the outcomes of experiments. Several generations of thorough trial leave physicists nearly unanimously confident that any numbers they compute using the Schrödinger equation will stand up to any experimental check -- at least to any experimental check involving only the degrees of freedom they have taken into account in their Schrödinger equation.

This leaves us few alternatives in resolving our dissatisfaction with the way randomness enters orthodox quantum mechanics. In these lectures we study the following alternative: We seek to embed the computational formalism of orthodox quantum mechanics, in particular the Schrödinger equation, into a more detailed description of nature -- so that ψ no longer fully describes the state -- in which the full time evolution is intrinsically random. This randomness in the dynamics is to be the only randomness present; in particular measurement is to have no special role in introducing randomness.

The most popular alternative is simply to ignore any dissatisfaction one has with the way randomness enters orthodox quantum mechanics, and to be satisfied with the predictive power of the Schrödinger equation. Since the Schrödinger equation, in principle at least, gives us a full set of predictions for any experiment we can perform today, it is possible to define physics so that only this second effortless alternative lies in the domain of physics.

There is an important caveat to this [1, p. 118-119]; namely there is no good evidence that quantum mechanics applies to gravitation. Perhaps string theory will provide the missing means to mate general relativity and quantum field theory, but this is not yet a foregone conclusion. So perhaps in trying to go beyond the wave function ψ to a more detailed description of nature we are pursuing physics and not only metaphysics. Perhaps we will arrive at a setting in which gravity and quantum mechanics fit smoothly together so that we can make fresh predictions. But certainly this is no foregone conclusion either.

Without attempting any further a priori justification of our endeavor to embed the computational formalism of orthodox quantum mechanics into a more detailed mechanics with stochastic time evolution, let us briefly describe how far one can get, and how naturally one can get that far.

The approach we discuss here was discovered by Nelson [2]. The stochastic mechanics developed by Nelson provides a description of nature in terms of diffusion processes, and it does so in a marvelously natural way. Nelson "randomized" classical mechanics, in some sense minimally, and found that out popped the Schrödinger equation in just such a way that his stochastic mechanics and orthodox quantum mechanics would predict the same results for experiments.

Nelson made his discovery in the sixties, but only in the eighties did it attract widespread attention. It is still by no means a complete theory, nor does the present development of it fully solve the problem posed in this introduction. But at the very least it affords us a very interesting start on the problem, and it suggests many specific questions inviting further mathematical and physical research.

We will discuss some of these problems later in these lectures. But before we spell out our agenda, we will give a rough sketch of Nelson's discovery.

Consider first a single particle of mass m moving in $\mathbb{R}^3$ under the influence of a potential V. For the moment, suppose V to be smooth. If we treat this system according to classical mechanics, we base our treatment on two assertions. First, that the motion of the particle is to be described by a smooth trajectory

$$t \longmapsto \xi(t) \tag{1.1}$$

in the configuration space $\mathbb{R}^3$. This is the kinematical law of classical mechanics. Second, the possible motions in any given time interval $[S,T]$ are exactly those which are critical points of the action functional

$$A(\xi,S,T) = \int_S^T (|\frac{d\xi(t)}{dt}|^2 - V(\xi(t)))dt \qquad (1.2)$$

under variations of the trajectory leaving the endpoints $\xi(S)$ and $\xi(T)$ fixed. This is the dynamical law of classical mechanics.

Nelson proposed the study of a stochastic mechanics in which one replaced the kinematical law of classical mechanics with a new one permitting a stochastic time evolution. Namely, the motion of the particle is to be described by a diffusion process $t \longmapsto \xi(t)$ in $\mathbb{R}^3$ satisfying a stochastic differential equation of the type

$$d\xi(t) = b(\xi(t),t)dt + \sqrt{\frac{\hbar}{m}}\, dw(t) \qquad (1.3)$$

where $b(x,t)$ is some time dependent vector field on $\mathbb{R}^3$, called the drift, and $dw(t)$ is the increment of a standard Brownian motion in $\mathbb{R}^3$. Of course $\hbar$ is the reduced Planck's constant.

While the first term on the right hand side of our stochastic differential equation has a rather general form, the second term is particularly simple. We restrict our attention to the class of diffusions on account of the following physical picture: The position of the particle at time t is taken to be random because of "quantum fluctuations". I would like to be more specific about the meaning of this term but I can't. In the nineteenth century physicists didn't know much about molecules, and so they couldn't be very specific about what "Brownian fluctuations" were. Perhaps there is a parallel here.

But even without a clear understanding of what quantum fluctuations might actually be, we can reasonably proceed on the assumptions that they are the manifestation of some isotropic, translation invariant phenomenon. That assumption made, it follows that the noise term in our stochastic differential equation must be a constant times the increment in of a Brownian motion. Since the mean of $|w(t)-w(s)|^2$ is $|t-s|$, it follows that $w(t)$ is measured in units of $(\text{time})^{1/2}$. Since $\xi(t)$ is to be measured in units of distance, it follows that the constant must have units of

$(\text{distance})/(\text{time})^{1/2}$. Of course, $(h/m)^{1/2}$ is such a constant, and we will soon explain how this choice fits into our program.

Having randomized the kinematical law of classical mechanics in this way, we now seek to adapt the dynamical law of classical mechanics to the new stochastic setting making as little further modification as possible. This was done in the Lagrangian setting we discuss here by Guerra and Morato [3]. Nelson had initially worked from Newton's second law, which is a consequence of the Lagrangian variational principle. There are several advantages to the Langrangian approach [3],[1]. In any case, given a time interval $[S,T]$ and a diffusion satisfying our kinematical requirements, we define its action to be

$$A(\xi,S,T) = E \int_S^T \left(\left|\frac{d\xi(t)}{dt}\right|^2 - V(\xi(t))\right)dt \tag{1.4}$$

and the dynamical law is that the possible diffusions are those which are critical points for the action functional under variations in the diffusion -- determined by a variation in the corresponding drift field -- under which the distributions of $\xi(S)$ and $\xi(T)$ are left unchanged.

Except for the E, denoting of course the expectation, in the action functional, the dynamical laws of classical mechanics and Nelson's stochastic mechanics have remarkably similar form. (In fact, if we identified the classical trajectory with the corresponding δ-measure on path space, we could include an expectation there too and they would have the <u>same</u> form.)

The only difficulty with the dynamical law of stochastic mechanics at this point is that it is highly formal. As is well known, the paths of a diffusion process are almost surely nowhere differentiable -- the difficulty stems from $dw^2 \approx dt$ -- so we have to be careful with the kinetic energy term in our action. However, Nelson has carefully examined a finite difference approximation to the action above and found that it converges to the sum of an infinite constant and a nice well defined functional of the diffusion. The way this works is that the terms in the approximate action which diverge are independent of the particular diffusion -- within the class specified by our kinematics -- being considered, hence our use of "constant". Since we are only interested in variations of the action, we need only consider the nice functional that results upon neglecting the divergent terms and then passing to the limit. We refer to this functional as the renormalized action functional.

Working with the renormalized action functional, Guerra and Morato then studied its variations and found that a diffusion is critical for their variational principle just when there is a solution $\psi(x,t)$ of the Schrödinger equation

$$i\hbar \frac{\partial}{\partial t} \psi(x,t) = (-\frac{\hbar^2}{m} \Delta + V(x))\psi(x,t) \tag{1.5}$$

so that the drift field $b(x,t)$ of the diffusion is given by

$$b(x,t) = \frac{\hbar}{m} (\mathrm{Re} \frac{\nabla\psi}{\psi} (x,t) + \mathrm{Im} \frac{\nabla\psi}{\psi} (x,t)) \tag{1.6}$$

and so that the probability that $\xi(t)$ is in a measurable set A is given by

$$\Pr\{\xi(t) \in A\} = \int_A |(x,t)|^2 dx . \tag{1.7}$$

The last equation can be expressed by saying that $\xi(t)$ has a density $\rho(x,t)$ given by

$$\rho(x,t) = |\psi(x,t)|^2 . \tag{1.8}$$

Note that it is the same potential which appears in (1.5) and which governs the motion of the diffusing particle. The choice of $(\hbar/m)^{1/2}$ as the diffusion constant in our kinematical law is reasonable for the fact that $\hbar$ and m appear in (1.5) in the usual places. This said, we henceforth put $\hbar = m = 1$.

The equation (1.7) is particularly encouraging in view of our goal. It says that stochastic mechanics and orthodox quantum mechanics make the same predictions for position measurements. Now the motion of a single spinless particle does not at all exhaust the scope of stochastic mechanics. Stochastic mechanics can be done for essentially any system of interest, and we find the natural generalization of the above result in all cases: Stochastic mechanics and orthodox quantum mechanics make the same predictions for the same configuration measurements. Insofar as all measurements are ultimately configuration measurements, the two descriptions cannot be distinguished experimentally.

The rest of the paper is devoted to a more detailed development of this approach to the problem of going beyond the orthodox quantum mechanical description of nature. It is organized in the following way. The next section, section 2, is a primer on diffusion theory. Here we briefly relate the mathematical lore the rest of the story will draw upon. In section 3 we discuss the variational principle and the dynamics of stochastic mechanics. Some unsolved problems are discussed at length, as well as, of course, the results just described. Having obtained a more detailed stochastic description of nature, we proceed in section 4 to study the new features. That is, we study the sample paths of our diffusions. In physical terms, we study the classical histories of our quantum system. Having worked hard to obtain them, it is a relief that they are by and large very well behaved. Finally in section 5 we discuss approaches to the stochastic mechanical treatment of quantum field theory. Our remarks here are necessarily quite tentative given the current state of development of this topic.

A DIFFUSION PRIMER

We now set about explaining what we mean by a solution of the stochastic differential equation (1.3). In the process we will fix our terminology and notation.

Let $\Omega = C(\mathbb{R},\mathbb{R}^n)$ be the space of continuous maps from $\mathbb{R}$ to $\mathbb{R}^n$ -- or in general into some Riemannian manifold M. Ω is equipped with the topology of uniform convergence on compacts; let $\mathcal{B}$ be its Borel σ-algebra. Define maps

$$\xi(t): \Omega \longrightarrow \mathbb{R}^n \quad \text{by} \quad \xi(t): \omega \longmapsto \omega(t) ;$$

these are just the evaluation maps.

The following σ-algebras are of special interest:

$$\mathcal{P}_t \equiv \sigma\{\xi(s) | s \leq t\} , \quad \mathcal{N}_t = \sigma\{\xi(t)\} , \quad \mathcal{F}_t = \sigma\{\xi(u) | u \geq t\} .$$

These are all subalgebras of $\mathcal{B}$ which happens to be $\sigma\{\xi(t) | t \in \mathbb{R}\}$. They are called the past, present and future fields at time t, respectively of course.

Now if we equip $(\Omega,\mathcal{B})$ with a probability measure Pr, the functions $\xi(t)$ become random variables, and $t \longmapsto \xi(t)$ becomes a stochastic process, the <u>configuration process</u>.

This process is Markovian in case under the law Pr it holds for all t that the past and the future are independent σ-algebras, given the present.

If $t \longmapsto X(t)$ is any square integrable process on $(\Omega,\mathcal{B},\mathrm{Pr})$, we define its mean forward derivative $DX(t)$ if it exists, by

$$DX(t) = \lim_{h \downarrow 0} \frac{1}{h} E\{X(t+h)-X(t) | \mathcal{P}_t\} \tag{2.1}$$

where the limit is taken in $L^2(\Omega,\mathcal{B},\mathrm{Pr})$. Similarly the mean backward derivative $D_*X(t)$ is given by

$$D_*X(t) = \lim_{h \downarrow 0} hE\{X(t)-X(t-h) | \mathcal{F}_t\} \tag{2.2}$$

under the same conditions. Here we use $E\{X(t+h)-X(t)\,|\,P_t\}$ to denote the conditional expectations of $X(t+h)-X(t)$ given the past at time t. The value of this conditional expectation at a particular trajectory ω is defined to be the average of $\tilde{\omega}(t+h)-\tilde{w}(t)$ over all trajectories $\tilde{\omega}$ satisfying $\tilde{\omega}(s) = \omega(s)$ for all $s \leq t$. We take the average according to the law Pr. The future conditional expectation may be similarly understood. For mathematical precision and further information, refer to [1].

Now let $f \in C_0^\infty(\mathbb{R}^n)$ be fixed, and consider $X(t) = f(\xi(t))$. Suppose that under Pr, $\xi(t)$ satisfies (1.3). Then making a Taylor expansion one has $f(\xi(t+h))-f(\xi(t)) = \nabla f(\xi(t))(b(\xi(t),t)dt+dw(t))$ $+\frac{1}{2}\frac{\partial^2 f}{\partial x^i \partial x^j}(b^i(\xi(t),t)dt+d^i w(t))(b^j(\xi(t),t)+dw^j(t))+o(d\xi^2(t))$ where of course we are being somewhat formal. Now since in each component the Brownian increments are independent of the past and guassian with mean zero and variance equal to the time increment, and since the different components are independent

$$E\{dw^i(t)\,|\,P_t\} = 0 \qquad E\{dw^i(t)dw^j(t)\,|\,P_t\} = \delta_{ij}dt$$

for all i, j. We should then have

$$Df(\xi(t)) = (\tfrac{1}{2}\Delta + b(\xi(t),t)\cdot\nabla)f(\xi(t)\,. \tag{2.3}$$

It is also easy to see that a solution of (1.3) should be Markovian -- again appealing to the independence of Brownian increments from the past, where the process goes next depends only on where it is, and not how it got there.

With this rough discussion as motivation, we now make the following precise definition.

<u>Def</u> Under the law Pr, the configuration process $t \longmapsto \xi(t)$ is a diffusion with forward generator

$$G = \frac{1}{2}\Delta + b(x,t)\cdot\nabla \tag{2.4}$$

and backward generator

$$G_* = -\frac{1}{2}\Delta + b_*(x,t)\cdot\nabla \tag{2.5}$$

and density $\rho(x,t)$ in case it is a Markov process and for all $f \in C_0^\infty(\mathbb{R}^n)$

$$Df(\xi(t)) = Gf(\xi(t)),\ D_*f(\xi(t)) = G_*f(\xi(t)),\ Ef(\xi(t)) = \int F(x)\rho(x,t)dx.$$

We have stated this definition in a somewhat redundant form (that is, some of the conditions imply the others), but one that is invariant under time reversal and suited to our purposes.

Using results in [1] one can see that this definition simply covers a special case -- the case relevant to stochastic mechanics -- of the Strook-Varadhan Martingale definition of diffusion. Usually reference to G_*, D_* and ρ is deleted from the definition. That is because whenever $Df(\xi(t)) = Gf(\xi(t))$ for all $f \in C_0^\infty(\mathbb{R}^n)$ and all $t \in \mathbb{R}$; and $b(x,t)$ is, say, smooth and bounded, then there exists a $b_*(x,t)$ and a $\rho(x,t)$ so that the rest of the definition -- including the Markov property -- holds. But the diffusions we consider will not usually have such nice drift fields.

As an example, consider Brownian motion in $\mathbb{R}^n$ started at the origin at time 0 (we restrict our attention to positive times here). This is a diffusion in our sense with

$$G = \frac{1}{2}\Delta \qquad G_* = \frac{1}{2}\Delta - \frac{x}{t}\cdot\nabla \qquad \rho(x,t) = (2\pi)^{-n/2}e^{-|x|^2/2t}.$$

In this example

$$b_* = b - \nabla\log\rho$$

and a fairly simple integration by parts argument shows that this equation holds in general; see [1, p. 30-32].

Let us define the current velocity $v(x,t)$ by

$$v(x,t) = \frac{1}{2}\,(b(x,t)+b_*(x,t)) \tag{2.6}$$

and the osmotic velocity by

$$u(x,t) = \frac{1}{2}\,(b(x,t)-b_*(x,t)) \;. \tag{2.7}$$

Then the general identity above relating b , b_* , and ρ may be expressed as

$$u = \frac{1}{2}\,\nabla \log \rho \;. \tag{2.8}$$

There is also an important relation between v and ρ .

$$\frac{d}{dt}\,Ef(\xi(t)) = EGf(\xi(t)) = EG_*f(\xi(t)) \quad \text{for all} \quad f \in C_0^\infty(\mathbb{R}^n)$$

and so averaging the last two expressions, the Laplacians cancel and

$$\frac{d}{dt}\int f(x)\rho(x,t)dt = \frac{d}{dt}\,Ef(\xi(t)) = \int (v(x,t)\cdot\nabla f(x))\rho(x,t)dx$$

which is a weak form of the continuity equation

$$\frac{\partial}{\partial t}\,\rho = -\nabla\cdot(\rho v) \;. \tag{2.9}$$

However when we refer to the weak continuity equation later, we will mean the still weaker equation

$$\int f(x,T)\rho(x,T)dx - \int f(x,S)\rho(x,S)dx = \int_S^T \int (v\cdot\nabla f(x,t))\rho(x,t)dxdt$$

for all $S < T$ and $f \in C_0^\infty(\mathbb{R}^n,[S,T])$.

Note that if we know the pair (v,ρ) we can compute G and G_* also by using (2.6), (2.7) and (2.8). We refer to the pair (v,ρ) as the infinitesimal characteristics of the diffusion $t \longmapsto \xi(t)$. They are of course completely determined by the diffusion. An important question at this point is the extent to which they uniquely determine a diffusion; that is a probability measure Pr on Ω under which they are the infinitesimal characteristics of the configuration process. Clearly we must require that v and ρ are compatible in that the weak continuity equation must be satisfied. It turns out we don't need much more.

In what follows, it is convenient to consider fixed a finite but otherwise arbitrary time interval $[S,T]$. We won't explicitly refer to dependence on $[S,T]$ in our terminology and notations, and when we say time dependent, we mean dependent on times in the interval $[S,T]$.

Def The space $\mathcal{P}$ of proper infinitesimal characteristics consists, as a set, of pairs (v,ρ) where v is a time dependent vector field on $\mathbb{R}^n$ and ρ is a time dependent probability density on $\mathbb{R}^n$ so that with $u = \frac{1}{2}\nabla\log\rho$

$$\int f(x,T_1)\rho(x,T_1)dx - \int f(x,S_1)\rho(x,S_1)dx = \int_{S_1}^{T_1}\int (v\cdot\nabla f)\rho(x,t)dxdt \tag{2.10}$$

and

$$\int_S^T\int (|u|^2+|v|^2)\rho(x,t)dxdt < \infty \tag{2.11}$$

for all $S \leq S' \leq T' \leq T$ and all $f \in C_0^\infty(\mathbb{R}^n\times[S,T])$.

We equip $\mathcal{P}$ with the metric

$$\begin{aligned} d_G((v_1,\rho_1),(v_2,\rho_2)) &= \sup_{t\in[S,T]} \|\sqrt{\rho_1}(\cdot,t) - \sqrt{\rho_2}(\cdot,t)\|_{L^2(\mathbb{R}^n)} \\ &+ \Big(\int_S^T \|v_1\sqrt{\rho_1}(\cdot,t)-v_2\sqrt{\rho_2}(\cdot,t)\|^2_{L^2(\mathbb{R}^n)} dt + \\ &\int_S^T \|u_1\sqrt{\rho_1}(\cdot,t)-u_2\sqrt{\rho_2}(\cdot,)\|^2_{L^2(\mathbb{R}^n)} dt\Big)^{1/2} . \end{aligned} \tag{2.12}$$

The fact that there is a unique law Pr on $(\Omega, \mathcal{B})$ for each pair (v,ρ) in $\mathcal{P}$ such that under Pr the configuration process is a diffusion with density ρ and current velocity v was proved by myself in [4], [5]. The above metric on $\mathcal{P}$ was introduced by Guerra in [6]. He proved that $\mathcal{P}$ is complete in this metric; in particular, the weak continuity equation is preserved on passing to a limit in the Guerra metric. We now state the following theorem, to which we will later return [4], [5], [7].

<u>Theorem</u> For each $(v,\rho) \in \mathcal{P}$, there exists a unique law Pr on $(\Omega, \mathcal{P})$ under which the configuration process is a diffusion with density ρ and current velocity v.

Since v and u are velocities (2.11) is a condition on the action. We may regard the above theorem as identifying $\mathcal{P}$ with the set of finite action diffusion laws, as clearly the map $(v,\rho) \longrightarrow Pr_{(v,\rho)}$ determined by the theorem is one to one. We may thus transfer the Guerra metric to this set of diffusion laws, and we do so. This leads to some interesting questions we will later raise, but we now have enough diffusion theory available to return to stochastic mechanics.

We also remark that the uniqueness under conditions (2.11) is a fairly simple consequence of Ershov's theorem [13]. Existence is what one must work to prove.

ACTION AND THE VARIATIONAL PRINCIPLE

We now turn to the Guerra-Morato variational principle. Our discussion is strongly influenced by the work of Lafferty [8]. As we have mentioned in the introduction, our first object is to make sense of the formal expression

$$E \int_S^T \left|\frac{d\xi(t)}{dt}\right|^2 dt \,.$$

Let us make a finite difference approximation and consider

$$\begin{aligned}\xi(t+h)-\xi(t) &= \int_t^{t+h} b(\xi(s),s)ds + w(t+h)-w(t)\\ &= \int_t^{t+h} (b(\xi(t),t)+\partial_i b(\xi(t),t)(\int_t^s b^i(\xi(r),r)dr+w^i(s)-w^i(t)))ds\\ &\qquad + w(t+h)-w(t) + O(h^{3/2})\end{aligned}$$

where we have been careful to take full account of the fact that $w(t+h)-w(t)$ is of order $\sqrt{h}$. The reasoning leading to (2.3) gives us

$$E\left(\frac{|\xi(t+h)-\xi(t)|^2}{h^2}\right) = \frac{n}{h} + E\,(|b(\xi(t),t)|^2+\nabla\cdot b(\xi(t),t))+O(h) \,.$$

Notice that the singular term $\frac{n}{h}$ -- the term that will cause trouble as h is decreased -- is completely independent of the particular diffusion under consideration, i.e. it is independent of b. This is because we have fixed the strength of the fluctuations in our kinematical law. Since we are only interested in variations of the action, we can quite reasonably neglect this singular term that doesn't vary.

We now replace $\left|\frac{d\xi(t)}{dt}\right|^2$ by $(|b(\xi(t),t)|^2+\nabla\cdot b(\xi(t),t))$ in the formal action and define

$$A(v,\rho),[S,T]) = E^{Pr_{(\rho,v)}} \int_S^T (\frac{1}{2} (|b(\xi(t),t)|^2 + \nabla\cdot b(\xi(t),t) - V(\xi(t)))dt \tag{3.1}$$

where $(v,\rho) \in \mathcal{P}$ specifies the diffusion law $Pr_{(\rho,v)}$ as discussed in the last section, under which the configuration process has density ρ, current velocity v and drift field $b = v + \frac{1}{2} \nabla \log \rho$.

Here we are thinking of the action functional as a functional on $\mathcal{P}$. Now clearly if we compute the expectation in terms of ρ we have

$$\begin{aligned} A((v,\rho),[S,T]) &= \int_S^T \int (\frac{1}{2} |b(x,t)|^2 + \nabla\cdot b(x,t)) - V(x))\rho(x,t)dxdt \\ &= \int_S^T \int \frac{1}{2} (|v(x,t)|^2 - |u(x,t)|^2) - V(x))\rho(x,t)dxdt \end{aligned} \tag{3.2}$$

where we have integrated by parts and again used $u = \frac{1}{2} \nabla \log \rho$.

<u>Def</u> $(v,\rho) \in \mathcal{P}$ is a <u>critical point</u> of the action function A in case for all sequences $(v_n, \rho_n) \longrightarrow (v,\rho)$ is the Guerra metric with $\rho_n(S,\cdot) = \rho(S,\cdot)$ and $\rho_n(T,\cdot) = \rho(T,\cdot)$ for all n

$$\left| \frac{A((v,\rho),[S,T]) - A((v_n,\rho_n),[S,T])}{d_G((v,\rho),(v_n,\rho_n))} \right| \longrightarrow 0 .$$

The dynamical law of stochastic mechanics expressed in the introduction is now properly formulated. A diffusion corresponding to $(v,\rho) \in \mathcal{P}$ is a possible motion under the influence of potential V in case it is a critical point of the above action functional under all variations leaving the initial and final distributions fixed.

Notice we do not assert the existence of critical points for arbitrary specification of $\rho(S,\cdot)$ and $\rho(T,\cdot)$.

Now let us find all critical points in $\mathcal{P}$. Suppose $(v,\rho) \in \mathcal{P}$ is critical. Let z be any time dependent vector field with $\nabla(\rho z) = 0$. Then for all $\lambda \in \mathbb{R}$, $(\rho, v+\lambda z) \in \mathcal{P}$ and we easily compute that

$$d_G((\rho,v+\lambda z),(\rho,v)) = \lambda(\int_S^T |z|^2\rho(x,t)dxdt)^{1/2}$$

and

$$\frac{d}{d\lambda} A((\rho,v+\lambda z),[S,T]])|_{\lambda=0} = \int_S^T (v\cdot z)\rho(x,t)dxdt \ .$$

Since (v,r) is critical, this must vanish and so v is orthogonal to the general divergence free vector field. Therefore $v(x,t) = \nabla S(x,t)$ for some function S. This is called the stochastic Hamilton-Jacobii condition. It is a necessary condition for (v,ρ) to be critical.

We now consider the general variation where v_λ and ρ_λ vary with λ. Using $v = \nabla S$, the continuity equation and several integrations by parts (to transfer all λ derivatives from v to ρ) we find

$$\frac{d}{d\lambda} P(v_\lambda,\rho_\lambda),[S,T])|_{\lambda=0} = \int_S^T\int(-\frac{\partial S}{\partial t} + \frac{1}{2}\nabla u - \frac{1}{2} v^2 + \frac{1}{2} u^2 - V)\left(\frac{\partial\rho_\lambda}{\partial\lambda}\bigg|_{\lambda=0}\right) dxdt \ .$$

Our Euler equation for the critical diffusion characteristics then is, putting $u = \nabla R$ (which we know holds with $R = \frac{1}{2}\log\rho$ for any diffusion).

$$\frac{\partial S}{\partial t} = \frac{1}{2}(|\Delta R|^2 - |\nabla S|^2 - V + \frac{1}{2}\Delta R \ . \qquad (3.3)$$

This is the <u>stochastic Hamilton-Jacobi equation.</u>

Simply rewriting the continuity equation in terms of R and S we have

$$\frac{\partial R}{\partial t} = -\nabla R\cdot\nabla S + \Delta \ . \qquad (3.4)$$

This gives us a coupled non-linear system of equations for R and S, and hence ρ and v.

If we now make the change of dependent variable $\psi = e^{R+iS}$, this coupled non-linear system is equivalent to

$$i \frac{\partial}{\partial t} \psi(x,t) = (-\frac{1}{2} \Delta+V)\psi(x,t) \tag{3.5}$$

where is the Schrödinger equation for the same potential V. Moreover $\rho(x,t) = |\psi(x,t)|^2$ follows from $R = \frac{1}{2} \log \rho$. So if $t \longmapsto \xi(t)$ is the diffusion corresponding to the critical pair (ρ,v)

$$Pr\{\xi(t) \in A\} = \int_A |\psi(x,t)|^2 dx \tag{3.6}$$

which is just the answer quantum mechanics gives for the probability of finding the particle in the set A when its quantum mechanical state is given by ψ.

So here we are with a mechanical system with stochastic time evolution which incorporates the computational formalism of orthodox quantum mechanics, but in which ψ is by no means the complete description of the state of the system. And we have arrived here by studying what is a beautifully natural and minimal "randomization" of classical mechanics.

However it has recently been pointed out by Goldstein [9] and Walstrom [10] that a careful analysis of the extent to which the computational formalism of orthodox quantum mechanics is absorbed into stochastic mechanics is required. The first point, observed independently by Goldstein and Walstrom at about the same time, is this: The argument sketched above shows only that every diffusion satisfying the dynamical law of stochastic mechanics corresponds to a solution of the Schrödinger equation and not vice-versa. And indeed with the dynamical law of stochastic mechanics formulated as it is here the reverse correspondence actually fails to hold. In particular Walstrom has shown [10] that a diffusion with time independent density can be a critical point for the variational principle only if the current velocity v vanishes identically. However solutions of the

Schrödinger equation which are simultaneous eigenvectors for the Hamiltonian and, say, the z-component of angular momentum will in general satisfy

$$\frac{\partial}{\partial t} |\psi(x,t)|^2 = 0 \qquad\qquad \operatorname{Im} \frac{\nabla\psi(x,t)}{\psi(x,t)} \neq 0$$

and hence cannot correspond to any stochastic mechanical diffusion. The culprit appears to be the stochastic Hamilton-Jacobi condition $v = \nabla S$. The sort of wave function we have just discussed cannot be written in the form $\psi = e^{R+iS}$ for any single valued function S on the configuration space. The diffusions critical for the dynamical law of stochastic mechanics always correspond to wave functions with such a single valued S; it is even given explicitly as a functional of the diffusion [3], [1, p. 70, eq. 14.7]. Ideally one would like to put constraints on the admissible variations, say, as a device to increase the number of critical points and then show that all physically meaningful wave functions correspond to critical diffusions.

Moreover as Goldstein pointed out, we cannot entirely abandon the stochastic Hamilton-Jacobi condition. To get rid of this condition, note that just as in classical mechanics, the Euler equation for our variational principle can be put in Newtonian form. If we write v for ∇S and u for ∇R and take one more gradient, our equations (3.3) and (3.4) become

$$\frac{\partial v}{\partial t} = -\nabla V + \frac{1}{2} \nabla(|u|^2 - |v|^2) + \frac{1}{2} \Delta u \tag{3.7}$$

$$\frac{\partial u}{\partial t} = -\nabla\cdot(u-v) + \frac{1}{2} \Delta v \tag{3.8}$$

Nelson's original approach to stochastic mechanics was his observation that the current and osmotic velocities of a diffusion in our kinematical class satisfy this coupled system just when the diffusion satisfies the stochastic Newton's equation

$$\frac{1}{2} (D_* D + D D_*) \xi(t) = -\nabla V(\xi(t)) \tag{3.9}$$

where the left hand side is a natural stochastic acceleration. If one then assumes $v = \nabla S$ for some S one passes to the Schrödinger equation by reversing the argument. What is encouraging is that there are other solutions besides these. In particular the diffusions corresponding to states of definite energy and angular momentum which we discussed above do satisfy the stochastic Newton's equation.

The reason we can't escape our plight by simply taking the stochastic Newton's equation as our dynamical law is that we would then have too many solutions the equations of motion. Here is the simplest example. We take as our state space the unit circle S^1 with its usual Riemannian structure. For any number s let $\xi_s(t)$ be Brownian motion on S^1 with uniform initial distribution and superimposed uniform angular speed s. The distribution will stay uniform at all times. Now, everything we have said about stochastic mechanics easily generalizes from to case where the configuration space in $\mathbb{R}^n$ to the case where it is a general Riemannian manifold [1]. The case of S^1 is particularly simple. One finds that

$$\frac{1}{2}\,(DD_*+D_*D)\xi_s(t) = 0 \tag{3.10}$$

for all s. So any angular velocity gives us a solution of the stochastic Newton's equation. However only for $s = n \in Z$ do these diffusions correspond to a solution of the free Schrödinger equation

$$i\,\frac{\partial}{\partial t}\,\psi(\theta,t) = -\,\frac{1}{2}\,\frac{\partial^2}{\partial\theta^2}\,\psi(\theta,t)$$

on S^1, namely

$$\psi_n(\theta,t) = e^{i(n\theta - \frac{n^2}{2}t)}\ .$$

Of these, only the $s = 0$ diffusion is critical for the zero potential Guerra-Morato variational principle on S^1. In fact taking the time interval for computing the action to be the unit interval one finds

$$A(\xi_s,[0,1]) = \frac{1}{2} s^2$$

and varying s is an admissible variation.

So the Guerra-Morato variational principle, which leads to the stochastic Hamilton-Jacobi condition, yields too few possible motions, while if we abandon this condition entirely and retain only the stochastic Newton's equation, we get too many possible motions. It is an important open problem to modify the dynamical law of stochastic mechanics so that we get just as many possible motions as there are physically meaningful solutions of the Schrodinger equation.

THE PATHWISE DESCRIPTION OF QUANTUM PHENOMENA

Given any solution $\psi(x,t)$ of the Schrödinger equation (3.5) we define

$$\rho(x,t) = |\psi(x,t)|^2 \quad \text{and} \quad v(x,t) = \begin{cases} \operatorname{Im} \dfrac{\nabla\psi(x,t)}{\psi(x,t)} & \psi(x,t) \neq 0 \\ 0 & \psi(x,t) = 0 \end{cases} \tag{4.1}$$

If the potential V is a form-small perturbation of the Laplacian -- as virtually everything of conceivable physical interest is -- and if $\int |\nabla\psi(x,0)|^2 dx < \infty$ then $\int |\nabla\psi(x,t)|^2 dx$ is continuous in t and hence bounded on any compact interval. Then with u defined in terms of ρ as before, a simple computation yields

$$\int (|u|^2+|v|^2)\rho(x,t)dx = \int |\nabla\psi(x,t)|^2 dx$$

and ρ and v satisfy the conditions (2.10) and (2.11) for membership in $\mathcal{P}$. Then by the theorem quoted in section 2, there is a unique diffusion $t \longmapsto \xi(t)$ with density ρ and current velocity v.

As discussed in the last section, at least some fraction of these diffusions arise as critical points of the Guerra-Morato variational principle. We now put aside the question as to what this fraction is an concentrate on the sample path properties of all these diffusions.

The point is that although

$$\Pr\{\xi(t) \in A\} = \int_A |\psi(x,t)|^2 dx$$

guarantees that we recover the correct quantum mechanical predictions for the probabilities of configurations, it says nothing about the behavior of the individual sample paths.

So two natural questions arise. First one might wonder whether the path behaves in anything like a classically reasonable fashion. Second, one might wonder how one sees the discrete -- quantal -- natural of quantum mechanics looking at the paths of a stochastic mechanical diffusion.

We can answer these questions by investigating scattering in stochastic mechanics. To the extent that all experiments are ultimately scattering experiments we get a fairly positive answer.

Consider the following typical scattering experiment. We have a projectile particle with fairly well determined initial momentum p_i . It is headed towards a target with some interesting internal structure. We will assume, as is often the case, that the target is fixed; i.e. it is much more massive than the projectile. After traveling freely for some time, the projectile enters the region where there is appreciable interaction with the target, and then complicated things happen until it again leaves this region. When it leaves, it moves off freely with a new momentum p_f , and the target is in a new internal state. Suppose the internal energy of target is E_i and the final energy E_f . Then by conservation of energy

$$\frac{m}{2} p_i^2 + E_i = \frac{m}{2} p_f^2 + E_f \tag{4.2}$$

or

$$(E_f - E_i) = \frac{m}{2} (p_i^2 - p_f^2) \ . \tag{4.3}$$

If E_i is the ground state energy, which may then be normalized to zero, a measurement of p_f yields a measurement of E_f And so the spectrum of possible values for p_f determines the spectrum of internal energies of the target.

The way one measures p_f is to place a detector at a great distance from the target -- so that on the way to the detector, the particle travels freely most of the way. One then measures the time it takes the particle to travel from the target to the detector. The momentum is m times the displacement of the detector from the target divided by the travel time.

Now let us consider this experiment in stochastic mechanical terms. Let $\xi(t)$ be the process for the position of the projectile. If the projectile does settle down to a final momentum pathwise, we should have

$$\lim_{t\to\infty} \frac{1}{t} m\xi(t,\omega) = p_f(\omega) \quad \text{a.s.} \qquad (4.4)$$

where the limit on the left defines the random variable on the right.

If this limit exists and so defines for us a final momentum random variable, we next enquire after its distribution. In particular, if we have a target with a discrete energy spectrum, and a fairly well defined initial momentum, we would expect to see a fairly discrete distribution for p_f.

This is exactly what does happen. If we model the effect of the interaction with the target by a potential $V(x)$, we have:

Theorem Let V be a nice potential. Let $t \longmapsto \xi(t)$ be a critical diffusion for V such that

$$\lim_{t\to\infty} \frac{1}{2T} \int_{-T}^{T} P_r\{|\xi(t) < R\}dt = 0 \quad \forall R > 0$$

and

$$E(u^2(\xi(0),0) + v^2(\xi(0),0) + |\xi(0)|^6) < \infty .$$

Then

$$\lim_{t\to\infty} \frac{m}{t} \xi(t,\omega) \equiv p_f(\omega)$$

exists almost surely and moreover p_f has the same distribution as does the final momentum in the quantum state corresponding to $t \longrightarrow \xi(t)$.

For the proof as well as the (reasonable) definition of nice, see [7] and [11]. The purpose of the first condition is to exclude diffusions describing bound, or partially bound, states. For bound states, it is easy to see that $\frac{1}{t}\xi(t)$ tends to zero almost surely. This quantity only defines a final momentum for scattering states.

The first result of this nature was obtained by Shucker [12] who considered the case $V = 0$. Shucker's conditions on the diffusion were also less natural than those above; he required an L^1 condition on the wave function.

The result presented here is not the most complete result to be found in our papers [7], [11], but certainly it gives the flavor. One further result is obtained in [7]; namely that -- under conditions to be found there -- p_f generates the tail field of the process. Physically, this means there is just enough information in the paths to define a final momentum pathwise, but not enough to define any extraneous asymptotic observables -- extraneous in the sense of lacking a counterpart in orthodox quantum mechanics.

Our proofs proceed by translating L^2 estimates on the wave function into L^2 estimates on the process which can be combined with martingale estimates and properties of Brownian motion to obtain the results. The details are to be found in the cited references.

To close this section, let us explain how we see the discreteness of quantum phenomena in stochastic mechanics. The answer is simple once we know p_f has the same distribution as does the orthodox quantum mechanical final momentum in the corresponding quantum state. If the target has a discrete spectrum, then the orthodox quantum final mechanical momentum will be discretely distributed; the possible values of p_f^2 will be fixed by (4.3). But the stochastic mechanical final momentum has the same discrete distribution. (We are speaking loosely; it is p_f^2 that is discretely distributed, not p_f itself; and it is only strictly discretely distributed in case p_i is. Actually there will always be some amount of spread.)

So the paths behave in a very reasonable manner, one which encourages me to take them seriously. After all, as we mentioned before, this is not at all a consequence of the fact that $\Pr\{\xi(t) \in A\} = \int_A |\psi(x,t)|^2 dx$. In fact, we can consider the Schrödinger equation in the momentum space representation. Then $|\hat{\psi}(k,t)|^2$ satisfies the Fokker-Plank equation for a certain type of jump process, and one can develop a momentum space version of stochastic mechanics in which one associates jump processes in momentum space to momentum space wave functions. This has many interesting properties, including of course, the fact that it does give the quantum mechanical answer for the probability to find the momentum in any given set A at time t . But no analog of the theorem of this section holds in momentum space stochastic mechanics: momenta keep jumping about forever at a residual constant rate -- even though the distribution tends to the right limit. There just is no final momentum in the momentum space representation! This is discussed in greater detail in [14]. Several miracles which did not have to occur have occurred in configuration space stochastic mechanics.

Having presented some positive results, we turn now to one which is more controversial. In [18] Nelson showed that stochastic mechanical diffusions can exhibit rather strange non-local behavior in their autocorrelation functions. As the sample paths, and hence the autocorrection functions have no direct analog in orthodox quantum mechanics, such a problem can't arise there. In this case it seems as if perhaps the extra details are embarrassing us, but there is little agreement here. While Nelson takes a strongly negative view of this non-locality, Goldstein actually takes a positive view of it in the interesting article [21]. Myself, I am not happy with this non-locality, but I am encouraged by the previously mentioned fact that at least there is no anamolous behavior observable in the tail field; that is, the asymptotic sample path observables display no such non-locality.

STOCHASTIC MECHANICS AND FIELD THEORY

The stochastic mechanics of relativistic systems deserves more attention than it has received. There are many interesting questions; but as yet, few results.

The problem, it seems to me, is that stochastic mechanics as we have developed it here is intimately dependent upon the Schrödinger representation of quantum dynamics. There is a Schrödinger representation of quantum field dynamics, but it is of course not Lorentz invariant. This causes no fundamental problem in orthodox quantum mechanics. One can use this non-invariant (can one say variant?) machinery to compute invariant quantities like scattering amplitudes. And indeed, all the quantities one computes which are of physical interest will be invariant.

However, if we associate a field configuration valued diffusion to a solution of the field theory Schrödinger equation -- and this can be done [15], [16] -- we encode a choice of frame into the sample paths. If we take the sample paths seriously as possessing a basic physical reality, we will then be embarrassed by the fact that Lorentz invariance is broken on a basic level. Even if the way in which the Lorentz invariance is broken is unobservable, the break is still there and is a source of dissatisfaction.

Nelson has discussed several possible modifications of stochastic mechanics which might lead to a way out of this quandry [17], [18]. At present, they are too little developed to make any assertion other than that they deserve close attention.

In the meantime, it is still interesting to consider the field configuration valued diffusions discussed in [16]. Several interesting questions about the non-relativistic limit and particle structure are developed there, and some answers given in [19]. Even here much remains to be done.

CONCLUDING REMARKS

I hope by now to have conveyed something of the goals and methods, of the accomplishments and open problems in stochastic mechanics. What I cannot hope to have conveyed is any indication of the enormous and fruitful activity in this field of my many colleagues. I am afraid I have stayed too close to my own papers in these lectures, though that is in the nature of them. Fortunately there is a marvelous book [21] by Blanchard, Combe and Zheng which is a clear guide to all the many frontiers of this subject. I recommend it together with references [1], [17], and [18] to anyone interested in the subject.

BIBLIOGRAPHY

[1] Nelson, E., Quantum Fluctuations, Princeton University Press, Princeton NJ, (1985).

[2] Nelson, E., "Derivation of the Schrödinger Equation from Newtonian Dynamics", Phys. Rev. 150, 1079-1085, (1966).

[3] Guerra, F., Morato, L., "Quantization of Dynamical Systems and Stochastic Control Theory", Phys. Rev. D27, 1774-1786, (1983).

[4] Carlen, E., "Conservative Diffusions", Comm. Math. Phys. 94, 293-315, (1983).

[5] Carlen, E., "Existence and Sample Path Properties of the Diffusions in Nelson's Stochastic Mechanics", in "Stochastic Processes in Mathematics and Physics", eds. Albeverio, S. et. al., Lecture Notes in Math. 1158, Springer Verlag, New York, 25-51, (1986).

[6] Guerra, F., "Carlen Processes: A New Class of Diffusions with Singular Drifts", in "Quantum Probability and Applications II", ed. Acardi, L. et. al., Lecture Notes in Math. 1136, Springer Verlag, New York, 259-267, (1985).

[7] Carlen, E., "The Pathwise Description of Quantum Scattering in Stochastic Mechanics", in "Stochastic Processes in Classical and Quantum Systems", Albeverio, S., et. al., eds., Lecture Notes in Physics, 262, Springer Verlag, New York, 139-147, (1986).

[8] Lafferty, J., "The Density Manifold and Configuration Space Quantization", to appear in T.A.M.S.

[9] Goldstein, S., Private communication, July 1987.

[10] Walstrom, T., Princeton thesis, (1988).

[11] Carlen, E., "Potential Scattering in Stochastic Mechanics", Ann. Inst. H. Poincaré, Physics Series, 42, 407-418, (1985).

[12] Shucker, D., "Stochastic Mechanics of Systems with Zero Potential", J. Funct. Anal. 38, 146-155, (1980).

[13] Ershov, M., "On the Absolute Continuity of Measures Corresponding to Diffusion Processes", Theory of Prob. and Appl. 17, 169-174, (1972).

[14] Blanchard, Ph., Carlen, E., "Sample Path Properties of the Jump Processes in Momentum Space Stochastic Mechanics, BiBoS Preprint, (1988).

[15] Guerra, F., Ruggiero, P., "A New Interpretation of the Euclidean Markov Field in the Framework of Physical Minkowski Space Time", Phys. Rev. Latt. 31, 1022-1025, (1973).

[16] Carlen, E., "Stochastic Mechanics of Free Quantum Fields", to appear in Lecture Notes in Math, Swansea conference proceedings, eds. Davies, I., et. al., Springer Verlag, New York, (1988).

[17] Nelson, E., "Stochastic Mechanics and Random Fields", Proceedings Ecole des Probabilités, Saint Flour, (1987). To appear in Lecture Notes in Math.

[18] Nelson, E., "Field Theory and the Future of Stochastic Mechanics", in "Stochastic Processes in Classical and Quantum Systems", Albeverio, S., et. a., eds., Lecture Notes in Physics 262, Springer Verlag, New York, 438-469, (1986).

[19] Blanchard, Ph., Carlen, E., Dell'Antonio, F.G., "Particles and Bumps in Quantum Field Configurations", BiBoS Preprint, (1988).

[20] Blanchard, Ph., Combe, Ph., Zheng, W., Mathematical and Physical Aspects of Stochastic Mechanics, Lecture Notes in Physics 281, Springer Verlag New York, (1987).

[21] Goldstein, S., "Stochastic Mechanics and Quantum Theory", to appear in J. Stat. Phys.

ON CAPTURE TIMES AND HITTING TIMES IN STOCHASTIC MECHANICS

by

Andrew Batchelor and Aubrey Truman
Department of Mathematics and Computer Science
University College of Swansea, Singleton Park
Swansea, SA2 8PP

ABSTRACT

New results are given for first hitting times, sojourn times and capture times in Nelson's stochastic mechanics. Possible applications in nuclear physics are discussed.

1. INTRODUCTION

The Schrödinger equation for a particle of mass m moving in $\mathbb{R}^d$, d-dimensional Euclidean configuration space, subject to a force $-\nabla V$, is equivalent to

$$i\hbar\,\psi^*(\underset{\sim}{x},t)\frac{\partial\psi}{\partial t}(\underset{\sim}{x},t) = -\frac{\hbar^2}{2m}\psi^*(\underset{\sim}{x},t)\Delta_x\psi(\underset{\sim}{x},t) + V(\underset{\sim}{x})\,|\psi(\underset{\sim}{x},t)|^2, \qquad (1)$$

where $\hbar$ is Planck's constant divided by 2π and $*$ denotes the complex conjugate. Here $\psi(\underset{\sim}{x},t)$ is the quantum mechanical wave function depending upon the time $t\in\mathbb{R}^+$, the positive reals, and the position vector $\underset{\sim}{x}\in\mathbb{R}^d$.

Following Nelson [9],[10], setting $\psi = e^{R+iS}$, for real-valued R and S, equating imaginary parts of the above equation gives

$$\frac{\partial\rho}{\partial t}(\underset{\sim}{y},t) = \mathrm{div}_y\{\frac{\hbar}{2m}\nabla_y\rho(\underset{\sim}{y},t) - \underset{\sim}{b}(\underset{\sim}{y},t)\rho(\underset{\sim}{y},t)\}, \qquad (2)$$

where $\rho(\underset{\sim}{y},t) = |\psi(\underset{\sim}{y},t)|^2$ is the quantum mechanical particle density and $\underset{\sim}{b} = \frac{\hbar}{m}\underset{\sim}{\nabla}(R+S)$. The last equation is just the forward Kolmogorov equation for the Nelson diffusion $\underset{\sim}{X}$, with

$$d\underset{\sim}{X}(t) = \underset{\sim}{b}(\underset{\sim}{X}(t),t)dt + \left(\frac{\hbar}{m}\right)^{\frac{1}{2}} d\underset{\sim}{B}(t), \tag{3}$$

$\underset{\sim}{B}$ being a $BM(\mathbb{R}^d)$ process with $\underset{\sim}{B}(s) = (B_1(s),\ldots,B_d(s))$ in cartesians and with covariance

$$\mathbb{E}\{B_i(t)B_j(s)\} = \delta_{ij}\min(s,t), \quad i,j = 1,2,\ldots,d.$$

Thus, if the parabolic forward Kolmogorov equation has unique solutions, denoting its fundamental solution by $\rho(\underset{\sim}{x},s;\underset{\sim}{y},t)$, $t > s$, with $\lim_{t\downarrow s} \rho(\underset{\sim}{x},s;\underset{\sim}{y},t) = \delta(\underset{\sim}{x}-\underset{\sim}{y})$, gives

$$\rho(\underset{\sim}{y},t) = \int \rho(\underset{\sim}{x},s)\rho(\underset{\sim}{x},s;\underset{\sim}{y},t)dx, \tag{4}$$

where $\rho(\underset{\sim}{x},s;\underset{\sim}{y},t)dy = \mathbb{P}(\underset{\sim}{X}(t) \in dy, \underset{\sim}{X}(s) = \underset{\sim}{x})$.

Nelson showed much more. Denoting the mean forward and backward time derivatives by $D_\pm$,

$$D_\pm f(\underset{\sim}{X}(t),t) = \lim_{h\downarrow 0} \mathbb{E}\left\{\frac{f(\underset{\sim}{X}(t\pm h),t\pm h) - f(\underset{\sim}{X}(t),t)}{\pm h}\middle| \underset{\sim}{X}(t)\right\}, \tag{5}$$

it follows from Eq. (3) that, if $\underset{\sim}{b}_+ = \underset{\sim}{b}$,

$$D_+\underset{\sim}{X}(t) = \underset{\sim}{b}_+(\underset{\sim}{X}(t),t) = \frac{\hbar}{m}\underset{\sim}{\nabla}(S+R)(\underset{\sim}{X}(t),t). \tag{6}$$

Nelson also showed that

$$D_-\underset{\sim}{X}(t) = \underset{\sim}{b}_-(\underset{\sim}{X}(t),t) = \frac{\hbar}{m}\underset{\sim}{\nabla}(S-R)(\underset{\sim}{X}(t),t). \tag{7}$$

Using Itô's formula, it is then straight-forward to show that

$$D_\pm f(\underset{\sim}{X}(t),t) = \left(\frac{\partial}{\partial t} + \underset{\sim}{b}_\pm \cdot \underset{\sim}{\nabla} \pm \frac{\hbar\Delta}{2m}\right) f(\underset{\sim}{X}(t),t). \tag{8}$$

A tedious calculation leads to

$$\frac{m}{2}(D_+D_- + D_-D_+)\underset{\sim}{X}(t) = \hbar\underset{\sim}{\nabla}\left(\frac{\partial S}{\partial t} - \frac{\hbar}{2m}(|\underset{\sim}{\nabla}R|^2 - |\underset{\sim}{\nabla}S|^2 + \Delta R)\right)(\underset{\sim}{X}(t),t). \tag{9}$$

If we now equate real parts of Eq. (1), we arrive at Nelson's amazing result:

$$\frac{m}{2}(D_+D_- + D_-D_+)\underset{\sim}{X}(t) = -\underset{\sim}{\nabla}V(\underset{\sim}{X}(t)). \tag{10}$$

This is just a stochastic version of Newton's second law of motion:

$$\text{Mass} \times \text{Acceleration} = \text{Force}. \tag{11}$$

Therefore, the net content of the Schrödinger equation is the forward Kolmogorov equation for the diffusion $\underset{\sim}{X}$ satisfying Eq. (3) and the dynamical principle for $\underset{\sim}{X}$ summarised in Eq. (10). This suggests that the sample paths of the diffusion $\underset{\sim}{X}$ have some physical significance. In this paper we investigate some of the properties of the hitting times and sojourn times for $\underset{\sim}{X}$ when the state ψ is a stationary state solution of the Schrödinger equation. We discuss whether or not these properties could be accessible to experiment in the context of nuclear physics.

2. FIRST HITTING TIMES FOR STATIONARY STATES FOR SPHERICALLY SYMMETRIC POTENTIALS

We consider Nelson diffusions corresponding to stationary state solutions of the Schrödinger equation for a quantum mechanical Hamiltonian $H = (-\frac{\hbar^2}{2m}\Delta + V)$, a self-adjoint linear operator on some appropriate domain in $L^2(\mathbb{R}^d)$. We specialize to the case $d = 3$ and further we assume that the potential V is spherically symmetric, $V = V(|\underset{\sim}{x}|)$, $\underset{\sim}{x} \in \mathbb{R}^3$, $|\cdot|$ being the Euclidean norm.

Let $\psi_{E_N}(\underset{\sim}{x},t)(\in L^2(\mathbb{R}^3))$ be such a stationary state, which for simplicity we take to be spherically symmetric.

$$\psi_{E_N}(\underset{\sim}{x},t) = \frac{1}{|\underset{\sim}{x}|} R_{E_N}(|\underset{\sim}{x}|)\, e^{-iE_N t/\hbar}, \tag{12}$$

where $\{E_N : E_N < E_{N+1},\ N = 0,1,2,\ldots\}$ is the discrete spectrum of H. Then, if the angular momentum is zero in the state ψ_{E_N}, as is well-known,

$$(-\frac{\hbar^2}{2m}\frac{d^2}{dx^2} + V(x) - E_N)\, R_{E_N}(x) = 0, \tag{13}$$

i.e. $(H_r - E_N)R_{E_N} = 0$, where $H_r = (-\frac{\hbar^2}{2m}\frac{d^2}{dx^2} + V(x))$ is the radial Hamiltonian for zero angular momentum. To simplify the analysis we assume that V is piecewise continuous with finite discontinuities on $(0,\infty)$ so that R_{E_N} is C^2 and, as usual, $R_{E_N}(0) = 0$.

Let us denote the ith zero of $R_{E_N}(x)$ by $a_{i,N} \geq 0$, $a_{0,N} \equiv 0$, $a_{i,N} \equiv \infty$ if $R_{E_N}(x) \neq 0$ on $(0,\infty)$. Obviously $|R_{E_N}(x)| > 0$ on $(a_{i,N}, a_{i+1,N})$. Moreover, using Eq. (12), a straight-forward application of Itô's formula yields for the Nelson diffusion $\underset{\sim}{X}(t)$ on $(a_{i,N}, a_{i+1,N})$, $i = 0,1,\ldots,$ corresponding to the state ψ_{E_N}:

$$d|\underset{\sim}{X}(t)| = \frac{\hbar}{m}\frac{d}{d|\underset{\sim}{X}|}\ln|R_{E_N}(|\underset{\sim}{X}(t)|)|dt + \left(\frac{\hbar}{m}\right)^{\frac{1}{2}}dB(t), \tag{14}$$

B being a BM($\mathbb{R}$) process. This is a one-dimensional time-homogeneous diffusion on $(a_{i,N}, a_{i+1,N})$, $i = 0,1,\ldots,$ with generator

$$L_x = \frac{\hbar}{2m}\frac{d^2}{dx^2} + \frac{\hbar}{m}\frac{d}{dx}(\ln|R_{E_N}(x)|)\frac{d}{dx}, \quad x \in (a_{i,N}, a_{i+1,N}). \tag{15}$$

From Eq. (13) it is clear that the radial diffusion $|\underset{\sim}{X}(t)|$ satisfies the Nelson-Newton law on $(a_{i,N}, a_{i+1,N})$:

$$\frac{m}{2}(D_+D_- + D_-D_+)|\underset{\sim}{X}(t)| = -\frac{d}{d|\underset{\sim}{X}|}V(|\underset{\sim}{X}(t)|). \tag{16}$$

Moreover, on each $(a_{i,N}, a_{i+1,N})$, for any function $f \in C^2$,

$$-R_{E_N}^{-1}(H_r - E_N)(R_{E_N}f) = -R_{E_N}^{-1}\{-\frac{\hbar^2}{2m}(R_{E_N}f'' + 2R'_{E_N}f' + R''_{E_N}f) + (V - E_N)R_{E_N}f\} = \hbar L_x f, \tag{17}$$

R_{E_N} being positive. The transition density for $|\underset{\sim}{X}(t)|, p(t,x,y)$, satisfies

$$L_x p(t,x,y) = 0, \quad p(0+,x,y) = \delta(x-y), \tag{18}$$

$x,y \in (a_{i,N}, a_{i+1,N})$, $\delta(x-y)$ being a delta function. For ground states R_{E_o}, where $a_{1,0} \equiv \infty$, Eq.'s (17) and (18) give

$$p(t,x,y) = R_{E_o}^{-1}(x)\exp\{-\frac{t}{\hbar}(H_r - E_o)\}(x,y)R_{E_o}(y), \tag{19}$$

$\exp\{-\frac{t}{\hbar}(H_r - E_o)\}(x,y)$ being the appropriate heat kernel for H_r on $\mathbb{R}^+ \times \mathbb{R}^+$. For simplicity of notation, we now set $\hbar = m = 1$.

Let us now introduce the speed measure, $M(x)$, and scale function,

$S(x)$, for our diffusion $|\underset{\sim}{X}(t)|$. We have

$$M(x) = \int_{x_o}^{x} R_{E_N}^2(u)du, \quad S(x) = \int_{x_o}^{x} R_{E_N}^{-2}(u)du, \quad L_x = \frac{1}{2}\frac{d}{dM(x)}\frac{d}{dS(x)}, \tag{20}$$

where x_o is an arbitrary point in $(a_{i,N}, a_{i+1,N})$. Clearly $M(a_{i,N})$ exists $\forall i$, since $R_{E_N} \in L^2(\mathbb{R})$, whilst $S(a_{i,N}) = -\infty$ and $S(a_{i+1,N}) = \infty \forall i$, as $R_{E_N}(a_{i,N}) = 0$. Further, define

$$u^1(x) = \int_{x_o}^{x} M(u)dS(u), \quad v^1(x) = \int_{x_o}^{x} S(u)dM(u). \tag{21}$$

Then $u^1(a_{i,N}) = \infty$. Thus, by the boundary classification scheme of Feller [7], the boundary points $a_{i,N}, a_{i+1,N}$ of the radial diffusion process $|\underset{\sim}{X}(t)|$ on $(a_{i,N}, a_{i+1,N})$ are inaccessible and therefore cannot be reached from an interior point of $(a_{i,N}, a_{i+1,N})$. Hence, the internodal regions $(a_{i,N}, a_{i+1,N})$ are non-communicating (see [1] e.g.) and on each of these regions $|\underset{\sim}{X}(t)|$ has a stationary distribution given by

$$\mathbb{P}\{|X(t)| \in A\} = \int_A R_{E_N}^2(u)du \Big/ \int_{a_{i,N}}^{a_{i+1,N}} R_{E_N}^2(u)du, \tag{22}$$

for each Borel set $A \subset (a_{i,N}, a_{i+1,N})$. We now work with the radial diffusion $|\underset{\sim}{X}(t)|$ on the first internodal region $(0, a_{1,N})$ $((0,\infty)$ for the ground state) to examine the first hitting time of the level a, $a \in (0, a_{1,N})$, defined by:

$$\tau_x(a) = \inf\{s > 0 : |\underset{\sim}{X}(s)| = a, |\underset{\sim}{X}(0)| = x\}. \tag{23}$$

The key result here is

$$\mathbb{E}\{\exp(-\lambda\tau_x(a))\} \equiv \hat{g}(x,a,\lambda) = \frac{\hat{p}(\lambda,x,a)}{\hat{p}(\lambda,a,a)}, \quad \lambda > 0, \tag{24}$$

where $\hat{p}(\lambda,x,y) = \text{L.T.}[p(t,x,y)](\lambda)$. The last identity follows because $|\underset{\sim}{X}(t)|$ is a one-dimensional time-homogeneous diffusion. For such a diffusion to have gone from x to y in time t the first hitting time of any intermediate point a must be less than t. Since the process starts afresh from a, for each fixed x,y and any intermediate a,

$$p(t,x,y) = \int_0^t g(x,a;u)p(t-u,a,y)du, \tag{25}$$

where $g(x,a;u)du = \mathbb{P}\{\tau_x(a) \in (u,u+du)\}$. The desired identity follows by taking Laplace transforms and letting $y \to a$. Writing $\hat{g}(x,\lambda)$ for $\hat{g}(x,a,\lambda)$, we see from Eqs. (18), (20) and (24) that

$$(L_x - \lambda)\hat{g}(x,\lambda) \equiv \left(\frac{1}{2}\frac{d}{dM}\frac{d}{dS} - \lambda\right)\hat{g}(x,\lambda) = 0, \quad \hat{g}(a,\lambda) = 1. \tag{26}$$

By definition, $0 \le \hat{g}(x,\lambda) \le 1$, thus from Eqs. (20) and (26)

$$\frac{d}{dx}\left(\frac{d}{dS}\hat{g}(x,\lambda)\right) = 2\lambda R_{E_N}^2 \hat{g}(x,\lambda) \ge 0. \tag{27}$$

This implies that $\frac{d}{dS}\hat{g}(x,\lambda)$ is increasing. For one-dimensional diffusions, continuity of sample paths ensures that as $|x-a|$ increases $\tau_x(a)$ increases. Thus, for $x < a$, $\hat{g}(x,\lambda)$ is increasing, and for $x > a$, $\hat{g}(x,\lambda)$ is decreasing. It follows that,

$$-1 \le \hat{g}(x,\lambda) - \hat{g}(a,\lambda) = \int_a^x \frac{d}{dS}\hat{g}(u,\lambda)dS(u) \le [S(x) - S(a)]\frac{d}{dS}\hat{g}(x,\lambda) \le 0, \quad (x > a), \tag{28}$$

$$1 \ge \hat{g}(a,\lambda) - \hat{g}(x,\lambda) = \int_x^a \frac{d}{dS}\hat{g}(u,\lambda)dS(u) \ge [S(a) - S(x)]\frac{d}{dS}\hat{g}(x,\lambda) \ge 0, \quad (x < a). \tag{29}$$

Letting $x \uparrow a_{1,N}$ in Eq. (28) and $x \downarrow 0$ in Eq. (29) yields:

<u>Proposition 1</u>

For the diffusion $|\underset{\sim}{X}(t)|$ on $(0,a_{1,N})$ satisfying Eq. (14), the Laplace transform of the distribution for the first hitting time of the level a starting from x is the solution of

$$(L_x - \lambda)\hat{g}(x,\lambda) \equiv (\frac{1}{2}\frac{d}{dM}\frac{d}{dS} - \lambda)\hat{g}(x,\lambda) = 0, \quad \lambda > 0, \tag{30}$$

with $g(a,\lambda) = 1$ and

$$\lim_{x \downarrow 0} \frac{d}{dS}\hat{g}(x,\lambda) = 0, \quad x < a; \qquad \lim_{x \uparrow a_{1,N}} \frac{d}{dS}\hat{g}(x,\lambda) = 0, \quad x > a. \tag{31}$$

In the case of ground states the following interesting result follows from Eq. (17) and the above proposition:

Proposition 2

Let $f_{E_o-\lambda}(x)$ and $h_{E_o-\lambda}(x) \in C^2$ satisfy, for each $\lambda > 0$,

$$-\frac{1}{2}\frac{d^2y}{dx^2} + (V(x) + \lambda - E_o)y = 0, \tag{32}$$

with $f_{E_o-\lambda}(0) = 0$, $f'_{E_o-\lambda}(0) = 1$, and $h_{E_o-\lambda}(x)$ the unique solution decreasing at infinity. Then:

$$\hat{g}(x,\lambda) = \frac{f_{E_o-\lambda}(x)R_{E_o}(a)}{f_{E_o-\lambda}(a)R_{E_o}(x)}, \; x < a; \quad \hat{g}(x,\lambda) = \frac{h_{E_o-\lambda}(x)R_{E_o}(a)}{h_{E_o-\lambda}(a)R_{E_o}(x)}, \; x > a. \tag{33}$$

We remark that the above hypotheses are met for piecewise continuous potentials $V(x)$ such that $x^2V(x)$ is analytic in a neighbourhood of the origin and $V(x) \to 0$ as $x \to \infty$.

The following is an easy corollary to Proposition 2:

Proposition 3

Let V have compact support with $\operatorname{supp} V \subset [0,a)$. Then, for $x > a$, the radial ground state process with energy $E_o(<0)$ has

$$\mathbb{E}\{\exp(-\lambda\tau_x(a))\} \equiv g(x,\lambda) = \exp(-\sqrt{(-2(E_o - \lambda))})(x - a)\exp(\sqrt{(-2E_o)}(x - a)), \tag{34}$$

leading to

$$\mathbb{P}\{\tau_x(a) \in (s, s + ds)\} = \frac{|a - x|}{\sqrt{(2\pi s^3)}}\exp\{E_o s - \frac{(x - a)^2}{2s} + \sqrt{(-2E_o)}(x - a)\}ds. \tag{35}$$

The hitting time density in Eq. (35) is easily seen to be the same as that for Brownian motion with constant drift $-\sqrt{(-2E_o)}$ (see [12] e.g.).

The preceding example is atypical in that one is able to invert

$\hat{g}(x,\lambda)$ and obtain an explicit formula for the density of $\tau_x(a)$. The Coulomb problem is more representative:

Proposition 4 (Coulomb Problem Ground State)

Let $V(x) = -Ze^2/|x|$. The ground state energy $E_o = -\frac{1}{2a_o^2}$, a_o being the Bohr radius $\frac{1}{Ze^2}$. The corresponding ground state wave-function is $R_{E_o}(x) = x\exp(-x/a_o)$. In this case we have

$$\mathbb{E}\{\exp(-\lambda\tau_x(a))\} \equiv \frac{a}{x}\exp\left(\frac{x-a}{a_o}\right)\frac{G_{\frac{1}{k},\frac{1}{2}}\left(\frac{2kx}{a_o}\right)}{G_{\frac{1}{k},\frac{1}{2}}\left(\frac{2ka}{a_o}\right)}, \tag{36}$$

$k = \sqrt{(1+2a_o^2\lambda)}$, $G \equiv W$ for $x \geq a$ and $G \equiv M$ for $x \leq a$, M and W being Whittaker functions [5]. (See [3]).

Needless to say we cannot invert the Laplace transforms in Eq. (36). For this reason we turn our attention to the first moment of the hitting time distribution, $\mathbb{E}\{\tau_x(a)\}$.

Due to its complex λ dependence, $\hat{g}(x,\lambda)$ proves ill-suited for the purpose of evaluating $\mathbb{E}\{(\tau_x(a))^N\}$. The following proposition is much more useful and may be deduced from Proposition 1. (For similar results see e.g. [8].)

Proposition 5

For the diffusion $|\underset{\sim}{X}(t)|$ on $(0,a_{1,N})$ of Proposition 1 let $V^j(x) = \mathbb{E}\{(\tau_x(a))^j\}$, $V^0(x) \equiv 1$. Then $V^N(x)$ satisfies, for $j = 1,2,\ldots$

$$\frac{1}{2}\frac{d}{dM}\frac{d}{dS}V^j(x) = -j\,V^{j-1}(x), \quad V(a) = 0, \tag{37}$$

$$\lim_{x\downarrow 0}\frac{d}{dS}V^j(x) = 0, \quad x < a; \qquad \lim_{x\uparrow a_{1,N}}\frac{d}{dS}V^j(x) = 0, \quad x > a.$$

Applying this to the Coulomb problem we find with $j = 1$:

Proposition 6 (Coulomb Problem Ground State and First Excited State.)

Let $\bar{\mathbb{E}}_N^+\{\tau_x(a)\} = \mathbb{E}\{\tau_x(a)\,|\,x \in (a,a_{1,N})\}$

and $\bar{\mathbb{E}}_N^-\{\tau_x(a)\} = \mathbb{E}\{\tau_x(a)\,|\,x \in (0,a)\}$

for the radial diffusion $|\underset{\sim}{X}(t)|$ on $(0,a_{1,N})$ corresponding to the state $R_{E_N}(x)$, x distributed according to the stationary distribution of Eq. (22). We have

$$\bar{\mathbb{E}}_0^+\{\tau_x(a)\} = a_o^2\int_{a/a_o}^{\infty} (x+1+\frac{1}{2x})^2 e^{-2x}\,dx, \tag{38}$$

$$\bar{\mathbb{E}}_0^-\{\tau_x(a)\} = a_o^2\int_{0}^{a/a_o} (\frac{\sinh x}{x} - (1+x)e^{-x})^2 dx, \tag{39}$$

$$\bar{\mathbb{E}}_1^+\{\tau_x(a)\} = \frac{8a_o^2}{(1-7e^{-2})}\int_{a/2a_o}^{1} \frac{1}{(1-x)^2}\{(x^3+x+1+\frac{1}{2x})e^{-x} + \frac{7e^{-(2-x)}}{2x}\}^2 dx, \tag{40}$$

$$\bar{\mathbb{E}}_1^-\{\tau_x(a)\} = \frac{8a_o^2}{(1-7e^{-2})}\int_{0}^{a/2a_o} \frac{1}{(1-x)^2}\{\frac{\sinh x}{x} - (x^3+x+1)e^{-x}\}^2 dx. \tag{41}$$

Further details of the above calculations may be found in Batchelor's Ph.D. thesis.

It is clear that the analogues of the above for the nth excited state would be formidable to evaluate. We therefore turn to asymptotic methods, as pioneered by Mandl.

3. ASYMPTOTIC RESULTS

The determination of the asymptotics of the first hitting time distribution as $a \sim 0$ in $\tau_x(a)$ hinges on some rather intricate analysis of a specially constructed solution to Eqs. (30) and (31). We simply catalogue the main steps here, the details may be found in [8].

Set for $N = 0,1,\ldots$ and $x \in (0,a_{1,N})$,

$$u^0(x) \equiv 1, \quad u^{N+1}(x) = \int_{x_o}^{x}\int_{x_o}^{y} u^N(u)\, dM(u)\, dS(y), \tag{42}$$

where M, S and x_o are as in Eq. (20). Define $U(x,\lambda)$ by

$U(x,\lambda) = \sum_{N=0}^{\infty} (2\lambda)^N u^N(x)$. It is easy to show

$$1 + 2\lambda u^1(x) \le U(x,\lambda) \le \sum_{N=0}^{\infty} \frac{(2\lambda)^N}{N!}(u^1(x))^N = \exp\{2\lambda u^1(x)\}, \tag{43}$$

$$2\lambda|M(x)| \le \left|\frac{dU(x,\lambda)}{dS}\right| \le 2\lambda|M(x)|\exp\{2\lambda v^1(x)\}, \tag{44}$$

where $v^1(x)$ is defined in Eq. (21). Further, introduce for $\lambda > 0$, the functions

$$U^+(x,\lambda) = U(x,\lambda)\int_x^{a_{1,N}} U(y,\lambda)^{-2} dS(y), \quad U^-(x,\lambda) = U(x,\lambda)\int_0^x U(y,\lambda)^{-2} dS(y). \tag{45}$$

It can be shown that $U^+(x,\lambda)$ and $U^-(x,\lambda)$ are well-defined. Moreover

$$\hat{g}(x,\lambda) = \frac{U^+(x,\lambda)}{U^+(a,\lambda)},\ x \ge a; \quad \hat{g}(x,\lambda) = \frac{U^-(x,\lambda)}{U^-(a,\lambda)},\ x \le a, \tag{46}$$

for our process $|\underset{\sim}{X}(t)|$ on $(0,a_{1,N})$. Let us take $x \ge a$. From Eqs. (45) and (46) we may write

$$\hat{g}(x,\lambda) = \frac{U(x,\lambda)\int_x^{a_{1,N}} U(y,\lambda)^{-2} dS(y)}{U(a,\lambda)\left\{\int_a^{x_o} U(y,\lambda)^{-2} dS(y) + \int_{x_o}^{a_{1,N}} U(y,\lambda)^{-2} dS(y)\right\}}. \tag{47}$$

Since 0 and $a_{1,N}$ are inaccessible and $M(0,a_{1,N}) \equiv M(a_{1,N}) - M(0) < \infty$ we can show that

$$\lim_{a \downarrow 0} \frac{u^2(a)}{S^2(a)} = 0. \tag{48}$$

Define $\gamma_N(a)$ by

$$\gamma_N(a) = -2M(0,a_{1,N})S(a) = -2\int_0^{a_{1,N}} dM(y)\int_{x_o}^{a} dS(y), \tag{49}$$

then using the definition of $U(x,\lambda)$ and Eq. (48) one can show:

(a) $\lim_{a\downarrow 0} U(x,\lambda/\gamma_N(a)) = 1$ while $\lim_{a\downarrow 0} U(a,\lambda/\gamma_N(a)) = 1 - \dfrac{2M(0)\lambda}{M(0,a_{1,N})}$;

(b) $\lim_{a\downarrow 0} \dfrac{\lambda}{\gamma_N(a)}\displaystyle\int_x^{a_{1,N}} U(y,\lambda/\gamma_N(a))^{-2}dS(y) = \dfrac{1}{2M(a_{1,N})}$;

(c) $\lim_{a\downarrow 0} \dfrac{\lambda}{\gamma_N(a)}\displaystyle\int_a^{x} U(y,\lambda/\gamma_N(a))^{-2}dS(y) = \dfrac{1}{M(0)}\left[\dfrac{2\lambda M(0)}{M(0,a_{1,N}) - 2\lambda M(0)}\right]$.

Substituting the above set of limits into the expression for $\lim_{a\downarrow 0} \hat{g}(x,\lambda/\gamma_N(a))$ obtained from Eq. (47) yields

$$\lim_{a\downarrow 0} \hat{g}(x,\lambda/\gamma_N(a)) = \frac{1}{1+\lambda}. \tag{50}$$

The same result holds for $x \le a$ as $x \uparrow a_{1,N}$, with $\gamma_N(a) = 2M(0,a_{1,N})S(a)$. This leads us to the following appealing result:

Proposition 7

For the diffusion $|\underset{\sim}{X}(t)|$ on $(0,a_{1,N})$ satisfying Eq. (14) we have:

$$\mathbb{P}\{\tau_x(a) < t\} \sim 1 - \exp(-t/\gamma_N(a)), \quad \bar{\mathbb{E}}_N^{\pm}\{\tau_x(a)\} \sim \gamma_N(a), \tag{51}$$

as $a \sim 0$ or $a \sim a_{1,N}$, where for each fixed x,

$$\gamma_N(a) = -2M(0,a_{1,N})S(a),\ x > a; \quad \gamma_N(a) = +2M(0,a_{1,N})S(a),\ x < a. \tag{52}$$

(An alternative method of proof can be found in [11].) We now return to the Coulomb problem.

The spherically symmetric radial wave-functions, $R_{E_N}(x)$, for the Coulomb problem are given by

$$R_{E_N}(x) = x\exp\left(\frac{-1}{(N+1)a_o}\right) L_N^1\left(\frac{2x}{(N+1)a_o}\right), \; N = 0,1,\ldots; \quad E_N = -\frac{Z^2e^4}{2(N+1)^2} \tag{53}$$

L_N^1 a Laguerre polynomial. For this system $a_{i,N} = \frac{(N+1)a_o}{2} x_{i,N}$, where $x_{i,N}$ are the zeros of L_N^1. From Proposition 7 and the asymptotics of L_N^1, we have:

Proposition 8

For the Coulomb problem

$$\lim_{a\downarrow 0} \frac{\gamma_N(a)}{\gamma_o(a)} = \frac{(N+1)}{2} \int_0^{x_{1,N}} u^2 e^{-u} (L_N^1(u))^2 du, \tag{54}$$

$$\lim_{N\uparrow\infty} \lim_{a\downarrow 0} \frac{\gamma_N(a)}{\gamma_o(a)} = \frac{1}{16} \int_0^{j_{1,1}} u^3 (J_1(u))^2 du, \tag{55}$$

$\gamma_o(a) \simeq \frac{ma_o^3}{2\hbar a}$ for $a \sim 0$, $j_{1,1}$ being the first zero of the Bessel function J_1. We remark that $u^{3/2} J_1(u)$ is the zero-energy wave-function for the Coulomb problem.

In the next section we discuss a possible application for some of the above results.

4. A POSSIBLE APPLICATION TO Π-MESIC HYDROGEN

If you replace the negatively charged electron in the hydrogen atom by a negatively charged pion π^-, the π^- feels only the Coulomb attraction due to the positively charged proton p^+ when at a distance exceeding $\hbar/m_\pi c$ (the pion Compton wave length) from the nuclear proton. The reason for this is the extremely short range $(\sim\hbar/m_\pi c)$ of the strong force which governs the decay process $p^+ + \pi^- \to n + \gamma$'s. It follows that the $p^+\pi^-$ system cannot decay until π^- first hits a sphere of radius $\hbar/m_\pi c$ centred at the nuclear proton. We will write:

$$\tau^N_{Decay} \equiv \text{Expected Decay Time for } p^+\pi^- \text{ after capture in } N^{th} \text{ excited state.} \tag{56}$$

Firstly, assume that the π^- has been captured into the ground state (N = 0). Then, if stochastic mechanics gives the correct first hitting time, we obtain from Proposition 6:

$$\tau^{o}_{Decay} > \frac{2\hbar^3}{m_\pi e^4}\int_{e^2/c\hbar}^{\infty} (x + 1 + \frac{1}{2x})^2 e^{-2x} dx \sim 7\times 10^{-18} \text{ secs.} \qquad (57)$$

Experimentalists suggest $\tau^{o}_{Decay} < 10^{-12}$ secs which is consistent.

Using Proposition 8 we can find approximations for $\bar{\mathbb{E}}^{+}_{N}\{\tau_x(a)\}$ and hence τ^{N}_{Decay}. Specifically we have, for $a = h/m_\pi c \sim 0$,

$$\tau^{N}_{Decay} > \frac{\bar{\mathbb{E}}^{+}_{N}\{\tau_x(a)\}}{\bar{\mathbb{E}}^{+}_{0}\{\tau_x(a)\}}\,\bar{\mathbb{E}}^{+}_{0}\{\tau_x(a)\} \simeq \lim_{a\downarrow 0}\frac{\gamma_N(a)}{\gamma_0(a)}\,\bar{\mathbb{E}}^{+}_{0}\{\tau_x(a)\}. \qquad (58)$$

Numerical calculations show that for the first few excited states the above approximation gives good agreement with exact values. For example, when N = 1, using Eq. (40), the error is less than 3%.

It will not surprise the reader that exactly the same arguments as used in the above may be employed to obtain the corresponding results for states with non-zero angular momentum. In the case of the Coulomb problem we now have:

$$R_{E_{N,\ell}}(x) = x^{\ell+1} L_N^{2\ell+1}\left(\frac{2x}{(N+\ell+1)a_o}\right)\exp\left(-\frac{x}{(N+\ell+1)a_o}\right); \quad E_{N,\ell} = -\frac{Z^2 e^4}{2(N+\ell+1)^2}, \qquad (59)$$

$N = 0,1,\ldots,\ \ell = 0,1,\ldots$. Generalising Proposition 8 to non-zero angular momentum gives:

<u>Proposition 9</u>

For the radial Nelson diffusion corresponding to the above radial wave-function, $R_{E_{N,\ell}}$ with energy $E_{N,\ell}$ on the first internodal region:

$$\lim_{a\downarrow 0}\frac{\gamma_{N,\ell}(a)}{\gamma_{o,\ell}(a)} = \frac{\frac{(2\ell+1)!}{(2\ell+2)}\left[\frac{N!}{(N+2\ell+1)!}\right]^2}{\left[\frac{(\ell+1)}{(N+\ell+1)}\right]^{2\ell+3}}\int_0^{x_{1,N,\ell}} u^{2\ell+2} e^{-u} (L_N^{2\ell+1}(u))^2 du, \qquad (60)$$

$$\lim_{N\uparrow\infty}\lim_{a\downarrow 0}\frac{\gamma_{N,\ell}(a)}{\gamma_{0,\ell}(a)} = \frac{(2\ell+1)!}{(\ell+1)^{2\ell+4}}\cdot\frac{1}{16}\int_0^{j_{1,2\ell+1}} u^3(J_{2\ell+1}(u))^2\,du, \quad (61)$$

$$\gamma_{0,\ell}(a) \sim \frac{8ma_o^2}{\hbar}\left(\frac{(\ell+1)}{2}\right)^{2\ell+4}(2\ell)!\left(\frac{a_o}{a}\right)^{2\ell+1}, \quad (62)$$

where $x_{1,N,\ell}$ is the first zero of $L_N^{2\ell+1}$ and $j_{1,2\ell+1}$ is the first zero of the Bessel function $J_{2\ell+1}$.

One can now approximate the decay time for the $p^+\pi^-$ system from states with angular momentum ℓ. For the $p^+\pi^-$ system with lowest possible energy and angular momentum unity $(N = 0, \ell = 1)$ we obtain an approximate lower bound of 4.2×10^{-12} secs. When compared with decay rates of 10^{-12} secs. this looks as though it might be accessible to experiment. However, one must guard against being too optimistic as the $P\to S$ transition rate may be too great (see [4]). Tables of $\lim_{a\downarrow 0}\frac{\gamma_{N,\ell}(a)}{\gamma_{0,\ell}(a)}$ for $\ell = 0,1,2$, $N = 1,\ldots,20$ are given at the end of this paper. We now turn our attention to sojourn times.

5. SOJOURN TIMES FOR GROUND STATES

Throughout this section we use units in which $\hbar = m = 1$ to simplify our algebraic expressions. Let $\underset{\sim}{X}_x$, with $\underset{\sim}{X}_x(0) = \underset{\sim}{x}$, be the Nelson diffusion corresponding to $\psi_E(>0)$ the ground state solution of the above Schrödinger equation in d-dimensional Euclidean space. Let Λ be an open subset of $\mathbb{R}^d$. Denote by $\tau_\Lambda(t)$ the time spent in Λ upto time t by the process $\underset{\sim}{X}_x$, so that

$$\tau_\Lambda(t) = \int_0^t \chi_\Lambda(\underset{\sim}{X}_x(s))ds, \quad (63)$$

where χ_Λ is the characteristic function of Λ. Then, using the usual probabilistic notation,

$$\mathbb{E}_x\{\tau_\Lambda(t)\} = \mathbb{E}_x\{\int_0^t \chi_\Lambda(\underset{\sim}{X}(s))ds\} = \int_0^t p(s,\underset{\sim}{x},\Lambda)ds, \tag{64}$$

with $p(s,\underset{\sim}{x},\Lambda) = \int p(s,\underset{\sim}{x},\underset{\sim}{y})\chi_\Lambda(\underset{\sim}{y})dy$, p being the transition density for the process $\underset{\sim}{X}$. Since by the above reasoning

$$p(s,\underset{\sim}{x},\underset{\sim}{y}) = \psi_E^{-1}(\underset{\sim}{x}) \exp\{-s(H-E)\}(\underset{\sim}{x},\underset{\sim}{y})\psi_E(\underset{\sim}{y}), \tag{65}$$

as expected, $|\psi_E(\underset{\sim}{x})|^2$ is the invariant density for $p(s,\underset{\sim}{x},\underset{\sim}{y})$, so averaging over the initial position $\underset{\sim}{x}$, gives

$$\bar{\mathbb{E}}\{\tau_\Lambda(t)\} = \int |\psi_E(\underset{\sim}{x})|^2\, \mathbb{E}_x(\tau_\Lambda(t))\,dx = t\int_\Lambda |\psi_E|^2(\underset{\sim}{x})\,dx. \tag{66}$$

By using the result

$$2^{-1}\bar{\mathbb{E}}\{(\tau_\Lambda(t))^2\} = \bar{\mathbb{E}}\{\int_0^t \chi_\Lambda(\underset{\sim}{X}(s_1))ds_1 \int_{s_1}^t \chi_\Lambda(\underset{\sim}{X}(s_2))ds_2\}, \tag{67}$$

one can show from the Markov property [12] that

$$2^{-1}\bar{\mathbb{E}}\{(\tau_\Lambda(t))^2\} = \tag{68}$$

$$\int_0^t ds_1 \int_{s_1}^t ds_2 \iint dx\,dy\,\psi_E(\underset{\sim}{x})\chi_\Lambda(\underset{\sim}{x}) \exp(-(s_2-s_1)(H-E))(\underset{\sim}{x},\underset{\sim}{y})\chi_\Lambda(\underset{\sim}{y})\psi_E(\underset{\sim}{y}).$$

This suggests:

<u>Proposition 10</u>

For operator valued α let $\mathcal{T}$ denote the time-ordered product:

$$\mathcal{T}\{\exp(\lambda\int_0^t \alpha(s)ds)\} = \sum_{n=0}^{\infty} \lambda^n \int_0^t ds_1 \int_{s_1}^t ds_2 \cdots \int_{s_{n-1}}^t ds_n\,\alpha(s_1)\ \ldots\ \alpha(s_n). \tag{69}$$

Then for the ground state process $\underset{\sim}{X}_x$, with a bounded locally Lipschitz drift $\psi_E^{-1}\underset{\sim}{\nabla}\psi_E$, for $\lambda > 0$,

$$\bar{\mathbb{E}}\{\exp(-\lambda\tau_\Lambda(t))\} = (\psi_E, \mathcal{T}\{\exp(-\lambda\int_0^t e^{-s(H-E)}P_\Lambda e^{s(H-E)}ds)\}\psi_E)_{L^2}, \tag{70}$$

where P_Λ is the projection defined by $(P_\Lambda f)(\underset{\sim}{x}) = \chi_\Lambda(\underset{\sim}{x})f(\underset{\sim}{x})$.

Further details of this are given in [6]. Observe that, if $\psi_E^{-1}\nabla\psi_E$ is bounded and measurable, by the Girsanov-Cameron-Martin theorem [12], we obtain for $\psi = e^R$

$$\mathbb{E}\{\exp -\lambda\int_0^t \chi_\Lambda(X_x(s))ds\} = \tag{71}$$

$$\mathbb{E}_x\{\exp(-\lambda\int_0^t \chi_\Lambda(B(s))ds + \int_0^t \nabla R(B(s)).dB(s) - 2^{-1}\int_0^t |\nabla R|^2(B(s))ds)\},$$

B being a BM($\mathbb{R}^d$) process.

Itô's formula now gives

$$R(B_x(t)) - R(B_x(0)) = \int_0^t \nabla R(B_x(s))\cdot dB_x(s) + \frac{1}{2}\int_0^t \Delta R(B_x(s))ds \tag{72}$$

and, since $R = \ln\psi_E$, we obtain

$$R(B_x(t)) - R(x) = \int_0^t \nabla R(B_x(s))\cdot dB_x(s) - \frac{1}{2}\int_0^t |\nabla R|^2(B_x(s))ds + \frac{1}{2}\int_0^t \psi_E^{-1}\Delta\psi_E(B(s))ds. \tag{73}$$

Combining these identities yields, using the Feynman-Kac formula,

$$\mathbb{E}_x\{\exp(-\lambda\int_0^t \chi_\Lambda(X(s))ds)\} = \psi_E^{-1}(x)\exp\{-t(H + \lambda P_\Lambda - E)\}\psi_E(x), \tag{74}$$

where we assume that the quantum mechanical Hamiltonian $H = (-2^{-1}\Delta + V)$ is self-adjoint on a domain containing ψ_E. Thus, we have shown:

<u>Proposition 11</u>

Let X be the Nelson diffusion corresponding to the ground state ψ_E of the self-adjoint quantum mechanical Hamiltonian $H = (-2^{-1}\Delta + V)$, with bounded Lipshitz $\psi_E^{-1}\nabla\psi_E$, and $E = \inf\operatorname{spec}(H)$. Then, defining u by

$$u(x,\alpha) = \int_0^\infty e^{-\alpha t}\,\mathbb{E}_x\{\exp(-\lambda\int_0^t \chi_\Lambda(X(s))ds)\}dt \quad (\alpha,\lambda > 0), \tag{75}$$

we obtain

$$u(\underset{\sim}{x},\alpha) = \psi_E^{-1}(\underset{\sim}{x})(H+\lambda P_\Lambda+\alpha-E)^{-1}\psi_E(\underset{\sim}{x}). \tag{76}$$

Hence, if $L = -\psi_E^{-1}(H-E)\psi_E$ is the generator for the Nelson diffusion $\underset{\sim}{X}$,

$$\{L_x-(\lambda\chi_\Lambda(\underset{\sim}{x})+\alpha)\}u(\underset{\sim}{x},\alpha) = -1, \tag{77}$$

with the boundary conditions implicit in the above i.e. $u(\underset{\sim}{x},\alpha)\in C^1$.

6. APPLICATION TO THE K-SHELL CAPTURE OF ELECTRONS

The inner K-shell electrons of heavy elements can be captured by the nucleus, decaying according to the weak decay $p^+ + e^- \to n+\nu$. We give a model of this decay using Nelson's stochastic mechanics.

The weak decay force is of short range so we shall assume:-

(i) Probability of decay/unit time $=\begin{cases} 0, \text{ if } e^- \text{ is outside nucleus } \Lambda, \\ \lambda, \text{ if } e^- \text{ is inside nucleus } \Lambda. \end{cases}$

(ii) K-shell electron e^- is in the ground state ψ_E.

(iii) Stochastic mechanics is valid at the level of sample paths.

It is simple to show, [6], that

$$\bar{\mathbb{P}}(\text{no decay up to time } t) = \bar{\mathbb{E}}\{\exp(-\lambda\tau_\Lambda(t))\}, \tag{78}$$

so that if τ_c denotes the capture (decay) time

$$\bar{\mathbb{P}}(\tau_c>t) = \bar{\mathbb{E}}\{\exp(-\lambda\tau_\Lambda(t))\}. \tag{79}$$

Now set

$$\bar{u}(\alpha) = \int|\psi_E(\underset{\sim}{x})|^2u(\underset{\sim}{x},\alpha)dx. \tag{80}$$

Using the result that $L_x^*|\psi_E(\underset{\sim}{x})|^2 = 0$, $|\psi_E(\underset{\sim}{x})|^2$ being the invariant density, from (77) we obtain

$$\lambda\int_\Lambda|\psi_E(\underset{\sim}{x})|^2u(\underset{\sim}{x},\alpha)dx + \alpha\bar{u}(\alpha) = 1. \tag{81}$$

But integration by parts gives

$$u(\underset{\sim}{x},\alpha) = \int_0^\infty e^{-\alpha t}\mathbb{P}_x(\tau_c>t)dt = \alpha^{-1}(1-\mathbb{E}_x\{\exp(-\alpha\tau_c)\}), \tag{82}$$

so that for the capture time τ_c we arrive at

$$\bar{\mathbb{E}}\{\exp(-\alpha\tau_c)\} = \lambda\int_\Lambda |\psi_E(\underset{\sim}{x})|^2 u(\underset{\sim}{x},\alpha)\,dx. \tag{83}$$

Further, from (82) and (77), we have for $\mathbb{E}_x\{\exp(-\alpha\tau_c)\} = v(\underset{\sim}{x},\alpha)$,

$$\{L_x - (\lambda\chi_\Lambda(\underset{\sim}{x}) + \alpha)\}v(\underset{\sim}{x},\alpha) = -\lambda\chi_\Lambda(\underset{\sim}{x}). \tag{84}$$

Thus, using the above we can prove:

<u>Proposition 12</u>

Let τ_c denote the capture time for the above model of the K-shell capture of electrons. Then for $\alpha > 0$

$$\begin{aligned}\bar{\mathbb{E}}\{\exp(-\alpha\tau_c)\} &= \lambda(P_\Lambda\psi_E,(H+\lambda P_\Lambda+\alpha-E)^{-1}\psi_E)_{L^2}\\ &= \lambda(\psi_E,(H+\lambda P_\Lambda+\alpha-E)^{-1}(P_\Lambda\psi_E))_{L^2},\end{aligned} \tag{85}$$

giving

$$\bar{\mathbb{P}}(\tau_c < \infty) = \lambda(\psi_E,(H+\lambda P_\Lambda-E)^{-1}(P_\Lambda\psi_E))_{L^2} = 1 \tag{86}$$

and for $N = 0,1,2,\ldots$

$$\begin{aligned}\bar{\mathbb{E}}\{(\tau_c)^N\} &= N!\lambda(\psi_E,(H+\lambda P_\Lambda-E)^{-(N+1)}(P_\Lambda\psi_E))_{L^2}\\ &= N!\lambda(P_\Lambda\psi_E,(H+\lambda P_\Lambda-E)^{-(N+1)}\psi_E)_{L^2}.\end{aligned} \tag{87}$$

The above result gives rise to exact formulae for the preceding expressions in the capture time when the Hamiltonian H is the Coulomb Hamiltonian $H = (-2^{-1}\Delta_x - \frac{ze^2}{|\underset{\sim}{x}|})$. [2].

ACKNOWLEDGEMENTS

It is a pleasure to thank Professors David Williams and John Taylor at Cambridge for helpful conversations concerning this paper.

REFERENCES

1. Albeverio, S., Blanchard, Ph., Hoegh-Krohn, R, "Newtonian Diffusions and Planets with a remark on non-standard Dirichlet forms and Polymers". In "Stochastic Analysis and Applications", Proceedings, Swansea 1983, editors A. Truman and D. Williams, 1-25. Lecture Notes in Maths. 1095, Springer-Verlag. (1984).

2. Batchelor, A., Truman, A., "Exact Formulae for Capture Times in Stochastic Mechanics", in preparation.

3. Batchelor, A., Truman, A., "First Hitting Times in Stochastic Mechanics", in preparation.

4. Bethe, H.A., Leon, M., "Negative Meson Absorption in Liquid Hydrogen", Phys. Rev. 127, 636-647. (1962).

5. Bucholz, H., "The Confluent Hypergeometric Function". New York: Springer-Verlag. (1969).

6. Carlen, E., Truman, A., "Sojourn Times and First Hitting Times in Stochastic Mechanics". In "Fundamental Aspects of Quantum Theory", editors Gorini, V. and Alberto, F., 153-161. Nato ASI Series. Series B: Physics, 144, Plenum. (1986).

7. Feller, W., "Diffusion Processes in One-Dimension", Trans. of Amer. Math. Soc., 97, 1-37. (1954).

8. Mandl, P., "Analytical Treatment of One-Dimensional Markov Processes". New York: Springer-Verlag. (1968).

9. Nelson, E., "Dynamical Theories of Brownian Motion". Mathematical Notes. Princeton: Princeton University Press. (1967).

10. Nelson, E., "Quantum Fluctuations". Princeton Series in Physics. Princeton: Princeton University Press. (1985).

11. Newell, G.F., "Asymptotic Extreme Value Distribution for One-Dimensional Diffusion Processes", J. of Math. and Mech., 11, 481-496. (1962).

12. Rogers, L.C.G., Williams, D., "Diffusions, Markov Processes and Martingales", Volume 2, Itô Calculus. Chichester: John Wiley. (1987).

ZERO ANGULAR MOMENTUM CASE

N	$\lim_{a \downarrow 0} \gamma_N(a)/\gamma_0(a)$
0	1.00000
1	0.42122
2	0.38735
3	0.37686
4	0.37221
5	0.36973
6	0.36826
7	0.36731
8	0.36666
9	0.36620
10	0.36586
11	0.36560
12	0.36540
13	0.36524
14	0.36511
15	0.36500
16	0.36492
17	0.36484
18	0.36478
19	0.36473

TABLE 1.

Note that:

$$\frac{1}{16}\int_0^{j_{1,1}} u^3[J_1(u)]^2 du = 0.36424 \quad 5 \text{ d.p.}$$

ANGULAR MOMENTUM ($\ell = 1$) CASE

N	$\lim_{a \downarrow 0} \gamma_{N,1}(a)/\gamma_{0,1}(a)$
0	1.00000
1	0.31516
2	0.25372
3	0.23203
4	0.22155
5	0.21562
6	0.21192
7	0.20944
8	0.20770
9	0.20643
10	0.20547
11	0.20473
12	0.20415
13	0.20368
14	0.20329
15	0.20298
16	0.20271
17	0.20250
18	0.20230
19	0.20214

TABLE 2.

Note that:

$$\frac{3}{512}\int_0^{j_{1,3}} u^3[J_3(u)]^2du = 0.20055 \quad 5 \text{ d.p.}$$

ANGULAR MOMENTUM ($\ell = 2$) CASE

N	$\lim_{a\downarrow 0} \gamma_{N,2}(a)/\gamma_{0,2}(a)$
0	1.00000
1	0.25432
2	0.17974
3	0.15269
4	0.13930
5	0.13155
6	0.12660
7	0.12324
8	0.12084
9	0.11906
10	0.11770
11	0.11664
12	0.11580
13	0.11511
14	0.11455
15	0.11408
16	0.11369
17	0.11335
18	0.11306
19	0.11277

TABLE 3.

Note that:

$$\frac{5}{4374}\int_0^{j_{1,5}} u^3[J_5(u)]^2 du = 0.11032 \quad 5 \text{ d.p.}$$

WEYL QUANTIZATION FOR $\mathbb{Z}_N \times \mathbb{Z}_N$ PHASE SPACE : STOCHASTIC ASPECT

Olivier COHENDET, Philippe COMBE* , Madeleine SIRUGUE-COLLIN**

Centre de Physique Théorique
CNRS - Luminy, Case 907
F - 13288, MARSEILLE Cedex 09 (France)

* et Université d'Aix Marseille II, ** et Université de Provence.

1. INTRODUCTION

Nelson's stochastic mechanics [1] [2] [3] provides a new interesting description of the non relativistic quantum dynamics based on the interpretation of the Schrödinger equation in terms of the Fokker - Planck equation of a Newtonian diffusion process. However the Nelson's program is widely based on the existence of a diffusion process in the configuration space and the extension of this program to equivalent descriptions of quantum mechanics makes some problems [4]. In fact in this context we cannot limit the class of processes to the diffusion one but rather to the class of infinitely divisible processes. Then it is very difficult to compare processes corresponding to different polarizations of phase space.

Another problem is the study of quantum systems described by mixed states [5]. A natural way of investigating such problems is to use Wigner functions rather than wave functions. In this framework the construction of a stochastic mechanics on the phase space leads to new problems which are clearly solved only in the case of systems with finite number of states. Nevertheless the case with an odd number of states has the flavour of the continuous case and can be seen as an approximation. In Section 2 we recall basic facts on Weyl quantization and Wigner functions in the continuous case, in Section 3 we develop this formalism in the case of cyclic groups. Finally in Section 4 we show the existence of a state dependent Maskov process, on an extended phase space, associated to the quantum evolution.

2. PHASE SPACE FORMULATION OF QUANTUM MECHANICS

Let us recall, in this section, the main feature of the Weyl quantization for usual quantum mechanics. In group theoretical approach a Weyl system is a unitary projective representation, on an Hilbert space $\mathcal{H}$, of the phase space translation group $\mathbb{R}^{2n}$ [see eg. 6,7]. It is a true unitary representation of a central extension of $\mathbb{R}^{2n}$ by U(1). An easy way to construct a Weyl system is to start with the usual commutation relations of quantum mechanics

$$[Q_i, P_j] = i\hbar\,\delta_{ij} \tag{2.1}$$

which defines the Lie algebra of the central extension labeled by $\hbar$, where $\hbar$ is the Planck constant divided by 2π. In exponential form (2.1) becomes

$$e^{\frac{i}{\hbar}p.Q}\, e^{\frac{i}{\hbar}q.P} = e^{-\frac{i}{\hbar}q.p}\, e^{\frac{i}{\hbar}q.P}\, e^{\frac{i}{\hbar}p.Q} \tag{2.2}$$

Moreover these unitary operator act on $L^2(\mathbb{R}^n, dx)$, the configuration representation, as follows

$$(e^{\frac{i}{\hbar}q.P}\psi)(x) = \psi(x+q) \tag{2.3}$$

$$(e^{\frac{i}{\hbar}p.Q}\psi)(x) = e^{\frac{i}{\hbar}p.x}\psi(x) \tag{2.4}$$

Now the family of unitary operators $\{W_{qp}, (q,p) \in \mathbb{R}^{2n}\}$ defined by

$$W^{\eta}_{qp} = \eta(q,p)\; e^{\frac{i}{\hbar}p.Q}\; e^{-\frac{i}{\hbar}q.P} \tag{2.5}$$

where $q,p \rightarrow \eta(q,p)$ is a complex function such that

$$|\eta(q,p)| = 1 \tag{2.6}$$

defines a Weyl system

$$W^{\eta}_{qp}\, W^{\eta}_{q'p'} = \eta(q,p)\,\eta(q',p')\,\bar{\eta}(q+q', p+p')\, e^{-\frac{i}{\hbar}q.p'}\; W_{q+q',p+p'} \tag{2.7}$$

$$(W^{\eta}_{qp})^{-1} = \bar{\eta}(q,p)\,\bar{\eta}(-q,-p)\,e^{-\frac{i}{\hbar}q.p}\,W_{-q-p} \tag{2.8}$$

$$(W^{\eta}_{qp})^{+} = (W^{\eta}_{qp})^{-1} \tag{2.9}$$

and the commutation relation

$$W^{\eta}_{qp}\,W^{\eta}_{q'p'} = e^{\frac{i}{\hbar}(p.q'-q.p')}\,W^{\eta}_{q'p'}\,W^{\eta}_{qp} \tag{2.10}$$

does not depend on η. Hence the Weyl system labeled by η is equivalent to the one labeled by η' in the sense that they are associated with the same commutation relations, and we can choose η to be the more convenient for our purpose. In other words the choice of η corresponds to a representative in the class cohomology associated to the bicharacter

$$b^{\hbar}(q,p\,;\,q',p') = e^{\frac{i}{\hbar}(p.q'-q.p')}\,.$$

In the following we choose

$$\eta(q,p) = e^{-\frac{i}{2\hbar}q.p} \tag{2.11}$$

and denote by $\{W_{qp}\}$ the Weyl system associated with this choice of η. This choice of η leads to the Weyl system

$$W_{qp} = e^{-\frac{i}{\hbar}(p.Q-q.P)} \tag{2.12}$$

which verify

$$W_{qp}\,W_{q'p'} = e^{\frac{i}{2\hbar}(p.q'-q.p')}\,W_{q+q',p+p'} \tag{2.13}$$

$$W_{qp}^{-1} = W_{qp}^{+} = W_{-q-p} \tag{2.14}$$

Moreover the action of W_{qp} on $L^2(\mathbb{R}^n, dx)$ is given by

$$(W_{qp}\psi)(x)\, e^{-\frac{i}{2\hbar} p.(q-2x)} \psi(x-q) \tag{2.15}$$

The Weyl quantization procedure associates to a function f (classical observable) on the phase space an operator Q(f) on the Hilbert space of states $\mathcal{H}$ by

$$Q(f) = \frac{1}{(2\pi\,\hbar)^n} \int_{\mathbb{R}^{2n}} \tilde{f}(q,p)\, W_{qp}\, dqdp \tag{2.16}$$

where $\tilde{f}$ is the symplectic Fourier transform of f

$$\tilde{f}(q,p) = \frac{1}{(2\pi\,\hbar)^n} \int_{\mathbb{R}^{2n}} e^{-\frac{i}{\hbar}(p.q'-q.p')} f(q',p')dq'dp' \ . \tag{2.17}$$

Usually (2.16) is defined in the weak sense.
The relation (2.16) can be rewritten in terms of the observable f [8] [9] [10] as follows

$$Q(f) = \frac{1}{(\pi\hbar)^n} \int_{\mathbb{R}^{2n}} f(q,p)\, \Delta_{qp}\, dq\, dp \tag{2.18}$$

where Δ_{qp} is the Fano operator

$$\Delta_{qp} = \frac{1}{(4\pi\hbar)^n} \int_{\mathbb{R}^{2n}} e^{\frac{i}{\hbar}(p.q'-q.p')} W_{q'p'}\, dq'dp' \tag{2.19}$$

We can remark that for the choice (2.11)

$$\Delta_{oo} = \frac{1}{(4\pi\hbar)^n} \int_{\mathbb{R}^{2n}} W_{qp}\, dqdp = S \tag{2.20}$$

where S is the parity operator defined by

$$W_{qp}\, S = S\, W_{-q-p} \tag{2.21}$$

which acts on $L^2\,(R^n, dx)$ by

$$(S\psi)\,(x) = \psi(-x) \tag{2.22}$$

and we have the relations

$$\Delta_{qp} = W_{qp}\;\, S\, W_{-q-p} = W_{2q\;\,2p}\;\, S \tag{2.23}$$

Moreover Δ_{qp} is a unitary hermitic operator

$$\Delta_{qp}{}^{+} = \Delta_{qp} \tag{2.24}$$

$$\Delta_{qp}{}^{2} = 1 \tag{2.25}$$

The mean value of Q(f) in the state defined by the density matrix ρ take the form

$$<Q(f)> = T_r\,(\rho Q(f)) = \int_{\mathbb{R}^{2n}} f(q,p)\;\; \mathcal{W}\,(q,p)\; dq\; dp \tag{2.26}$$

where the function

$$\mathcal{W}(q,p) = T_r\,(\rho\; \Delta_{qp}) \tag{2.27}$$

is the Wigner function. This function which contains all information about the state of the quantum system is real and bounded, moreover

$$\int_{\mathbb{R}^{2n}} \mathcal{W}\,(q,p)\; dq\; dp = 1 \tag{2.28}$$

but it can assume negative values and cannot be considered as a probability density.
In the case of a pure state ψ, the density matrix is just the projector on normalized vector ψ and the Wigner function rewrites

$$\mathcal{W}_\psi(q,p) = \frac{1}{(2\pi)^n} \int_{\mathbb{R}^n} e^{-ip.x} \; \bar{\psi}(q-\frac{\hbar}{2}x)\,\psi\,(q+\frac{\hbar}{2}x)\,)dx \qquad (2.29)$$

and the "marginal distributions" verify

$$\int_{\mathbb{R}^n} \mathcal{W}_\psi(q,p)dp \; = \; |\psi(q)|^2 \qquad (2.30)$$

$$\int_{\mathbb{R}^n} \mathcal{W}_\psi(q,p)dq \; = \; |\,\mathcal{F}\psi(p)|^2 \qquad (2.31)$$

where $\mathcal{F}$ is a usual Fourier transform on $\mathbb{R}^n$.
The time evolution of Wigner function is associated to the evolution of the density matrix. If we assume that the Hamiltonian of the problem is such that $H = H_o + V$ we obtain easily the equation of the time evolution of the Wigner function :

$$\frac{\partial}{\partial t}\mathcal{W}_t(q,p) + \frac{p}{m}\nabla_q \mathcal{W}_t(q,p) + \int_{\mathbb{R}^{2n}} \mathcal{W}_t(q,p-u)J(u,q)\,du = 0 \qquad (2.32)$$

where

$$J(u,q) = \frac{1}{\hbar(2\pi)^n} \int_{\mathbb{R}^n} e^{-i\,u.v}\left[V\,(q+\frac{\hbar}{2}v) - V\,(q-\frac{\hbar}{2}v)\right]dv$$

Identifying this equation with the backward Kolmogorov equation of a stochastic process in an extended phase space leads to a probabilistic interpretation of the evolution of the Wigner function in terms of the mean value of a function (tied to the initial condition) of the stochastic process (which is state independent).
But to define a phase space stochastic mechanics "à la Nelson" we have to interpret this equation as a Fokker Planck equation (of a state dependent process) which is the aspect we would like to develop in the following special case of a quantum system with finite number of states.

3. WEYL SYSTEM AND WIGNER FUNCTIONS ON $\mathbb{Z}_N \times \mathbb{Z}_N$.

3.1 Some properties of the cyclic groups

Let us consider the abstract cyclic group of order N generated by a, $a^N = \varepsilon$ where ε

denotes the neutral element. This group is isomorphic to $\mathbb{Z}_N$, the additive group of integers modulo N. In the following we denote always by $\mathbb{Z}_N$ the cyclic group of order N. It is well known that the irreducible representation of $\mathbb{Z}_N$ are of dimension 1 and that there are N non equivalent unitary irreducible representations, the characters of $\mathbb{Z}_N$, denoted by

$$\chi_p(q) = e^{i \frac{2\pi}{N} q.p} \qquad q \in \mathbb{Z}_N , p \in \mathbb{Z}_N \tag{3.1}$$

The characters verify the orthogonality relation

$$\sum_{q \in \mathbb{Z}_N} \chi_p (q) \chi_{p'}(q) = N \delta_{pp'} \tag{3.2}$$

Moreover the characters form a group $\hat{\mathbb{Z}}_N$, the dual of $\mathbb{Z}_N$, isomorphic to $\mathbb{Z}_N$. Let us remark that χ_p defines an isomorphism from $\mathbb{Z}_N$ in the Nth roots of the identity only if p and N are prime numbers one respect to the other . Indeed,

<u>Proposition</u> [see eg 12]

Let $p \in \mathbb{Z}_N$, the mapping $\mathbb{Z}_N \to \mathbb{Z}_N$ defined by $q \to pq \bmod N$, $q \in \mathbb{Z}_N$, is an isomorphism if N and p are prime numbers one respect to the other.

Let us now consider the Hilbert space $\mathbb{C}^{\mathbb{Z}_N}$ isomorphic to $\mathbb{C}^N$, of the complex functions on $\mathbb{Z}_N$. The characters define an orthogonal basis of this space

$$\psi = \frac{1}{\sqrt{N}} \sum_{q \in \mathbb{Z}_N} (\chi_p , \psi)\chi_p \tag{3.3}$$

where (,) denote the ordinary scalar product in $\mathbb{C}^{\mathbb{Z}_N}$. Then we have the natural notion of Fourier transform

$$(\mathcal{F}\psi) (p) = \frac{1}{\sqrt{N}} \sum_{q \in \mathbb{Z}_N} e^{-i \frac{2\pi}{N} qp} \psi(q) \tag{3.4}$$

and

$$\psi(q) = \frac{1}{\sqrt{N}} \sum_{p\in \mathbb{Z}_N} e^{i\frac{2\pi}{N}qp} (\mathcal{F}\psi)(p) \tag{3.5}$$

$\mathcal{F}$ is an unitary operation on $\mathbb{C}^{\mathbb{Z}_N}$. Moreover

$$(\mathcal{F}\chi_p)(q) = \frac{1}{\sqrt{N}} \delta_p(q) \tag{3.6}$$

and

$$\mathcal{F}^4 = \mathbb{1} \tag{3.7}$$

Then $\mathcal{F}$ have, at most, four eigenvalues $e^{i\alpha\pi/2}$ $\alpha=(0,1,2,3)$. The projectors on eigenspaces associated to a value of α have the following expression

$$P_\alpha = \frac{1}{4} \sum_{\gamma=0}^{3} \left(e^{-i\alpha\frac{2\pi}{N}} \mathcal{F}\right)^\gamma \tag{3.8}$$

and the dimension of E_α is given by

$$\dim E_\alpha = T_r\, P_\alpha = \left[\frac{n+4}{4} - \frac{\alpha}{4}\left(\frac{19}{6} - \frac{3}{2}\alpha + \frac{1}{3}\alpha^2\right)\right] \tag{3.9}$$

where [.] denotes the integer part.

On $\mathbb{C}^{\mathbb{Z}_N}$ there exists a parity operator S define by

$$(S\psi)(q) = \psi(-q) \tag{3.10}$$

where -q denotes the inverse in $\mathbb{Z}_N$. Then S have the following properties

$$S^2 = 1 \tag{3.11}$$

$$S = S^+ = S^{-1} \tag{3.12}$$

Moreover

$$\mathrm{Tr}\, S = \begin{cases} 2 & \text{if } N = 2n \\ 1 & \text{if } N = 2n+1 \end{cases} \tag{3.13}$$

and

$$\mathrm{Det}\ S = (-1)^{\frac{(N-1)(N-2)}{2}} \tag{3.14}$$

To conclude this section let us recall that the Hilbert space $\mathbb{C}^{\mathbb{Z}_N}$ is isomorphic to the space of periodic complex functions on $\mathbb{Z}$ with period N.

3.2. Weyl system on $\mathbb{Z}_N \times \mathbb{Z}_N$

To study all possible Weyl systems we have to consider the second cohomology group Z_2 ($\mathbb{Z}_N \times \mathbb{Z}_N$, U(1)). However our aim in this lecture is to construct a Weyl system which in some way recalls the Weyl system of usual quantum mechanics. To do this a natural way is to define some equivalent formulation of the space and momentum translation (2,3) and (2,4). More precisely let U and V be two unitary operators in $\mathcal{L}(\mathbb{C}^{\mathbb{Z}_N})$

$$(U\psi)(x) = \psi(x-1) \quad x \in \mathbb{Z}_N \tag{3.15}$$

and

$$(V\psi)(x) = e^{i\frac{2\pi}{N}x}\ \psi(x) \quad x \in \mathbb{Z}_N \tag{3.16}$$

Then for q and p in $\mathbb{Z}_N$

$$(U^q\psi)(x) = \psi(x-q) \tag{3.17}$$

$$(V^p\psi)(x) = e^{i\frac{2\pi}{N}px}\ \psi(x) \tag{3.18}$$

Moreover

$$(U^q)^+ = U^{-q} \tag{3.19}$$

$$(V^p)^+ = V^{-p} \tag{3.20}$$

and we have the commutation relation

$$U^q V^p = e^{-i\frac{2\pi}{N}qp} V^p U^q \qquad (3.21)$$

Let us remark that U^q and V^p define dual representations of $\mathbb{Z}_N$

$$\mathcal{F} V^p = U^p \mathcal{F} \qquad (3.22)$$

$$\mathcal{F} U^q = V^{-q} \mathcal{F} \qquad (3.23)$$

where $\mathcal{F}$ is the Fourier transform on $\mathbb{Z}_N$.

To these unitary representations of $\mathbb{Z}_N$ we can associate the Weyl system $\{W_{qp}, (q,p) \in \mathbb{Z}_N \times \mathbb{Z}_N\}$, defined by

$$W_{qp} = \eta(q,p)\, V^p U^q \qquad (3.24)$$

with

$$|\eta(q,p)| = 1 \qquad (q,p) \in \mathbb{Z}_N \times \mathbb{Z}_N \qquad (3.25)$$

Then W_{qp} defines a unitary projective representation of $\mathbb{Z}_N \times \mathbb{Z}_N$

$$W_{qp}\, W_{q'p'} = \omega(q,p\,;\, q',p')\, W_{q+q'\ p+p'} \qquad (3.26)$$

with

$$\omega(q,p\,;\, q',p') = \eta(q,p)\, \eta(q',p')\, \overline{\eta}(q+q',p+p') \quad e^{-i\frac{2\pi}{N}qp} \qquad (3.27)$$

and

$$W^{-1}_{qp} = \overline{\eta}(q,p)\, \overline{\eta}(-q,-p) \quad e^{-i\frac{2\pi}{N}qp}\, W_{-q-p} \qquad (3.28)$$

$$W^{+}_{qp} = W^{-1}_{qp} \qquad (3.29)$$

Hence the commutation relations, in exponential form, are given by

$$W_{qp}\, W_{q'p'} = b(q,p;q',p')\; W_{q'p'}\, W_{qp'} \qquad (3.30)$$

where the bicharacter does not depend on η

$$b(q,p;q',p') = e^{i\frac{2\pi}{N}(pq'-qp')} \qquad (3.31)$$

In our approach by (3.16) we define a class of cohomology, by (3.24) we specify an element of this class. More precisely, choosing $(V\psi)(x) = \chi_k(x)\psi(x)$ with $k \neq 1$ we obtain other elements of the second cohomology group, but if k and N are prime numbers one respect to the other, the Weyl systems corresponding to different values of k are isomorphic.

Let us remark that the family W_{qp}, $(q,p)\in \mathbb{Z}_N \times \mathbb{Z}_N$ defines an orthonormal basis of $\mathcal{L}(\mathbb{C}^{\mathbb{Z}_N})$.

$$(W_{qp}|W_{q'p'}) = \frac{1}{N}\, T_r\,(W^{+}_{qp}\, W_{q'p'}) = \delta_q(q')\, \delta_p(p') \qquad (3.32)$$

Another orthonormal basis is given by the Fano operators Δ_{qp}

$$\Delta_{qp} = \frac{1}{N} \sum_{(q',p')\in \mathbb{Z}_N^2} e^{i\frac{2\pi}{N}(pq'-qp')}\, W_{q'p'} = W_{qp}\, \Delta_{oo}\, W^{+}_{qp} \qquad (3.33)$$

which verifies

$$(\Delta_{qp}|\Delta_{q'p'}) = \frac{1}{N}\, T_r\,(\Delta^{+}_{qp}\, \Delta_{qp}) = \delta_q(q')\, \delta_p(p') \qquad (3.34)$$

3.3. Wigner functions

As for the usual quantum mechanics, by Weyl quantization we associate to a function f on $\mathbb{Z}_N \times \mathbb{Z}_N$ an operator in $\mathcal{L}(\mathbb{C}^{\mathbb{Z}_N})$ by the procedure

$$Q(f) = \frac{1}{N} \sum_{(q,p)\in \mathbb{Z}_N^2} \tilde{f}(q,p)\, W_{qp} \qquad (3.35)$$

where $\tilde{f}$ is the "symplectic" Fourier transform of f :

$$\tilde{f}(q,p) = \frac{1}{N} \sum_{(q',p')\in \mathbb{Z}_N^2} e^{-i\frac{2\pi}{N}(pq'-qp')} f(q',p') \tag{3.36}$$

or in terms of Fano operators

$$Q(f) = \frac{1}{N} \sum_{(q,p)\in \mathbb{Z}_N^2} f(q,p)\ \Delta_{qp} \tag{3.37}$$

We want to associate hermitic operators to real functions, then η has to be such that

$$\eta(qp)\ \eta(-q,-p) = e^{-i\frac{2\pi}{N}qp} \tag{3.38}$$

Under such condition the Fano operators are hermitian

$$\Delta^{+}_{qp} = \Delta_{qp} \tag{3.39}$$

Then the Wigner function associated to the density matrix ρ

$$\mathcal{W}(q,p) = \frac{1}{N} T_r(\rho\Delta_{qp}) \tag{3.40}$$

is real and bounded, moreover

$$\sum_{(q,p)\in \mathbb{Z}_N^2} \mathcal{W}(q,p) = 1 \tag{3.41}$$

In case of a pure state ψ, under the conditions

$$\eta(o,o) = \eta(q,o) = \eta(o,p) = 1 \tag{3.42}$$

we have the "marginal destributions"

$$\sum_{p \in \mathbb{Z}_N} \mathcal{W}_\psi\ (q,p) = |\ \psi(q)|^2.$$

$$\sum_{q \in \mathbb{Z}_N} \mathcal{W}_\psi\ (q,p) = |\mathcal{F}\ \psi(p)|^2. \tag{3.43}$$

Now we can ask if it is possible to have bounds for the Wigner function, which are independent of the number of states. In the case where $N = 2s\text{-}1,\ s \in \mathbb{N}$, it is possible to choose

$$\eta(q,p) = e^{i \frac{2\pi}{2s-1} sqp} \tag{3.44}$$

For such a choice the Fano operators becomes unitary

$$\Delta_{qp}^{2} = 1 \tag{3.45}$$

as in the continuous case, and it is possible to choose a bound on $\mathcal{W}(q,p)$ which does not depend on N.

Let us remark that the choice (3.44) is possible because $\mathbb{Z}_{2s-1}$ is a divisible group as $\mathbb{R}^n$ that is for each $q \in \mathbb{Z}_N$ there exists $y \in \mathbb{Z}_{2s-1}$ such that $q = 2y$. This property is no longer verified in the case of $N = 2s$, $s \in \mathbb{N}$. Indeed in this case $\mathbb{Z}_{2s}$ is isomorphic to $\mathbb{Z}_2 \times \mathbb{Z}_s$ and $\mathbb{Z}_{2s}$ is not a divisible group. Moreover there exists for $\mathbb{Z}_2$ no hermitian and unitaryFano operators .

In order to conclude this section let us remark that if we can expect that the finite quantum system described by $\mathbb{Z}_{2s-1}$ provides an approximation of the usual quantum mechanics [12] it is certainly not the case for $\mathbb{Z}_{2s}$. Indeed the sequence of quantum systems described in terms of $\mathbb{Z}_{2}s$ converges towards the U.H.F. algebra which is associated to infinitely many countable spins one half on a lattice [see eg.13].

4. STOCHASTIC MECHANICS IN EXTENDED PHASE-SPACE

In this section we limit ourselves to the case of odd N, the case of even N can be developed in a similar way.

4.1. Dynamics

The dynamics of a quantum system associated to $\mathbb{Z}_N \times \mathbb{Z}_N$ is given by an Hamiltonian operator $H \in B(l_2(\mathbb{Z}_N))$. It defines the evolution of any density matrix ρ_t through the Heisenberg equation :

$$\frac{d}{dt}\rho_t = -i\,[H,\rho_t]\ . \tag{4.1}$$

The Hamiltonian H has a decomposition both in terms of Fano operators and Weyl operators

$$H = \sum_{(q,p)\in \mathbb{Z}_N^2} \mathcal{H}(q,p)\ \Delta_{qp} \tag{4.2}$$

and

$$H = \sum_{(q,p)\in \mathbb{Z}_N^2} \hat{\mathcal{H}}(q,p)\ W_{qp} \tag{4.3}$$

where $\hat{\mathcal{H}}_{(q,p)} = \tilde{\mathcal{H}}_{(2q,2p)}$. $\tilde{\mathcal{H}}$ is, as previously, the "sympletic" Fourier transform of $\mathcal{H}$. In order to define a conservative dynamics $\mathcal{H}$ has to be chosen real. In fact to avoid an unessential difficulty we take $\tilde{\mathcal{H}}$ positive in the following. Under this condition the evolution of Wigner function is given by the equation

$$\frac{d}{dt}\ \mathcal{W}_t(q,p) = \sum_{q',p'} i\ \hat{\mathcal{H}}\ (q',p')\ e^{-i\frac{2\pi}{N}s(pq'-qp')}\ \{\mathcal{W}_t(q+q',p+p') - \mathcal{W}_t(q-q',p-p')\} \tag{4.4}$$

Let us remark that this equation looks like the forward Kolmogorov equation of a jump process in phase-space but

i) transition probabilities could be negative

ii) $\mathcal{W}_t$ is not a probability density because it can assume negative values.

To avoid these problems let us introduce the function

$$g_t(q,p,\sigma) = \frac{1}{4N}\ \left(\frac{2}{N} + \sigma\ \mathcal{W}(q,p)\right) \tag{4.5}$$

where σ is a dichotomic variable.

Then $g_t(q,p,\sigma)$ is a positive real function on $\mathbb{Z}_N \times \mathbb{Z}_N \times \{-1,1\}$ such that

$$\frac{1}{4N^2} \leq g_t(q,p,\sigma) \leq \frac{3}{4N^2} \tag{4.6}$$

and

$$\sum_{\substack{(q,p)\in Z_N^2 \\ \sigma=\pm 1}} g_t(q,p,\sigma) \;\; f(q,p) \tag{4.7}$$

Moreover if f is a function on $\mathbb{Z}_N \times \mathbb{Z}_N$ then the quantum expectation of Q(f) is just

$$<Q(f)> = 2N \sum_{\substack{(q,p)\in Z_N^2 \\ \sigma=\pm 1}} \sigma \; g(q,p,\sigma) \;\; f(q,p) \tag{4.8}$$

Now we can interpret g as a probability density on the extended phase space $\mathbb{Z}_N \times \mathbb{Z}_N \times \{-1,1\}$. It is easy to see that $g_t(q,p,\sigma)$ verifies the same evolution equation than $\mathcal{W}_t(q,p)$ and we can rewrite this equation in the following form

$$\frac{d}{dt} g_t(q,p,\sigma) = \sum_{(q',p')\in Z_N^2} 2 \sin\left[\frac{4\pi}{N}(pq' - qp')\right] \hat{\mathcal{H}}(q'-q,\, p'-p) \;\; g_t(q',p',\sigma) \tag{4.9}$$

Using the same technics as in [14] we can transform this equation into a forward Kolmogorov equation which is of course state dependent. We have the following proposition [12] :

Proposition

There exists a stochastic Markov process X_t on $\mathbb{Z}_N \times \mathbb{Z}_N \times \{-1,1\}$ whose density is given at each time by $g_t(q,p,\sigma)$ and which verifies the forward Kolmogorov equation

$$\frac{d}{dt} g_t(q,p,\sigma) = \sum_{\substack{(q',p')\in Z_N^2 \\ \sigma'=\pm 1}} A_t\;(q,p,\sigma;q',p',\sigma') \;\; g_t(q',p',\sigma') \tag{4.10}$$

where

$$A_t\;(q,p,\sigma;q',p',\sigma) = -\frac{4}{g_t(q,p,\sigma)} \sum_{(q',p')\neq(0,0)} \hat{\mathcal{H}}\;(q',\, p')$$

$$A_t(q,p,\sigma;q,p,\sigma') = 0 \quad \text{if } \sigma \neq \sigma'$$

$$A_t\,(q,p,\sigma;q',p',\sigma') = 2\,\frac{1}{g_t(q',p',\sigma')} - \delta_{\sigma\sigma'}\,\sin\left[\frac{-4\pi}{N}(pq'-qp')\right]\,\hat{\mathcal{H}}\,(q-q',\,p-p')$$

$$\text{if } (q',p') \neq (q,p).$$

This completes what we wanted to show, how to construct a stochastic mechanics on a phase space, at least for a finite phase space.

REFERENCES

1. Nelson, E., "Dynamical Theories of Brownian Motion" Princeton University Press 1967.

2. Nelson, E., "Quantum Fluctuations", Princeton University Press, 1985.

3. Blanchard, Ph., Combe, Ph., Zeng, W., "Mathematical and Physical Aspect of Stochastic Mechanics". Lecture Notes in Physics, 281, Springer Verlag, 1987.

4. De Angelis, G.F., "A Route to Stochastic Mechanics" in "Stochastics Processes in Classical and Quantum Systems", Lecture Notes in Physics, 262, Springer Verlag, 1986.

5. Jaekel, M.T., Pignon, D., "Stochastic Mechanics of Mixed States". J. Phys. A 17 (1984), 131-140.

6. Manuceau, J., Sirugue, M., Testard, D., Verbeure, A., "The Smaller C*-algebra for Canonical Commutation Relations". Comm. Math. Phys. 32 (1973) 231-243.

7. Combe, Ph., Rodriguez, R., Sirugue-Collin, M., Sirugue, M., "A Uniqueness Theorem for Anticommutation Relations and Commutation Relations of Quantum Spin Systems". Comm. Math. Phys. 63 (1978), 219-235.

8. Fano, U., "Description of States in Quantum Mechanics by Density Matrix and Operator Techniques". Rev. Mod. Phys. 29 (1957) 79-93.

9. Grossman, A., "Parity Operators and Quantization of δ functions" Comm. Math. Phys. 48 (1976), 191-194.

10. Combe, Ph., Rodriguez, R., Sirugue, M., Sirugue-Collin, M. , "High Temperature Behaviour of Quantum Mechanical Thermal Functionals". Pub. Rims, Kyoto Univ. 19 (1983)355-365.

11. Combe, Ph., Guerra, F., Rodriguez, R., Sirugue, M., Sirugue-Collin, M., "Quantum Dynamical Time Evolutions as Stochastic Flows on Phase Space". Physica, 124A (1984) 561-574.

12. Cohendet, O., "Etude d'une Mécanique Stochastique dans l'espace des Phases". Thèse, Marseille, 1987.

13. Combe, Ph., Rodriguez, R. Sirugue-Collin, M., Sirugue M., "On the Quantization of Spin Systems and Fermi Systems", J.M.P. 20 (1979) 611-616.

EUCLIDEAN QUANTUM MECHANICS

J.C. Zambrini

Mathematics Institute
University of Warwick
Coventry CV4 7AL
ENGLAND

ABSTRACT

This is a short review of a new Euclidean approach to quantum phenomena, founded on a new probabilistic interpretation of the classical heat (or diffusion) equation.

1. INTRODUCTION

The Schrödinger equation on $L^2(\mathbb{R}^d)$, with appropriate initial condition χ, and for a scalar potential V,

$$i\frac{\partial\psi}{\partial t} = -\tfrac{1}{2}\Delta\psi + V\psi \equiv H\psi \qquad \psi(0) = \chi \tag{1}$$

has been formally interpreted by R. Feynman in the fifties [1] in a way suggesting the existence of an underlying configuration diffusion process. In particular, any solution of (1) was represented as a formal integral over the space of all paths, involving exclusively the action functional of the associated classical system. It has been proved, however, that no stochastic process with all the properties imposed by Feynman exists in the mathematical sense, in spite of the amazing efficiency of this approach, also in quantum field theory. The term "Euclidean Quantum Mechanics" was referring exclusively, till recently, to Kac's interpretation of Feynman's idea, involving the heat equation associated to (1) that I shall write in a slightly Byzantine way as

$$-\frac{\partial\eta^*}{\partial t} = -\tfrac{1}{2}\Delta\eta^* + V\eta^* \qquad \eta^*(0) = \chi \tag{2}$$
$$= H\eta^*$$

(* is *not* a complex conjugation. This one will be denoted by a line, $^-$)
Kac's analog of Feynman's path integral is a well defined mathematical object, underlying a well defined stochastic process, and is called *Feynman-Kac formula* [2]. This is the starting point of Euclidean field theory, initiated in the sixties by Schwinger, Nakano and Symanzik.

The success of this programme has been very partial, from the physical point of view, and a whole generation of mathematical physicists seems to have given up the hope to come closer to what Feynman did.

On the other hand, the theoretical physicists object that (2) is not a natural starting point for studying quantum phenomena because

a) The physical phenomena described by (2) are, qualitatively, completely different from quantum phenomena. For example they are irreversible in time.

b) In Feynman-Kac formula, the potential V is interpreted as a "Killing rate" for the Brownian trajectories. This has no physical meaning: V has the same interpretation in classical and quantum dynamics.

Theoretical physicists could be right. There is another "Euclidean" starting point for quantum physics, founded on a new probabilistic interpretation of (2) and such that the objections a) and b) disappear. This is what I call Euclidean Quantum Mechanics (EQM) and what I am going to summarize here.

2. BERNSTEIN PROCESSES

The usual approach involves the theory of Markovian (diffusion) processes X_t, i.e. of random variables on a probability space (Ω,σ_I,P), indexed by an interval $I = \left[-\frac{T}{2},\frac{T}{2}\right]$ and such that for any bounded measurable g and $0 < t$ in I,

$$E[g(X_t) \mid \mathcal{P}_0] = E[g(X_t) \mid X_0] \tag{3}$$

where $\mathcal{P}_0$ is the past filtration generated by X(s), for $s \leq 0$ in I. Historically, this Markov property (3) has been introduced as a probabilistic analogue of the classical causality principle according to which the knowledge of a single state in the past is sufficient to describe the future state. It follows that the data of an initial probability and of the (forward) transition probability is sufficient to obtain all consistent finite dimensional distributions and therefore, from the Kolmogorov theorem, the process X_t itself.

There is another, more subtle, concept of causality involved, for example, in the least action principle of classical mechanics. It is implicit in the following problem put forward by E. Schrödinger (1931–32) for the one-dimensional free (i.e. $V = 0$) heat equation (2) [3]: let $p_{-T/2}(x)\,dx$ and $p_{T/2}(y)\,dy$ two arbitrary probabilities. Is it possible to construct a unique probabilistic evolution of (free) diffusion particles for any t in I? S.N. Bernstein, the precursor of the theory of stochastic differential equations suggested that this problem is to investigate processes Z_t, t in I, such that, instead of (3),

$$E[g(Z_t) \mid \mathcal{P}_s \cup \mathcal{F}_u] = E[g(Z_t) \mid Z_s, Z_u] \tag{4}$$

holds for any $s < t < u$ in I, where $\mathcal{F}_u$ is the future filtration generated by $Z(v)$ for $v \geq u$ in I [3].

I call such processes "Bernstein processes" since he discovered the relevance of this "local Markov property" (4) for Schrödinger's problem and suggested how to construct them. Euclidean Quantum Mechanics is the theory founded on the solution of Schrödinger's problem.

For a Borel set B in $\mathbb{R}^d$, x,y in $\mathbb{R}^d$ and $s \leq t \leq u$ in I, a Bernstein transition $H(s,x\,;\,t,B\,;\,u,y)$ is such that $B \mapsto H(s,x\,;\,t,B\,;\,u,y)$ is a probability measure on the Borel sigma-algebra of $\mathbb{R}^d$, $\mathcal{B}(\mathbb{R}^d)$, and satisfies a consistency relation analog to the Chapman Kolmogorov equation. Suppose that a "joint" probability measure on $\mathcal{B}(\mathbb{R}^d) \times \mathcal{B}(\mathbb{R}^d)$, denoted by m, is given. Then we have the

<u>Theorem</u> (Jamison [4])

There is a unique probability measure P_m such that, with respect to (Ω, G_I, P_m), Z_t is a Bernstein process, for t in I, and

a) $P_m(Z_{-T/2} \in B_S, Z_{T/2} \in B_E) = m(B_S \times B_E)$, $B_S, B_E \in \mathcal{B}(\mathbb{R}^d)$

b) $P_m(Z_t \in B \mid Z_s, Z_u) = H(s, Z_s, t, B, u, Z_u)$

c) $P_m(Z_{-T/2} \in B_S\,,\, Z_{t1} \in B_1, \dots, Z_{tn} \in B_n\,,\, Z_{T/2} \in B_E) =$

$$\int_{B_S\times B_E} dm(x,y) \int_{B_1} H(-\frac{T}{2},x,t_1,dx_1,\frac{T}{2},y)\ldots \int_{B_n} H(t_{n-1},x_{n-1},t_n,dx_n,\frac{T}{2},y)$$

Now let $h(s,x,s+t,y) \equiv h(x,t,y)$ be the integral kernel of the semigroup associated with (2), e^{-tH}, as an integral operator on $L^2(\mathbb{R}^d)$, with V such that h is strictly positive (for ex. the harmonic potential or the Coulomb potential). Then

$$h(s,x,t,\xi,u,y) = \frac{h(s,x,t,\xi)\, h(t,\xi,u,y)}{h(s,x,u,y)} \tag{5}$$

is the density of a Bernstein transition (by the semigroup property). For an arbitrary choice of joint probability m, the Bernstein process Z_t resulting from the Theorem is generally not Markovian but a particular choice characterizes a Markovian Bernstein:

<u>Proposition.</u>

Let $H(s,x,t,B,u,y)$ be the Bernstein transition with density (5), and m a probability measure on $\mathcal{B}(\mathbb{R}^d) \times \mathcal{B}(\mathbb{R}^d)$. The resulting Bernstein process is also Markovian iff $m \equiv M$ is of the form

$$M(B_S \times B_E) = \int_{B_S\times B_E} \theta^*_{-T/2}(x)\, h(x,T,y)\theta_{T/2}(y)\, dxdy$$

for some bounded measurable functions $\theta^*_{-T/2}$ and $\theta_{T/2}$ on $\mathbb{R}^d$.

After substitution of $m = M$ and (5) in the finite dimensional distribution c) of the Theorem, we get, for $-\frac{T}{2}<t_1<\ldots<t_n<\frac{T}{2}$,

$$p_M(dx_1,t_1,\ldots,dx_n,t_n) = \int \theta^*_{-T/2}(x)\, h(x,t_1+\frac{T}{2},dx_1) \ldots h(dx_n,\frac{T}{2}-t_n,y)\, \theta_{T/2}(y)dxdy$$

$$= \int Q_*(t_1,dx_1,t_2,x_2) \ldots Q_*(t_n,dx_n,\frac{T}{2},y)p(y,\frac{T}{2})\, dy$$

$$= \int dx\, p(x,-\frac{T}{2})\, Q\,(-\frac{T}{2},x,t_1,dx_1) \ldots Q(t_{n-1},x_{n-1},t_n,dx_n)$$

where all the integrals are on $\mathbb{R}^d$ and the densities of Q_* and Q are, respectively,

$$q_*(s,x,t,y) = \frac{\int \mathcal{O}^*_{-T/2}(\xi)\, h(\xi, s+\frac{T}{2}, x)\, d\xi}{\int \mathcal{O}^*_{-T/2}(\xi)\, h(\xi, t+\frac{T}{2}, y)\, d\xi}\, h(x,t-s,y) \tag{6}$$

and

$$q(s,x,t,y) = h(x,t-s,y)\, \frac{\int h(y, \frac{T}{2}-t, \xi)\, \mathcal{O}_{T/2}(\xi)\, d\xi}{\int h(x, \frac{T}{2}-s, \xi)\, \mathcal{O}_{T/2}(\xi)\, d\xi} \tag{6'}$$

Also, the definitions

$$\begin{cases} p(y, \frac{T}{2}) = \mathcal{O}_{T/2}(y) \int \mathcal{O}^*_{-T/2}(x)\, h(x,T,y)\, dx \\ p(x, -\frac{T}{2}) = \mathcal{O}^*_{-T/2}(x) \int h(x,T,y)\, \mathcal{O}_{T/2}(y)\, dy \end{cases} \tag{7}$$

have been introduced.

Notice that q_* involves a solution of (2) at two different times in I, for an initial conditions $\mathcal{O}^*_{-T/2}$, and a solution of the time reversed equation $\frac{\partial \eta}{\partial t} = H\eta$ at two different times in I, for a final condition $\mathcal{O}_{T/2}$. Formally, $q*$ and q are the densities of the backward and forward transition probabilities for the Markov process $Z_t, t \in I$, and $p(y,\frac{T}{2})$, $p(x,-\frac{T}{2})$ are its final and initial probability densities. In Schrödinger's original problem, these two probability densities $p_{-T/2}$ and $p_{T/2}$ are given. So the condition (7) on the marginals of M can also be regarded as a system of equations for $\mathcal{O}^*_{-T/2}$ and $\mathcal{O}_{T/2}$, for $p_{-T/2}$, $p_{T/2}$ and the kernel h given. I call this the system of Schrödinger [3 and 5].

Theorem. (Beurling [4])

Let $p_{-T/2}(dx)$ and $p_{T/2}(dy)$ two probability measures on $\mathbb{R}^d$ with strictly positive densities. Let $h(x,T,y) \equiv h(x,y)$ be a bounded and strictly positive integral kernel on $\mathbb{R}^d \times \mathbb{R}^d$. Then positive (not necessarily integrable) solutions of Schrödinger's system exist and are unique.

If $\eta^*_\chi(x,t)$ denotes $(e^{-tH}\chi)(x)$ and $\eta_{\bar{\chi}}(y,t)$ is $(e^{tH}\bar{\chi})(y)$, for t in I and χ in $\mathcal{D}(e^{\frac{T}{2}H})$, where $\mathcal{D}(e^{\frac{T}{2}H})$ is a set, to be specified, of very regular vectors in $L^2(\mathbb{R}^d)$, it is clear that $\eta^*_\chi(x,-\frac{T}{2})$ and $\eta_{\bar{\chi}}(y,\frac{T}{2})$ solves Schrödinger's system by construction. When χ and $\bar{\chi}$ can be regarded as positive, the probability distribution of the associated Markovian Bernstein Z_t, $t \in I$, is

$$p_M(dx,t) = \eta^*_\chi(x,t)\ \eta_{\bar{\chi}}(x,t)\ dx \tag{8}$$

in striking analogy with Born interpretation in Quantum Mechanics. One shows easily that Z_t is a diffusion process, without killing, time reversible, with local characteristic (drift and diffusion coefficient) given by [5].

$$B_*(x,t) = -\frac{\nabla\eta^*_\chi}{\eta^*_\chi}(x,t) \text{ and } C_*(x,t) = \text{Id (the } d \times d \text{ identity matrix)} \tag{9}$$

Notice that the drift B_* can be regarded as a mean velocity, namely

$$D_*Z(s) \equiv \lim_{\Delta s \downarrow 0} E_s\left[\frac{Z(s) - Z(s-\Delta s)}{\Delta s}\right]$$

where E_s denotes $E[\ldots \mid \mathcal{P}_s]$. Then the following Least Action principle holds:

Theorem [5]

If $\bar{L}(\omega,\dot{\omega}) = \frac{1}{2}|\dot{\omega}|^2 + V(\omega)$ is the classical Euclidean Lagrangian of the system, a smooth Markovian Bernstein $Z(s) = Z(s)$, $s \in I$, solves the stochastic Newton equation

$$D_*D_*Z(s) = \nabla V(X(s))$$

and minimizes the action function with initial condition

$$J[Z(\cdot)] = E_t \int_{-T/2}^{t} \bar{L}(Z(S),D_*Z(S))ds + E_t A^*(Z(-\tfrac{T}{2}), -\tfrac{T}{2})$$

on the one-parameter family of diffusion processes Z^{ε} such that $D_* Z^{\varepsilon}(-\frac{T}{2}) = \nabla A^*(Z^{\varepsilon}(-\frac{T}{2}), -\frac{T}{2})$ and $Z^{\varepsilon}(t) = y$.

One of the nontrivial by-products of EQM is the development of a very natural probabilistic extension of the classical calculus of variations whose the last Theorem is an illustration. This idea is originated in Yasue [6].

3. HILBERT SPACE FORMULATION

If the self-adjoint Hamiltonian H of (1) is lower bounded, the analytical continuation $\tau \to it$ of the solution is possible by the functional calculus and, according to the Spectral theorem,

$$\eta^*_{\chi}(x,t) = (e^{-tH}\chi)(x) = \Big(\int_{-\infty}^{\infty} e^{-t\lambda}\, dE^H(\lambda)\chi\Big)(x) \tag{10}$$

solves (2). Let $\mathcal{D}(e^{\frac{T}{2}H})$ the set of analytic vectors for H [7] with radius $\frac{T}{2}$. For χ in $\mathcal{D}(e^{\frac{T}{2}H})$ and t in I,

$$\eta_{\chi}(x,t) = (e^{tH}\chi)(x) = \Big(\int_{-\infty}^{\infty} e^{t\lambda}\, dE^H(\lambda)\chi\Big)(x) \tag{11}$$

is also well defined. Moreover, one has

$$\int_{\mathbb{R}^d} \eta_{\bar{\chi}}\eta^*_{\chi}(x,t)dx = \|\chi\|_2^2.$$

A linear subspace of $L^2(\mathbb{R}^d)$ is defined for each t in I by

$$\tilde{\mathcal{V}}^*_t = \{\eta^*_{\chi}(t), \chi \text{ in } \mathcal{D}(e^{\frac{T}{2}H})\}.$$

If U_t^{-1} denotes the linear operator $U_t^{-1} : \tilde{\mathcal{V}}^*_t \to \mathcal{D}(e^{\frac{T}{2}H})$ such that $\eta^*_{\chi}(t) \mapsto \chi$,

$$(\eta^*\chi_1(t) \mid \eta^*\chi_2(t))_t \equiv \langle U_t^{-1}\eta^*\chi_1(t) U_t^{-1}\eta^*\chi_2(t)\rangle_2 = \langle \chi_1 \mid \chi_2\rangle_2 \tag{12}$$

is clearly a scalar product. The forward Hilbert space $\mathcal{V}^*_t$ is the completion of $\tilde{\mathcal{V}}^*_t$ with respect to $(.|.)_t$.

The position, momentum and Hamiltonian operators on $\mathcal{V}^*_o = L^2(\mathbb{R}^d)$ are defined by

$$Q\chi = x\chi\ ,\ P\chi = -\nabla\chi \text{ and } H\chi = -\tfrac{1}{2}\Delta\chi + V\chi \tag{13}$$

on the usual quantum mechanical domains.

Clearly, Q,iP and H are self-adjoint operators on $\mathcal{V}^*_o$. They can be defined as symmetric operators with self-adjoint extensions in $\mathcal{V}^*_t$ $t \neq 0$ in I, as well [5].

For any operator A in $\mathcal{V}^*_t$ with domain $\mathcal{D}_A \supset \mathcal{V}^*_t$ and appropriate χ,φ,

$$(\eta^*\chi(t) \mid A\, \eta^*\varphi(t))_t = \langle \chi \mid U_t^{-1} A\, U_t\, \varphi\rangle_2 \tag{14}$$

is a bilinear form. When it is closable, $A_H(t) = U_t^{-1} A\, U_t, t \in I$, denotes the associated operator on $\mathcal{D}(e^{\frac{T}{2}H})$. The expected value at time t of A corresponds to the case $\varphi = \chi$, namely,

$$\langle A\rangle_t = (e^{-tH}\chi \mid Ae^{-tH}\chi)_t = \langle \chi \mid A_H(t)\chi\rangle_2 \tag{15}$$

In the sense of quadratic form, the Euclidean Heisenberg equation clearly holds,

$$\frac{d}{dt}A_H(t) = [H, A_H(t)]. \tag{16}$$

For example, for the (Euclidean) Hamilton operator on $\mathcal{V}^*_o$, $H = -\frac{1}{2}P^2 + V(Q)$, using the definitions of $Q(t) = Q_H(t)$ and $P(t) = P_H(t)$ one gets the Hamiltonian equations

$$\frac{d}{dt}\, Q(t) = P(t)\quad ,\quad \frac{d}{dt}P(t) = \nabla V(Q(t)) \tag{17}$$

If the basic operators of the theory are regarded as observables, they have a natural probabilistic interpretation in terms of absolute expectations of the Bernstein Z_t, t in I,

$$\langle Q\rangle_t = E[Z(t)]\quad ,\quad \langle P\rangle_t = E[B_*(Z(t),t)]$$
$$\langle H\rangle_t = E[-\tfrac{1}{2}BB_*(Z(t),t) + V(Z(t))] \tag{18}$$

where B is the drift of the Bernstein with respect to the (growing) filtration $\mathcal{P}_t$.

4. COMPARISON WITH OTHER APPROACHES

There is an explicit procedure to associate a Bernstein process $Z_t, t \in I$, to any solution of the Schrödinger equation [5], so it is really possible to use them in concrete situations. The moments of these processes, for $t_1 < t_2 <\ldots < t_n$ in I, are given by

$$E[Z(t_1)\, Z(t_2)\ldots Z(t_n)] =$$

$$\int_{\mathbb{R}^d} \eta^*_\chi(x,-\tfrac{T}{2})\, h(x,t_1+\tfrac{T}{2},x_1)\, x_1\, h(x_1,t_2-t_1,x_2)\, x_2 \ldots x_n\, h(x_n,\tfrac{T}{2}-t_n,y)\, \eta_{\bar{\chi}}(y,\tfrac{T}{2})\; dx\, dx_1 \ldots dx_n\, dy \tag{19}$$

Setting $\tau = -it$ everywhere, this becomes, in terms of the integral kernel of the Schrödinger equation (1),

$$\int_{\mathbb{R}^d} \psi_\chi(x,-\tfrac{s}{2})\, e^{-i(\tau_1+\frac{s}{2})H}(x,x_1)\, x_1\, e^{-i(\tau_2-\tau_1)H}(x_1,x_2)\, x_2 \ldots x_n\, e^{-i(\frac{s}{2}-\tau_n)H}(x_n,y)\, \overline{\psi_\chi(y,\tfrac{s}{2})}\; dx\, dx_1 \ldots dx_n\, dy \tag{20}$$

This expression has been used formally by Feynman as the moments of a configuration diffusion process. Such process does not exist but (20) indeed describes quantum dynamics. This dynamical content is entirely preserved in (19), i.e. in Euclidean quantum mechanics.

If $\tau_1 = \tau_2 = .. = \tau_n$, however, (20) indeed reduces to the single time moments of a diffusion process, $E[X^n(\tau)]$. This is Nelson's process, whose properties have been thoroughly described by E. Carlen [8]. When several times are involved, the relation between Bernstein processes and Nelson processes is not so straightforward, because the dynamical structure of the associated theories is quite different.

5. CONCLUSIONS

One whispers sometimes that Quantum Field Theory requires new ideas to come out of the doldrums. Euclidean Quantum Mechanics should be regarded as an alternative starting point to investigate physical fields, and could be useful in this respect.

It is a pleasure to thank the organizing committee of the Karpacz Winter School for their kindness and their remarkable efficiency.

REFERENCES

1. Feynman, R.P., Rev. Mod. Phys, 20, 267 (1948).

2. Kac, M., Proc. of 2nd Berkeley Symp. on Probability and Statistics, J. Neyman, Ed, Berkeley, Univ. of California Press (1951).

3. Schrödinger, E., Ann. Inst. Henri Poincaré, 11, 300 (1932); Bernstein, S., "Sur les liaisons entre les grandeur aléatoires", Verh. des intern. Mathematikerkong., Zurich, Band 1 (1932)

4. Jamison, B. and Wahrscheinlich, Z., Gebiete, 30, 65 (1974); Beurling, A., Annals of Math., 72, 1, 189 (1960).

5. Zambrini, J.C., J. Math. Phys., 27, 9, 2307 (1986); Phys, Rev. A, 35, 9, 3631 (1987); with Albeverio, S. and Yasue, K., "Euclidean Quantum Mechanics: Analytical Approach", to appear.

6. Yasue, K., J. Funct. Anal., 41, 327 (1981).

7. Nelson, E., Ann. Math., 70, 572 (1959).

8. Nelson, E., "Dynamical theories of Brownian motion", Princeton, Princeton University Press (1967); "Quantum fluctuations", Princeton University Press (1985); Carlen, E., Comm. in Math. Physics, 94, 293 (1984).

SEMICLASSICAL (STOCHASTIC) QUANTUM MECHANICS

Piotr Garbaczewski[1]

Institute of Theoretical Physics, University
of Wrocław, PL-50-205 Wrocław, Poland

Dariusz Prorok[2]

Mathematics Department King's College,
Strand, London, United Kingdom

Summary:

We explore the probabilistic aspects of the quantized Coulomb-Kepler problem in the (extremally) semiclassical regime.

---

(1) Delivered by P.G.

(2) On leave from the ITP, University of Wrocław

1. Statement of purpose: Departing from the stochastic mechanics model for the condensation of planets out of a protosolar nebula one arrives at the Schrödinger equation for the central problem. To explain how elliptic Kepler trajectories arise in this setting, we investigate the semiclassical regime of appropriate wave packets. The stochastic mechanics motivations follow from [1-3,34-36,28,29], while the relevant semiclassical analysis is related to [4-33].

2. Paradigm case, the Coulomb-Kepler problem and Gaussians.

The Coulomb-Kepler problem both in its classical and quantum versions is a standard text-book companion of the harmonic oscillator in any descriptiom of basic physical concepts. It is a part of folk-lore that everything relevant has been said and nothing relevant can be added to the issue. The situation is however not that obvious.

As is well known[20] the Wilson-Sommerfeld quantization rule, in case of the Coulomb-Kepler problem incorporates contributions from both circular and elliptic orbits (although the circular orbits suffice for the understanding of the spectrum). It is however somehow overlooked that Schrödinger's claim "wave groups can be constructed which move around highly quantized Kepler ellipses and are the representation by wave mechanics of the hydrogen electron", until recently has not received an adequate attention. This gap in the understanding of the fundamental model was the reason of a renaissance in the study of semiclassical features of the quantum Coulomb-Kepler problem[21-34].

Our own interest in the subject can be motivated as follows:

(i) Nelson's stochastic mechanics links the time development of quantum states with this for the appropriate stochastic (diffusion) processes. A better understanding of the physical meaning of this stochasticity should be arrived at while passing to the semiclassical regime. The analysis of the issue is far from being complete, apart from the numerical study of[2,3]. Recently the correspondence limit of the sample path of Nelson's mechanics for the $\psi_{n,n-1,n-1}$ orbital was shown[34] to converge in the L^2 sense to a classical trajectory (circular Kepler orbit). There is a close relationship of this investigation with our own[28-30] search for elliptic Kepler orbits in the framework of stochastic mechanics.

From another standpoint[32,33] the classical limits of quantal probability distributions ($\hbar \to 0$ at constant energy) were studied and the corresponding classical ensembles introduced. In particular classical orbits for a statistical beam undergoing classical Coulomb scattering were derived in this way, see also[31].

(ii) The stochastic model for the orbits of planets and satellites[35,36,34] is based on the Schrödinger type equation for the central problem. Being in principle capable of providing the segregation of matter mechanisms, with respect to the implementation of realistic classical motions, it could be linked with the circular orbit prediction of[22] only.

However, one must be able to explain how stochastic mechanics leads to the coplanar elliptic orbits, consistent with Kepler's third law of planetary motion.

The discussion of the circular case can be found in[34] with the qualitative argument that the accretion of matter by the diffusing planetesimal makes the diffusion coefficient $\varepsilon^2=\hbar/m(t)$ to diminish with the growth of time. As a consequence the Nelson diffusion should converge to a Keplerian orbit.

(iii) Since Gaussian wave packets are particularly useful for the study of semiclassical features of the quantum motion (6-11,22,26-30), see also(37,38), we get confronted with the problem of how to reconcile the classical time development of coherent state labels

$$\dot{\alpha} = \{\alpha, H_c\} \qquad \dot{\bar{\alpha}} = \{\bar{\alpha}, H_c\}. \tag{2.1}$$

with the time dependence of coherent states. In general[6]

$$\begin{aligned} |\alpha,t) &= \exp(-\tfrac{i}{\hbar} Ht)|\alpha) \neq |\alpha(t)) \\ H_{cl} &= (\alpha|:H:|\alpha) \end{aligned} \tag{2.2}$$

where $\alpha(t)$ is determined by (2.1).
This problem we encountered before[28-30] but it is implicit as well in[27]
The comparison of the time evolution of coherent states to this a classical particle would have in the same potential, has been the subject of analytic estimates ($\hbar \to 0$ regime) in[7,10] while a numerical analysis of the issue was attempted in[38] in slightly different context. Since generally the coherent states of interest fail to obey the exact Schrödinger equation, the main goal of the recent paper[30] was to investigate the accuracy with which they can be viewed to <u>approximate</u> true solutions. Since we refer to the central problem, let us mention that we use the Gaussian states, although the $|x| \to 0$ singularity would apparently lead to difficulties. We are motivated by the semiclassical analysis of[21,22], and we account for the fact that the semi-

classical wave must be concentrated around a classical trajectory, which never crosses the origin. It is a reasonable working assumption to exclude the ball of the size of the nucleus surrounding the singularity of the potential from considerations.

In the next section, motivated by[30] we shall discuss this approximation in more detail. Albeit disregarding the spreading effects (which may be significant in the narrow tube along the Kepler orbit[22]) we find the approximation satisfactory in the proper parameter regime (mass and minimal distance from the central body, while on the orbit). We emphasize at this point that a mathematical correctness of the standard $\hbar \to 0$ prescription not always is physically clear : it is inevitable to identify proper physical quantities, with numerical values sufficiently large compared to $\hbar$ (which is kept fixed as a universal constant) to arrive at a physically sound picture.

The parameter implementing the semiclassical regime in below will not be $\hbar$ but the (large) mass parameter. In case of realistic masses spreading effects do not matter on time scales equating the age of the Universe .

3. Non-spreading ("frozen") coherent states as approximate solutions of the Schrödinger equation in case of the Coulomb-Kepler problem

Inserting the wave function

$$\psi(\vec{x},t) = \phi(\vec{x},t) \exp \frac{i}{\hbar} S(\vec{x},t) \tag{3.1}$$

to the equation

$$i\hbar \frac{\partial\psi}{\partial t} = \left(- \frac{\hbar^2}{2m} \Delta + V(\vec{x})\right)\psi \tag{3.2}$$

we arrive at:

$$\left[\frac{\partial S}{\partial t} + H(\frac{\partial S}{\partial x_i}, x_i, t)\right]\phi + (-i\hbar)\left[\frac{\partial \phi}{\partial t} + \frac{1}{m}\nabla S \nabla \phi + \frac{1}{2m}\phi \Delta S\right] - \frac{\hbar}{2m}\Delta\phi = 0 \tag{3.3}$$

Let us investigate what would have happen if we choose:

$$\phi(\vec{x},t) = (2\pi\sigma)^{-3/4} \exp\left[-\frac{1}{4\sigma}(\vec{x} - \vec{Q}(t))^2\right] \tag{3.4}$$

$$S(\vec{x},t) = S_o + \vec{P}(t)\cdot(\vec{x} - \vec{Q}(t) + $$

$$+ \int_0^t \{\vec{P}(\tau)\dot{\vec{Q}}(\tau) - H(\vec{P}(\tau),\vec{Q}(\tau))\}d\tau - \frac{3}{2}\hbar\omega t$$

where $\vec{Q}(t)$, $\vec{P}(t)$ are classical solutions of Hamilton equations generated by the Hamiltonian

$$H(\vec{Q},\vec{P}) = \frac{1}{2m}\vec{P}^2 + V(\vec{Q}) \tag{3.5}$$

We denote $\vec{Q}(t=0)=\vec{Q}_o$, $\vec{P}(t=0)=\vec{P}_o$.

(3.4) implies that:

$$\frac{\partial \phi}{\partial t} = \frac{1}{2\sigma}(\vec{x} - \vec{Q})\cdot\dot{\vec{Q}}\phi$$

$$\frac{\partial \phi}{\partial x_i} = -\frac{1}{2\sigma}(x_i - Q_i)\phi \tag{3.6}$$

$$\frac{\partial S}{\partial x_i} = P_i \qquad \frac{\partial^2 S}{\partial x_i^2} = 0$$

and because of $\dot{Q}_i = \frac{\partial H}{\partial P_i} = \frac{1}{m}P_i$ there holds:

$$\frac{\partial \phi}{\partial t} + \frac{1}{m}\nabla S\cdot\nabla\phi + \frac{1}{2m}\phi\Delta S = 0 \tag{3.7}$$

Therefore the imaginary part of (3.3) disappears.

Taking into account the formula:

$$\frac{\partial S}{\partial t} = \dot{\vec{P}}(\vec{x} - \vec{Q}) - \vec{P}\dot{\vec{Q}} + \vec{P}\dot{\vec{Q}} - H(\vec{P},\vec{Q},t) - \frac{3}{2}h\omega =$$

$$= \dot{\vec{P}}(\vec{x} - \vec{Q}) - H(\vec{P},\vec{Q},t) - \frac{3}{2}h\omega \qquad (3.8)$$

$$H\left(\frac{\partial S}{\partial x},x,t\right) = H(\vec{P},\vec{Q} + (\vec{x}-\vec{Q}),t) =$$

$$= H(\vec{P},\vec{Q},t) - \dot{\vec{P}}(\vec{x}-\vec{Q}) + O((\vec{x}-\vec{Q})^2)$$

where

$$O((\vec{x}-\vec{Q})^2) = V(\vec{x}) - V(\vec{Q}) - \sum_{i=1}^{3}\frac{\partial V(Q)}{\partial Q_i}(x_i-Q_i) =$$

$$= \frac{1}{2}\sum_{i,j=1}^{3}\frac{\partial^2 V(Q)}{\partial Q_i \partial Q_j}(x_i-Q_i)(x_j-Q_j) + \qquad (3.9)$$

$$+ \frac{1}{3!}\sum_{i,j,k=1}^{3}\frac{\partial^3 V(Q)}{\partial Q_i \partial Q_j \partial Q_k}(x_i-Q_i)(x_j-Q_j)(x_k-Q_k) +$$

$$+ \ldots$$

while

$$\Delta\phi = \frac{1}{2\sigma}\left(\frac{1}{2\sigma}(\vec{x} - \vec{Q})^2 - 3\right)\phi(\vec{x},t) \qquad (3.10)$$

we find that the left-hand-side of (3.3) produces the term

$$L(\vec{x},t) = \left[O((\vec{x}-\vec{Q})^2) - \frac{m\omega^2}{2}(\vec{x} - \vec{Q})^2\right]\phi(\vec{x},t) \qquad (3.11)$$

which is the measure of inaccuracy with which $\psi(\vec{x},t)$ solves the

Schrödinger equation.

Let us investigate the case of the central potential $V(x) = -\frac{A}{|\vec{x}|}$ where $A=Ze^2$ in case of Coulomb problem, while $A = \gamma mM$ for the Kepler case.

For all $|\vec{x}| \geqq a > 0$, a finite it is easy to verify that:

$$O((\vec{x} - \vec{Q})^2) = -\frac{A}{x}\frac{1}{Q^3}\left[x\vec{Q}(\vec{x} - \vec{Q}) - (x - Q)Q^2\right] \tag{3.12}$$

$$x = |\vec{x}| \qquad Q = |\vec{Q}|$$

Consequently (3.11) reads:

$$L(\vec{x},t) = -(2\pi\sigma)^{-3/4}\left\{\frac{A}{x}\frac{1}{Q^3}\left[x\vec{Q}(\vec{x}-\vec{Q}) - (x-Q)Q^2\right]\right. \tag{3.13}$$

$$\left. + \frac{1}{2}m\omega^2(\vec{x}-\vec{Q})^2\right\} \exp\left[-\frac{1}{4\sigma}(\vec{x} - \vec{Q})^2\right]$$

Our inaccuracy estimate will pertain to the semiclassical regime. For say $|\vec{x}-\vec{Q}| \leqq 0{,}1\ Q$ we have:

$$(\vec{x} - \vec{Q})^2 = x^2 + Q^2 - 2Qx\cos(\vec{x},\vec{Q}) \cong$$

$$\cong x^2 + Q^2 - 2Qx = (x - Q)^2 \tag{3.14}$$

so that:

$$L(\vec{x},t) \cong -(2\pi\sigma)^{-3/4}\left[\frac{A}{x}\frac{1}{Q^2}(x - Q)^2 + \frac{1}{2}m\omega^2(x - Q)^2\right]\cdot \tag{3.15}$$

$$\exp\left[-\frac{1}{4\sigma}(x - Q)^2\right]$$

Because the maximum of the function $(x-Q)^2 \exp\left[-\frac{1}{4\sigma}(x-Q)^2\right]$ at its points $x_{1,2}=Q\pm 2\sqrt{\sigma}$ equals $4\sigma e^{-1}$, upon $\sqrt{\sigma}<<Q$ we can write $x_{1,2} \cong Q$. Furthermore if $|x-Q| \cong 0,1\ Q$ then there holds

$$- L(\vec{x},t) \leq (2\pi\sigma)^{-3/4} \left(\frac{A}{x} 10^{-2} + \frac{1}{2} m\omega^2 Q^2 10^{-2} \right) \exp\left(- \frac{1}{4} Q^2 10^{-2} \right) \tag{3.16}$$

We are interested in the semiclassical features of the system, hence admitting Q to be of the macroscopic size we realize that (3.16) is very small (the exponential damping). Denoting $x=x_{1,2}$ we arrive at:

$$- L(\vec{x}_{1,2},t) \cong (2\pi\sigma)^{-3/4} \left(\frac{A}{Q^3} + \frac{1}{2} m\omega \right)^2 4\sigma e^{-1} =$$

$$= 4(2\pi)^{-3/4} e^{-1} \sigma^{1/4} \left(\frac{A}{Q^3} + \frac{1}{2} m\omega^2 \right) \tag{3.17}$$

The frequency ω is a free parameter, which can be appropriately adjusted. Minimizing (3.17) with respect to ω and accounting for $\sigma=\hbar/2m\omega$ we find:

$$u(\omega) = \frac{A}{Q^3} \omega^{-1/4} + \frac{1}{2} m\omega$$

$$u'(\omega) = \frac{1}{4} \omega^{3/4} \left(- \frac{A}{Q^3} \omega^{-2} + \frac{7}{2} m \right) = 0 \tag{3.18}$$

$$u''(\omega) = \frac{5}{16} \omega^{-9/4} \frac{A}{Q^3} + \frac{21}{32} m\,\omega^{-1/4} \geq 0$$

the minimum being reached at $\omega = \left(\frac{A}{m}\right)^{1/2} Q^{-3/2}$ when

$$- L(\vec{x}_{1,2},t) = \left(\frac{\hbar}{m}\right)^{1/4}\left[\frac{A}{Q^3}\left(\frac{A}{m}\right)^{-1/2} Q^{3/8} + \right.$$

$$\left. + \frac{1}{2}m\left(\frac{A}{m}\right)^{7/8} Q^{-21/8}\right] \cong \hbar^{1/4} m^{-1/8} A^{7/8} Q^{-21/8} \tag{3.19}$$

Hence

$$|L(\vec{x},t)| \leq h^{1/4}\, m^{-1/8}\, A^{7/8}\, Q^{-21/8} \tag{3.20}$$

If we consider $|x-Q|>>4\sigma$, then $L(\vec{x},t)$ is negligibly small in virtue of the damping factor $\exp\left[-\frac{1}{4\sigma}(x-Q)^2\right]$ n the above estimate ω depends on time through the classical trajectory $Q(t)$. To be a proper (time independent) frequency parameter it needs a proper fit of Q_{min} (we equate it to the major semi-axis of the ellipse) instead of $Q(t)$. Then $\omega = \left(\frac{A}{m}\right)^{1/2} Q_{min}^{-3/2}$, and the Kepler period for motions on elliptic orbits does arise. Now:

$$|L(\vec{x},t)| \leq \hbar^{1/4}\, m^{-1/8}\, A^{7/8}\, Q_{min}^{-21/8} \tag{3.21}$$

$$A_{Coul} = Ze^2 \qquad\qquad A_{Kepl} = \gamma m M$$

After accounting for the exponential term, we realize that our inaccuracy measure makes essential contributions in a close surrounding of the classical trajectory. It is amusing to observe that the estimate (3.21) gives account of the three regimes exploited in the study of the semiclassical features of the hydrogen problem. Namely the $\hbar^{1/4}$ factor refers to the $\hbar \to 0$ regime, the $m^{-1/8}$ factor refers to the $m \to \infty$ case, while $Q_{min}^{-21/8}$ factor is related to the $l,n \to \infty$ regime studied in[21,22].

Both in the Coulomb and Kepler case the combined effect of m large, Q_{min} large (while keeping the very small $\hbar$ fixed), is capable of giving rise to essentially classical features of the quantum motion.

Let us emphasize that apart from the discussed inaccuracy measure, it is the width of the wave packet $2\sqrt{2\sigma}$ which is a proper measure of how classical the quantum motion is.

Let us examine the meaning of our estimates for the realistic examples of the electrostatic and gravitational potentials. In case of the hydrogen atom, for the choice of the classical orbit (Rydberg atoms) $Q(t) \geqq 10^{-2}$m there holds

$$|L(\vec{x},t)| \leqq 10^{-24}[J/m^{3/2}] \ , \qquad \omega = 10^{4}[s^{-1}] \tag{3.22}$$

while the wave packet width in each of the coordinates

$$\Delta x_i = 2\sqrt{2\sigma} = 2 \cdot 10^{-4}[m] \tag{3.23}$$

proves that on a short time scale only (see e.g. at [22] for the analysis of spreading times, while at [24] for the description of Rydberg atoms and their life-times) the semiclassical picture is reliable.

Notice that the gravitational motion of the Earth around the Sun corresponds to the Gaussian with

$$|L(\vec{x},t)| \leqq 2 \cdot 10^{-2}[J/m^{3/4}] \ , \qquad \omega = 10^{-6}[s^{-1}] \tag{3.24}$$

where however:

$$\Delta x_i = 2\sqrt{2\sigma} \cong 10^{-26}[m] \tag{3.25}$$

to be compared with (3.23).

Consequently we are capable of constructing approximate solutions of the Schrödinger equation, which display basic features of the classical motion of the mass point representing the Earth around the point-like Sun.

The corresponding coherent state certainly does not reflect any realistic triggering (fluctuations) of the Earth trajectory, but rather an extremely narrow tube of nearby classical orbits.

Notice that the characterization of the pure Coulomb case

$$
\begin{aligned}
&|L(\vec{x},t)| \leq \hbar^{1/4}\, m^{-1/2}\, A^{7/8}\, Q_{min}^{-21/8} \\
&\sqrt{2\sigma} = \left(\frac{\hbar}{m\omega}\right)^{1/2} = h^{1/2}\, m^{-1/4}\, A^{-1/4}\, Q_{min}^{3/4} \quad , \; A = Ze^2
\end{aligned}
\tag{3.26}
$$

is different from this of the pure Kepler problem:

$$
\begin{aligned}
&|L(\vec{x},t)| \leq \hbar^{1/2}\, m^{3/4} (\gamma M)^{7/8}\, Q_{min}^{-21/8} \\
&\sqrt{2\sigma} = \hbar^{1/2}\, m^{-1/2} (\gamma M)^{-1/4}\, Q_{min}^{3/4}
\end{aligned}
\tag{3.27}
$$

It is amusing to observe a striking correlation between m and Q_{min} for planets of the solar system which is reflected by the numerical values of the upper bound for $|L(\vec{x},t)|$ and the width $2\sqrt{2\sigma}$.

These values read respectively:

Mercury	$\|L(\vec{x},t)\| \leq 2 \; 10^{-2}$	$\sqrt{2\sigma} =$	$2{,}2 \; 10^{-26}$
Venus	$4{,}5 \; 10^{-2}$		10^{-26}
Earth	$1{,}7 \; 10^{-2}$		10^{-26}
Mars	10^{-3}		$4.3 \; 10^{-26}$

Jupiter	$|L(x,t)| \leq 1{,}7\ 10^{-2}$	$1{,}9\ 10^{-27}$
Saturn	$1{,}44\ 10^{-3}$	$0{,}56\ 10^{-26}$
Uranus	$5{,}7\ 10^{-5}$	$2{,}4\ 10^{-26}$
Neptune	$2\ 10^{-5}$	$3\ 10^{-26}$
Pluto	$4{,}8\ 10^{-7}$	$4{,}7\ 10^{-27}$

Even in this extremal regime we still deal with Gaussians, and Nelson´s approach induces sample paths of the (classically controlled)Wiener noise.Albeit mathematically acceptable , this picture does not seem to be physically correct:it was the planetary dust which was the reason of randomness in the time evolution of the planetesimal. The accretion of matter cannot last indefinitely, since there is a finite amount of dust accessible. Moreover, with the time passing,the no-where differentiable sample trajectories should be in principle replaced by piecewise differentiable with a finite number of random disturbances in a finite time interval , which get eventually transformed into everywhere differentiable ones,when there is practically no dust around the planet.

This problem needs further investigations, see also (41). Would it amount to the jump process approximation of the diffusion process ?

References:

1.E.Nelson, Quantum fluctuations,Princeton Univ.Press,1985
2.K.Yasue,Computer stochastic mechanics, BiBoS 240-86
3.D.de Falco,E.Pulignano,Sample paths in stochastic mechanics,BiBoS 251-87
4.L.E.Ballentine,Rev.Mod.Phys.42,(1970),358
5.R.H.Young, Found.Phys.10, (1980), 33
6.C.L.Mehta,P.Chand, E.C.G.Sudarshan, Phys.Rev.157,(1967), 1198
7.G.A.Hagedorn,Commun.Math.Phys.71, (1980), 77
8.E.J.Heller, J.Chem.Phys.62, (1975), 1544

9.E.J.Heller, J.Chem.Phys.65, (1976),4979
10.S.L.Robinson, The semiclassical limit of quantum dynamics, Blacksburg preprints 1987
11.R.G.Littlejohn, Physics Reports,138, (1986), 193
12.P.A.M.Dirac, Quantum mechanics, Oxford Univ.Press,1958
13.R.D.Karlitz,D.A.Nicole, Ann.Phys (NY), 164,(1985), 411
14.M.V.Berry, N.L.Balazs, Am.J.Phys.47, (1979), 264
15.M.V.Berry, M.Tabor, Proc.Roy.Soc. A 349, (1976), 101
16.H.M.Bradford, Am.J.Phys. 44, (1976), 1058
17.W.H.Louisell, in: Quantum optics, ed.S.M.Kay,A.Maitlamd, Academic Press,London, 1970
18.M.Lax, Phys.Rev.172, (1968), 350
19.R.Azencott, H.Doss,in:Lecture Notes in Mathematics, vol 1109, Springer, Berlin, 1985
20.J.L.Powell. B.Crasemann, Quantum mechanics, Addison-Wesley, Reading, 1961
21.P.J.Brussard, H,A.Tolhoek,Physica, 23, (1957), 955
22.L.S.Brown, Am.J.Phys.41, (1973), 525
23.J.Mostowski,Lett.Math.Phys. 2, (1977), 1
24.D.Kleppner,M.G.Littman, M.L.Zimmermann, Sci.Am.,May 1981
25.D.R.Snider,Am.J.Phys.51, (1983), 801
26.C.C.Gerry, Phys.Rev. A33, (1986), 6
27.D.Bhaumik,B.Dutta-Roy, G.Ghosh,J.Phys. A 19, (1986),1355
28.P.Garbaczewski, Phys. Rev. D 33, (1986), 2916
29.P.Garbaczewski,D.Prorok, Fortschr.Phys,,35, (1987), 771
30.E.J.Heller, J.Chem.Phys.75, (1981), 2923
31.S.Qian, X.Huang, Phys.Lett. A 115, (1986), 319
32.E.G.P.Rowe, J.Phys. A 20, (1987), 1419
33.E.G.P.Rowe, Eur.J.Phys. 8, (1987), 81
34.R.Durran, A.Truman, Planetesimal diffusions, Swansea preprint, 1987
35.S.Albeverio, Ph.Blanchard, R.Hoegh-Krohn, in:Lecture Notes in Mathematics, vol.1109, Springer, Berlin, 1985
36.S.Albeverio ,Ph.Blanchard, R.Hoegh-Krohn,Exp.Math. 4, (1983), 365

37.J.R.Klauder, B.S.Skagerstam, Coherent states, World Scientific, Singapore, 1985

38.P.Garbaczewski, Classical and quantum field theory of exactly soluble nonlinear systems, World Scientific, Singapore, 1985

39.V.P.Gutschick, M.M.Nieto, Phys.Rev.D22, (1980), 403

40.C.C.Gerry, J.Kiefer, Radial coherent states for the Coulomb problem ,Phys.Rev A , to appear

41.A.D.Ventzel,M.I.Frejdlin,Fluctuations in the dynamical systems under the action of small random perturbations, (in Russian), Nauka, Moscow, 1979

ON REGULARITY OF MARKOV JUMP PROCESSES IN A POLISH SPACE AND SOME APPLICATIONS TO SCHRÖDINGER EQUATION

A.M. Chebotarev
Moscow Institute for Electronic Engeneering
109028 Moscow
USSR

ABSTRACT

In order to treat by means of stochastic processes quantum mechanical problems with vector-valued potential and variable metric we consider regular Markov jump processes with unbounded intensity of jumps, which is associated with Hamiltonian of the problem. Necessary and sufficient regularity conditions are obtained for abelian and non-abelian cases. Some applications are given for Schrödinger and Lindblad equations with unbounded interactions.

1. INTRODUCTION

Several years ago a class of absolutely continuous transforms of probability measures was proposed which contains measures representing solutions of some quantum mechanical problems as Lebesque integrals. From physical point of view such integrals are Feynman integrals in momentum representation. A number of publications on this subject characterize possible applications and generalizations of the method [1-7].

Absolute continuous transforms used in [1-7] are combinations of two types of transformations. The first one is the absolutely continuous transform associated with continuous parts of the process. This transform is well known both in mathematics and in physics and is given by Feynman -Kac expression. The second one is associated with jumps of the process. This transform was not very popular among mathematicians and was not used for physical studies up to 1978. It consists of multiplications of probability measure by unitary multiplicators depending on the jumps value and on the value of the process before it jumps. Such a transforms have a good sense if the jump process is regular. It means that the process performs almost surely any finite number of jumps. Regular Markov processes arised in [1-6] are of bounded intensity. The class of all regular processes is much wider. It contains in particular the processes associated with Hamiltonians of Schrödinger equation with vector-valued potentials and variable metrics. We shall present such application later and now we shall discuss the problem of regularity.

2. REGULARITY CONDITIONS FOR KOLMOGOROV-FELLER EQUATION

Let us consider a Markov jump process in a measurable Polish space $(\mathbf{X},\mathbf{B})$ performing with intensity $w(x)$ random jumps from point x, distributed in space $\mathbf{X}$ with probability $m_x(B)$. Here w is any positive measurable function and m_x is a measurable family of probability measures. In what follows $(w;m_x)$ will be called parameters of the jump process and we shall use the left continuous modification of this process in backward time. The set of all left continuous functions performing any finite number of jumps in a Polish space is measurable with respect on the standart σ-algebra generated by all cylindrical sets of functions. If the intensity of jumps is an unbounded function then it may happen that the process performs an infinite number of jumps on a finite time interval with a positive probability and all absolutely continuous transformations we need further will break down. So we have to exclude nonregular process from our considerations.

There exist two equivalent ways to obtain regularity conditions. The first uses conditional probability measure

P_x of the Markov chain with the same transition probability $m_q(B)$ starting from the point x. This one gives the stationary regularity conditions. The second gives nonstationary conditions and uses so called minimal solution of Kolmogorov-Feller equation for transition probability of this process

$$\frac{d}{dt} P(B|x,t) = w(x) \int_X m_x(dq)(P(B|x+q,t) - P(B|x,t)) \tag{2.1}$$

<u>Theorem 2.1.</u> If X is a Polish space then following conditions are equivalent:

1. Markov jump process with parameters (w,m_x) is regular

2. $\sup_N P_x(\sum_1^N (w(x_n))^{-1} \geq R) = 1$ for every $R>0$, $x \in X$.

3. $\inf_N M_x f(\sum_1^N (w(x_n))^{-1}) = 0$ for every $x \in X$ and any bounded measurable decreasing function f such that $\lim_{z\to\infty} f(z)=0$.

4. Minimal solution of (2.1) is the unique solution preserving complete probability.

5. $\inf_N R_N(x,t)=0$, for every $x \in X$, $t \in R_t$.

Here P_x and M_x are the conditional probability measure and the mathematical expectation for the Markov chain starting from the point $x \in X$, $\{x_n\}_1^\infty$ is the random sequence of Markov chain coordinates and $\{R_n\}_1^\infty$ is the recursive sequence of functions

$$R_1 \equiv 1,$$

$$R_{n+1}(x,t) = \int_{[0,t)} \int_X \exp(-w(x)\ (t-s))w(x)R_n(x+q,t)w_x(dq)ds.$$

The following lemma makes constructive the third condition mentioned above.

Lemma 2.2. Let f(r) be such bounded measurable decreasing function on R_+ that $F(z)=f(r^{-1/a})$ is a concave function for any positive number a. Then for any sequence of positive random variables $\{r_n\}_1^N$,which may be dependent, the following inequality for mathematical expectation value holds:

$$Mf(z_1^{-1}+ \dots +z_N^{-1}) \leq f((Mz_1^a)^{-1/a} + \dots + (Mz_N^a)^{-1/a}).$$

If we apply the third condition from Theorem 2.1. to function $f(z)=(1+r^a)^{-1}$, $a\in(0,1]$, then lemma 2.2. gives sufficient regularity condition.

Theorem 2.3. If $\sup_{a\in(0,1]} \left[\sum_1^\infty (M_x w^a(x_n)^{-1/a} \right]^a = \infty$, then the Markov jump process with parameters (w,m_x) in a Polish space is regular. Trivial necessary condition follows from the second condition of Theorem 2.1.

$$\sum_1^\infty M_x(w(x_n))^{-1} = \infty$$

Example 2.1. Finite dimensions

Jumps distribution in R^4	Upper bounds for intensities of regular process	Lower bounds for intensities of nonregular process
Normal	$C(1+\|x\|^2)\ln(2+\|x\|)$	$C(1+\|x\|^2)\ln^{1+a}(2+\|x\|)$
Cauchy	$C(1+\|x\|)\ln(2+\|x\|)$	$C(1+\|x\|)\ln^{1+a}(2+\|x\|)$

for any a>0. This example shows that the gap between the necessary and sufficient conditions seems to be narrow and contains only slowly growing multipliers.

Example 2.2. Sufficient regularity conditions in infinite dimensions.

Upper bounds for intensity in X	Upper bounds for moments of jumps distribution
$C(1+\|x\|)$	$\sup_x \int \|q\| m_x(dq) < \infty$
$C(1+(x,Ax))$	$\sup_x \int (q,Ax) m_x(dq) <\infty$, $\sup_x \int (x,Aq) m_x(dq) <\infty$ $\sup_x \int (q,Aq) m_x(dq) <\infty$

where $(x,Ax)\geq 0$ for every $x \in X$.

3. APPLICATIONS TO SCHRÖDINGER EQUATION

Let p_s, $s\in[0,t]$ be a left continuous path of a regular Markov jump process in backward time and $M_{p,t}$ be a conditional mathematical expectation with respect to all paths starting from the point $p \in X$ at a moment $s=t$. In what follows p_o denotes random point where the process terminates at moment $s=0$. Let us consider absolutely continuous transform of the probability measure of this process, which is generated by the pair of real measurable functions $H_o(p)$ and $Q(p,q)$:

$$\Psi(p,t) = M_{p,t} \exp(iA[p_s])\Psi_o(p_o) \tag{3.1}$$

where Ψ_o is a complex measurable bounded function on X_-, A - - a stochastic "action" in the momentum representation:

$$A[p_s] = \int_{[0,t]} (H_o(p_s)ds + \int_X (Q(p_s,q) + \frac{\pi}{2}) r(ds,dq|p_s))$$

Here r is the whole numbered random Poisson measure with the average value

$$Mr([s,u),B|p) = (u-s)w(p)m_p(B).$$

It is so called Lèvy measure of the process p_s, which gives integral expression for the number of the jumps with values in subset B on time interval [u,s) along the trajectory p_s:

$$N_{[u,s)}(B) = \int_{[u,s)} \int_B r(dv,dq|p_v).$$

It is an easy thing to verify that the function (3.1) satisfies Schrödinger-like equation with dissipative extra term.

$$\frac{d}{dt}\psi(p,t) = (iH - w)\psi(p,t) \tag{3.2}$$

where $H = H_o + H_{int}$, $(H_{int}\psi)(p) = w(p)\int e^{iQ(p,q)} m_p(dq)\psi(p+q)$ and $w\psi(p) = w(p)\psi(p)$. We can eliminate this dissipative extraterm on every mesurable set $S_L = \{p: w(p) \leqq L\}$ by means of the next absolute continuous transform:

$$\psi_L(p,t) = M_{p,t} \exp(iA_L[p_s])\,\psi_o(p_o) ,$$

where $iA_L[p_s] = iA[p_s] + \int_{[0,t)} w_L(p_s)ds$, and $w_L(p) = \min(L,w(p))$, $L>0$.

The Markov property of the process p_s and additivity of the exponential $A_L[p_s]$ let us to define one parameter semigroup of operators in the space of all bounded measurable complex functions with dissipative generator $iH_L = iH-(w-w_L)$. Operator of multiplication by difference $w(p)-w_L(p)$ appears here since we eliminate the dissipative term in (3.2) on S_L only.

Let us suppose that there exists a Hilbert space $h=h(X)$ of complex functions on X with inner product $(\cdot,\cdot)$ and let operators iH and iH_L are generators of strongly continuous one parameter semigroups $U(t)$ and $U_L(t)$. Then the following theorem holds.

<u>Theorem 3.1.</u> Let $D(H) \subseteq D(H_L) \subseteq D(W) \subseteq D(W-W_L)$ and $\|(W-W_L)\psi\|_h \to 0$ on $D(W)$ when $L\to\infty$. Then $\|U_L(t)\psi_o\|_h \uparrow \|U(t)\psi_o\|_h$ and $U_L(t) \underset{s}{\to} U(t)$.

This kind of convergence motivates Monte-Carlo computation algorithms in the case when ψ_o is a continuous function from $h=h(X)$ and $\phi \in h$ acts as integral with respect to any probability measure l:

$$(\psi,\phi) = \int \psi(q) e^{iQ_o(q)} \, l(dq).$$

<u>Theorem 3.2.</u> Let the conditions mentioned above hold, then

$$(\phi | e^{iHt} | \psi_o) = \lim_{N\to\infty} \frac{1}{N} \sum_{j=1}^{N} \exp(iQ_o(p^j) + iA_{L(N)}[p_s^j])\psi_o(p_o^j),$$

where p^j is random initial point of trajectory p_s^j, $s\in[0,t]$ of regular Markov jump process with parameters (w,m_p), $P(p^j \in B)=l(B)$ and $L(N)=0\,(\ln N)$ for fixed values of t.

We are finishing this section with examples of Hamiltonian functions for which the conditions of theorem 3.1. hold

<u>Example 3.1.</u>

Space h(X)	Hamiltonian function H(p,x)	Fourier structure of coefficients	Upper bounds for moments
$L_2(R^n)$	$\frac{1}{2}(p+A(x))^2+A_o(x)$ $A=(A_1,\dots,A_n)$	$A_j(x) =$ $=\int e^{ipx+iQ_j(p)} m_j(dp)$	$\int(1+\|p\|)m_j(dp)<\infty$ $0\leqq j\leqq n$
$L_2(R^1)$	$\frac{1}{2}(p+A_1(x))^2 G(x)+$ $+ A_o(x)$ $\inf_x G(x) > 0$	$(x) =$ $=\int e^{ipx+iQ_2(p)} m_2(dp)$	$(1+\|p\|^2)m_j(dp)<\infty$ $0\leqq j\leqq 2$

4. REGULARITY CONDITIONS FOR LINDBLAD EQUATION

Time dependent evolution of open quantum systems is described by Lindblad equation [10]. In Heisenberg picture this equation is similar to Kolmogorov-Feller equation:

$$\frac{d}{dt} X(t) = L[X(t)], \quad L[X]=P[X]-P[I]\circ X + i[H_1 X] \qquad (4.1)$$

where X is an observable, H-Hamiltonian operator of the quantum system, P - completely positive linear mapping which describes interaction between the quantum system and devices, $[\cdot,\cdot]$ is a commutator and $A\circ B=\frac{1}{2}(AB+BA)$-Jordan product. A mapping L of such type is called absolutely dissipative. One of the physical principles imposed on equation (4.1) is the identity conservation principles for one parameter semigroup of operators generated by mapping L. This condition is similar to regularity condition for Kolmogorov-Feller equation and as we will see later the sufficient regularity condition for (4.1) will be also similar to condition of Theorem 2.3.

Let again h be a Hilbert space with inner product $(\cdot,\cdot)$ and D a dense linear subspace in h, A be any C*-algebra of bounded operators in h, I-identity from A, A^+ be the set of positive operators from A. Let us denote by $F^s(A)$ and $F^s_D(A)$ the linear subspaces of strongly continuous linear mappings from A into A and from A into A_D correspondingly, where A_D is a linear topological space of linear operators from D into h. Strong topology in A_D is given by the system of seminorms $n_\psi(A)=\|A\psi\|_h$, $\psi\in D$. One can regard A_D as completion of A in this topology.

Completely dissipative operator L will be called regular if one parameter strongly continuous semigroup of operators generated by L preserve unity. Regularity of L is rather simple for bounded interactions.

Let operators H and P[I] are hermitian unbounded operators and

$$D = D(H) \subseteq D(P[I]) \subseteq D(P[X]) \qquad (4.2)$$

for any $X \in A$, $0 \leq X \leq I$ and let operators

$$G = iH + \frac{1}{2}P[I] \quad , \quad G^* = -iH + \frac{1}{2}P[I]$$

are closable in h and for any $c>0$ operators $G+cI$, G^*+cI have dense ranges in h:

$$\overline{R(G+cI)} = \overline{R(G^*+cI)} = h \, . \qquad (4.3)$$

In this case operator G and G^* generate one parameter strongly continuous semigroups of operators in h which will be denoted $W_t = \exp(-Gt)$ and $W_t^* = \exp(-G^*t)$.

Theorem 4.1. If the hypotheses (4.2-4.3) hold, then the following statements are equivalent:

1 Completely dissipative operator L is regular.

2. There exists the unique completely positive strongly continuous dynamical semigroup of normal operators generated by L and preserving the unity

3. $\inf_N \langle\psi, R_t^{(N)}\psi\rangle = 0$ for every $\psi \in D$ and all $t \in R_+$, where

$$R_t^{(0)} = I \; , \; R_t^{(n+1)} = \int_{[0,t]} ds\, W_{t-s}^* P[R_s^{(n)}] W_{t-s} \, .$$

The last condition become constructive due to the following lemma which generalizes inequality of lemma 2.2 in nonabelian case when $a=1$, $f(r) = (1+r)^{-1}$.

Lemma 4.2. Let $r_o = I \in A$ and Q-completely positive mapping from $F(A)$, $\{r_i\}_1^n$ and $\{Q[r_i]\}_o^n$ are positive bounded operators in h and $\{r_i^{-1}\}_1^n$, $\{Q[r_i]^{-1}\}_o^n$ exist ,then

$$Q[(\sum_0^n r_i^{-1})^{-1}] \leq (\sum_0^n (Q[r_i])^{-1})^{-1} \, .$$

The third regularity condition and inequality of this lemma lead to the sufficient regularity condition.

Theorem 4.3. Completely dissipative operator L is regular if

$$\sup_{c\in R_+, N>0} e^{-ct} \sup_{a\in[0,1]} \sum_{n=0}^{N} \frac{1}{\langle\psi, Q_c^n[B_{cN}^{(a)}]\psi\rangle} = \infty$$

for every $\psi \in D$, where $Q_c^n[X]=\underbrace{Q_c[\ldots Q_c}_{n}[X]\ldots]$,

$$Q_c[X] = \int_0^\infty e^{-ct} W_t^* P[X] W_t dt \ , \quad A_c[X] = \int_0^\infty e^{-ct} W_t^* W_t dt$$

$$B_c = A_c^{-1} - cI, \quad B_{cN}^{(a)} = (N+1) B_c ((N+1) + aB_c)^{-1} \in A$$

Acknowledgments

I would like to thank Professor S.Albeverio and Members of Organizing Committee of the 24th Winter School of Theoretical Physics in Karpacz, for inviting me to talk in "Stochastic Methods in Mathematics and Physics".

REFERENCES

1. Chebotarev,A.M. Mat.Zametki (in Russian),vol.24,No 5,pp. 699-705(1978)

2. Chebotarev,A.M. Maslov,V.P. Lecture Notes in Physics, vol.106,pp.58-72,Springer Verlag (1979)

3. Combe,Ph.,Rodriguez,R.,Høegh-Krohn,R.,Sirugue M., Sirugue-Collin M., Phys.Reports, vol.77,No 3,pp.221-233(1987)

4 Combe,Ph.,Høegh-Krohn,R.,Rodriguez,R.,Sirugue,M., Sirugue-Collin,M., Journal of Math.Phys.,vol.23,No 3, pp. 405-411,(1982)

5. Combe,Ph.,Rodriguez,R.,Sirugue,M., Sirugue-Collin,M., Publications of RIMS Kyoto Univ.vol.19,No 1,pp.355-365 (1983)

6. Bertrand,J.,Gaveau,B.,Rideau,G. Lecture Notes in Math. vol.1109,pp.74-80, Springer Verlag (1985)

7. Blanchard,Ph.,Combe,Ph.,Sirugue,M.,Sirugue-Collin,M., Lecture Notes in Math.,vol.1136,pp.104-111;Springer Verlag,(1985)

8. Chebotarev,A.M.,Teoria verojatnostei i primenenija (in Russian) vol.33,No 1,pp.25-39(1988)

9. Chebotarev,A.M.,Dokladi Akademii Nauk SSSR (in Russian) vol.298,No 2,pp.51-55,(1988)

10.Lindblad,G., Commun.Math.Phys.,vol.48,pp.119-130,(1976)

STATISTICAL MECHANICS

CHALLENGING THE CONVENTIONAL WISDOM ON FERROMAGNETS AND LATTICE GAUGE THEORIES*

E. SEILER

Max-Planck-Institut für Physik und Astrophysik
– Werner-Heisenberg-Institut für Physik –
P.O.Box 40 12 12, Munich (Fed. Rep. Germany)

ABSTRACT

It is generally taken for granted that two dimensional $O(N)$ spin models and nonabelian lattice gauge models in four-dimensions share the properties of asymptotic freedom and spontaneous mass generation. A simple energy-entropy principle that is quite successful in predicting known phase transitions, predicts, however, that these models also show phase transitions, presumably leading to massless low temperature phases. We examine the numerical evidence for the conventional picture and find it inconclusive; furthermore, we report on our own simulations that support our "heretic" view.

---

*Extended version of a lecture given at the Karpacz Winter School, January 1988

1. Introduction

Over the years a certain consensus has developed in the physics community concerning certain systems of equilibrium statistical mechanics. These systems include two-dimensional continuous spin ferromagnets and four-dimensional lattice gauge theories. According to this "conventional wisdom" there is a drastic difference in these systems depending on whether their symmetry group is abelian or non-abelian (without any continuous center).

While it is rigorously known through the works of [1] and Fröhlich & Spencer [2, 3] that the abelian models, namely the two-dimensional plane rotator and four-dimensional (compact) lattice QED ($CQED_4$) undergo a phase transition from a massive, i.e. exponentially clustering, strong coupling phase to a "soft", massless (and in the case of $CQED_4$ nonconfining) phase at weak coupling, the common belief is that the nonabelian models remain massive (and confining in the case of $4D$ lattice gauge theory) at all couplings.

This second part of the conventional wisdom has resisted all attempts at proving it (several years ago an announcement of a proof appeared [4], but by now I think the statute of limitations should be applied to it).

Patrascioiu, Stamatescu and the author have challenged this picture in [5], based on very simple arguments and in [6] they presented an extensive numerical study of some two-dimensional spin models to test their conclusions. In this lecture I want to describe briefly the ideas presented in [5], some tests of the heuristic principle that is formulated there, our own numerical studies contained in [6] as well as preliminary results of an investigation of four-dimensional lattice gauge models that is under way [7].

2. The Heuristic Principle

All the models we are concerned with here share the property that they can be represented as gases of certain objects called polymers that interact by a hard core repulsion and are dilute either at strong or at weak coupling (high or low temperature). These polymer representations are the basis of cluster expansions that converge in the respective dilute regimes.

The simplest example is, not surprisingly, the Ising model. At low temperatures it can be represented as a dilute gas of "Peierls contours" or domain walls between subsets of different spin orientation. By using Kramers-Wannier duality one obtains a dual representation of the Ising model as a gas of dual domain walls (in 2D) or defects (in general) that are dilute at high temperature. These dual domain walls can also be related to the polymers of the well-known high temperature expansion of the Ising model. To avoid misunderstanding we should stress here that these polymer representations are not limited to the domain of convergence of the respective expansion; for instance one can describe any configuration of the Ising model by giving the domain walls (and one spin value).

Our central hypothesis is that these polymer systems undergo a phase transition, which we dubbed "polymerization" for lack of a better term, once the activities of the polymers do not decrease sufficiently with their size. This phase transition should be closely related to percolation (formation of infinite polymers), and in $2D$ we may identify the notions of percolation and polymerization. In $D \geq 3$, however, an infinite cluster (think of a domain wall of the Ising model) can be an essentially one-dimensional tube-like object; if we look at a two-dimensional slice we would see a finite object, or possibly many such objects, like the holes in a Swiss cheese. If in the Ising model we think of the holes of the Swiss cheese as filled by "down" spins and the complement by "up" spins we would still have a net magnetization in the "up" direction. What we mean by polymerization is the formation of a truly space-filling, sponge-like cluster that is characterized by the property that any two-dimensional slice of it is still percolating. I think that such a sponge should also be duality invariant in the following sense: If we replace the p-cells making up the polymer by their dual $(d-p)$-cells we should still get an object that shows percolation in all $2D$ slices.

The origin of these ideas is, of course, an attempt to "invert" the Peierls argument [8] that shows that in the low temperature phase the domain walls of an Ising model are both small and sparse. This is so because a domain wall of size N has an entropy factor of the order C^N (fixing one point) and an energy (Gibbs) factor $e^{-2\beta N}$ where β is the inverse temperature. Increasing the temperature will increase the average density and size of the domain walls, until at a certain $\beta = \beta_p$ percolation occurs (an infinite domain wall forms). In $2D$ β_p is equal to the critical point β_c of the Ising model at which the spontaneous magnetization disappears. In $D \geq 3$ the situation is expected to be more complicated as explained above; in fact it has been shown by Aizenman, Bricmont and Lebowitz [9] that in large D $\beta_p < \beta_c$. According to our heuristic picture β_c should be the polymerization threshold at which a space-filling "sponge" of domain walls is formed that shows percolation also in each 2-dimensional slice. The energy-entropy balance for polymerization should have the same form as for percolation, but we expect in $D \geq 3$ that the entropy per size of a "sponge", due to the stronger connectivity requirements, is smaller than for general polymers. This explains why $\beta_p < \beta_c$ in $d \geq 3$.

It is clear how this picture generalizes to the low temperature (weak coupling) phase and its disappearance in any discrete spin model. The polymers are simply the domain walls separating the different spin orientations.

Likewise the picture can be applied to discrete lattice gauge models: The polymers are now co-connected sets of plaquettes p (=2-cells) that are excited, i.e. for which the ordered product $\prod_{\langle xy\rangle \epsilon \partial p} g_{xy}$ differs from the unit element of the discrete group G. These polymers (called defect networks in [10] are again small and dilute at low temperature (weak coupling) and should undergo a polymerization transition once the temperature (coupling constant) reaches a certain critical value.

There is a different polymer representation of these spin and gauge models that is normally used as a starting point of the high temperature expansion (and related to the defects of the dual model): For each bond b (nearest neighbor link for ferromagnets, plaquette for gauge models) there is a Gibbs factor

$$e^{-\beta S_b} \tag{2.1}$$

which can be written as

$$e^{-\beta S_b} = 1 + \rho_b \tag{2.2}$$

We assume that S_b is normalized in such a way that the average of the Gibbs factor in the a priori measure is 1; this has the following group theoretic meaning: If we expand the Gibbs factor according to irreducible representations of the symmetry group, ρ_b is the sum of all terms corresponding to nontrivial representations.

Expanding the full Gibbs factor

$$\prod_b e^{-\beta S_b} = \prod_b (1 + \rho_b) \tag{2.3}$$

in the ρ's and integrating over the basic variables with the a priori measure produces again a polymer system (see for instance [10]), the polymers consisting of connected sets of bonds (links, plaquettes).

For small β large polymers will have an activity decreasing exponentially with their size. A closer look reveals that the activity of large polymers will in fact be essentially determined by the largest coefficient of the expansion of ρ_b according to irreducible representations of the symmetry group (see below).

According to our heuristic principle there should again be a "critical" β_c at which this system undergoes polymerization and this β_c should mark the end of the high temperature phase. It is not excluded that β_c lies outside the domain of convergence of the high temperature expansion: The polymers exist not only inside this domain of convergence just as the molecules of a gas also exist outside the domain of convergence of the virial expansion. The rewriting of the system as a system of polymers sketched here is valid for any β and should be viewed as the appropriate replacement of Kramers-Wannier duality for nonabelian systems.

We end this section by remarking that our heuristic principle passes the simplest test: There is no polymerization of polymers made of links in $D = 1$ or of plaquettes in $D = 2$, in accordance with the fact that $1D$ (short range) spin models and $2D$ gauge models have no phase transition; their high temperature expansion, however, has a finite radius of convergence.

3. Testing the Heuristic Principle (I): Abelian Systems

In general it is difficult to obtain precise estimates of the entropy of large polymers. We can, however, use the known critical temperatures of some models to estimate the critical temperatures of other models that have a similar polymer structure but different activities. Some such estimates are given in [5]. While not being rigorous, they are in general surprisingly successful numerically.

Consider first the $Z(N)$ clock model described by the action

$$S = \sum_{\langle xy \rangle} (1 - \cos(\phi_x - \phi_y)) \tag{3.1}$$

where ϕ_x is an angle taking the values $\frac{2\pi n}{N}$, $n = 0, 1, \ldots, N-1$ and the sum is over the nearest neighbors $\langle xy \rangle$ of a square or simple cubic (hypercubic) lattice. As discussed above, at low temperatures we have a dilute gas of domain walls much like in the Ising model (which is the special case $N = 2$). The predominant domain walls are the ones where the angle jumps by one unit $\frac{2\pi}{N}$; such a domain wall of length (hypersurface area) K has a Gibbs factor

$$e^{-\beta K(1-\cos\frac{2\pi}{N})} \tag{3.2}$$

In general, there is the possibility of branching of domain walls (that is absent in the Ising model) which gives the domain walls an increased effective entropy per size. So we predict a critical $\beta_M(N)$ for the "melting" of the ordered low temperature phase that satisfies

$$\beta_M(N) \geq \beta_M(2) \frac{2}{1 - \cos\frac{2\pi}{N}} \tag{3.3}$$

where $\beta_M(2)$ is the β_c of the Ising model.

The situation is very similar for $Z(N)$ gauge models described by the action

$$S = \sum_p (1 - \cos \phi_{\partial p}) \tag{3.4}$$

where ϕ_{xy} is now an angle taking one of the values $\frac{2\pi u}{N}, n = 0, 1 \ldots, N-1$ for each link $\langle xy \rangle$ and $\phi_{\partial p}$ is the sum (mod 2π) of these angles around the boundary of a plaquette p.

We predict that the critical $\beta_M(N)$ for the "melting" of the "frozen" low temperature phase again satisfies (3.3), where $\beta_M(2)$ now of course refers to the $Z(2)$ gauge model. In $D = 3$ this model is dual to the Ising model which has $\beta_c \sim .22$, which implies

$$\begin{aligned} \beta_M(2) &= -\frac{1}{2} \, ln \; th \; .22 \\ &= .76 \end{aligned} \tag{3.5}$$

Bhanot and Creutz [11] determined $\beta_M(N)$ for $3D$ gauge models and N up to 20; their data show that (3.3) is satisfied as a near equality (they fit their data with $\beta_M(N) = .75 \frac{2}{1-\cos\frac{2\pi}{N}}$.

Let us now return to $Z(N)$ spin models in $2D$. These models are essentially self-dual, and we can exploit duality to estimate the end of the high temperature phase by our heuristic principle. Together with (3.3) this yields a lot of information that can be checked.

The model dual to the $Z(N)$ clock model in $2D$ is again a $Z(N)$ spin model, but with a more complicated Gibbs factor

$$g_N(\phi_x - \phi_y, \beta) = \frac{\sum_{l=0}^{N-1} e^{il(\phi_x - \phi_y)} \; e^{-\beta(1-\cos\frac{2\pi l}{N})}}{\sum_{l=0}^{N-1} \; e^{-\beta(1-\cos\frac{2\pi l}{N})}} \tag{3.6}$$

$$(\phi_x, \phi_y = \frac{2\pi n}{N}, \quad n = 0, 1, \ldots, N-1)$$

By the same argument that gave (3.3) we obtain then that the high temperature phase (low temperature phase of the dual model) should "melt" at

$$\tilde{\beta}_M(N) \leq b(N) \tag{3.7}$$

where $b(N)$ is given by

$$\begin{aligned} g_N(\frac{2\pi}{N}, b) &= e^{-2\beta_c(2)} = th\beta_c(2) \\ &= \sqrt{2} - 1 \end{aligned} \tag{3.8}$$

Let us look at the implications of (3.3) and (3.7) for various N.

a) $N = 3$: This is the Potts model that is known to have only one transition at the selfdual point

$$\begin{aligned} \beta_M(3) = \tilde{\beta}_M(3) &= \frac{2}{3} ln \frac{\sqrt{3}-1}{2} \\ &= .67 \end{aligned} \tag{3.9}$$

This satisfies both (3.3) and (3.7)

b) $N = 4$: It is known that the $Z(4)$ clock model at β is equivalent to two uncoupled Ising models at $\frac{\beta}{2}$ [12]. Therefore, it has again only one transition at

$$\begin{aligned} \beta_c(4) &= 2\beta_c(2) = -\ln(\sqrt{2} - 1) \\ &= .88 \end{aligned} \tag{3.10}$$

The Gibbs factor (3.6) is

(3.11)

which satisfies the identity

$$g_4(\frac{2\pi}{4}, \beta) = g_2(\frac{2\pi}{2}, \frac{\beta}{2}) \tag{3.12}$$

Both (3.3) and (3.7) are satisfied as equalities.

c) $N \geq 5$: Here the inequalities (3.3) and (3.7) imply

$$\tilde{\beta}_M(N) < \beta_M(N) \tag{3.13}$$

so there are at least two transitions in accordance with the predictions of Elitzur, Pearson and Shigemitsu [13] and the results of Alcaraz and Köberle [14]. For $N = 5$ the lower bound (3.3) is about 1.27 and the upper bound (3.7) about 0.9. For $N \to \infty$ we obtain the plane rotator $O(2)$ model; the upper bound (3.3) diverges like N^2, whereas the lower bound goes to a finite value in accordance with having only one Kosterlitz-Thouless transition in the $O(2)$ model near $\beta = 1$. (It is known rigorously that $\beta_M(\infty) \geq 2\beta(2)$ [15].)

All this is in agreement with the known phase structure of these models, including the rigorous results of Fröhlich and Spencer [2]. Alcaraz and Köberle [14] also study more general actions giving rise to more complex phase diagrams; these results are still in agreement with our heuristic principle, as discussed in [5].

$Z(N)$ gauge models with the standard action (3.4) in $D = 4$ can be analyzed in exactly the same way as $2D$ spin models with action (3.1). Again one obtains the estimates (3.3) and (3.7) which for $N \geq 5$ require the existence of two separate transitions, in accordance with [13]. Creutz, Jacobs and Rebbi in their review [16] report numerical simulations of these models which are in general agreement with our predictions. They give a fit for $\beta_M(N)$, however, that has the form

$$\beta_M(N) \approx .78/(1 - \cos\frac{2\pi}{N})$$

whereas eq. (3.3) says

$$\beta_M(N) \geq .88/(1 - \cos\frac{2\pi}{N})$$

I think that the discrepancy is due to the prescription used in [16] to locate the phase transition. (They say that they define the location of the transition as the β at which $|\frac{dF}{d\beta}|$ takes its maximum, where F is the free energy. Presumably they do not mean that, but the maximum of $\frac{dU}{d\beta}$ where $U = -\frac{d}{d\beta}(\beta F)$ is the internal energy or average action. Taking instead the maximum of the specific heat which is $\beta^2\frac{dU}{d\beta}$ would already give a larger value for β_M).

If we now turn to $Z(N)$ spin models in $D = 3$ we are for the first time led to predictions that are in conflict with general beliefs. Our heuristic principle predicts again inequalities of the form (3.3) and (3.7), which again imply that for large enough N there will be two distinct transitions! Conventional wisdom, as for instance expressed in [11] asserts that there is only one such transition, corresponding to the melting of the low temperature phase of the dual $Z(N)$ gauge model. The melting of the ordered phase of the $Z(N)$ spin model is supposed to occur right there, in conflict with (3.7). We find this view highly implausible and in [11], in spite of their claim to the contrary, there is even numerical evidence in our favor. They find a sharp drop in the string tension of the gauge models (= the surface tension of the spin models) for $n \geq 8$ in a place that is in agreement with (3.7) (provided we take into account that there spin models have a somewhat different action). See Fig. of [11], reproduced as Fig.7 in [5].

If we send $N \to \infty$ in an $Z(N)$ spin model we normally obtain an $O(2)$ (plane rotator) model. But making β N-dependent in the right way we can obtain another type of model, an "integer spin model" with spins taking any value $n_x \epsilon Z$. The simplest models of this type are the "discrete Gaussian" models (DGM) described by the action

$$S = \frac{1}{2} \sum_{\langle xy \rangle} (n_x - n_y)^2 \tag{3.14}$$

By well-known transformations this model is in $3D$ equivalent to a Coulomb gas (the so-called Villain Coulomb gas) and a $U(1)$ lattice gauge model (the Villain form of compact QED_3. For reasons of convenience, that is in order to have the right inverse temperature for the Coulomb gas, we use for the DGM (3.14) the Gibbs factor

$$e^{-\frac{1}{\beta} S} \tag{3.15}$$

The DGM obviously has a low temperature (small β) phase that is ordered and describable as a dilute gas of domain walls. By our principle for $D \geq 2$, by increasing β we should eventually hit a phase transition at a value β_c (DGM). Comparing to the Ising model we predict

$$\beta_c(DGM) \leq \frac{1}{4\beta_c(2)} \tag{3.16}$$

In $D = 2$ this transition is again the well-known Kosterlitz-Thouless transition whose existence was proven in [2]. Inserting $\beta_c(2) \approx .44$ for $D = 2$ resp. $\beta_c(2) \approx .22$ for $D = 3$ we obtain

$$\beta_c(DGM) \leq .55 \quad (D = 2) \tag{3.17}$$

$$\beta_c(DGM) \leq 1.13 \quad (D = 3) \tag{3.18}$$

Note that in $D = 3$ the transition according to conventional wisdom does not exist!

We simulated the DGM in $D = 2, 3$ and measured in particular the quantity

$$\langle (\Delta n)^2 \rangle_{\mathrm{DGM}} \tag{3.19}$$

which is related to the average expectation value of the square of the charge density of the Coulomb gas, $\langle \rho^2 \rangle_{\mathrm{Coul}}$, by

$$\langle \rho^2 \rangle_{\mathrm{Coul}} = \frac{1}{4\pi^2 \beta}(2D - \frac{1}{\beta}\langle (\Delta n)^2 \rangle_{\mathrm{DGM}}) \tag{3.20}$$

We saw clear signs of transitions, apparently first order in $D = 3$, possibly second order in $D = 2$, at β_c values more or less saturating the bounds (3.17), (3.18).

The transition in $3D$, which by conventional wisdom is denied its right to exist, but which we think we have seen, is a very physical transition in the Coulomb language: It is the transition from a gas to a liquid. More on this can be found in [17].

4. Testing the Heuristic Principle (II): Two-Dimensional $O(N)$ Spin Models

Contrary to the conventional wisdom our principle predicts that all two-dimensional $O(N)$ models have a phase transition caused by polymerization of links carrying non-trivial representations of $O(N)$. It is natural to expect that this transition is connected with the change from massive to massless behavior (from exponential to power-like clustering) just as it happens for $N = 2$. Actually, this seemed to be the prevalent view up to the seventies, cf. [18], until the work of Polyakov [19] and Brézin & Zinn-Justin [20] appeared.

In order to check this, we did an extensive numerical study [6].

Instead of simulating continuous $O(N)$ spin models we replaced them by discrete ones; concretely we looked at the $Z(10)$-clock model and a discretized version of the $O(4)$ model in which the spins take their values in the 120 element subgroup $\tilde{Y} \subset SU(2) \cong S_3$. This has the additional advantage that we can test our principle even further by modifying the action in such a way that the "freezing" transition to the magnetized discrete phase stays in place but the high temperature transition predicted by our principle moves.

For the $Z(10)$ model we used the standard action (3.1); we used it mainly to have something to compare the nonabelian models with.

For the $\tilde{Y}$ models we used the one-parameter family of actions

$$\begin{aligned} S_\lambda = & \sum_{\langle xy \rangle} (1 - S(x) \cdot S(y))\theta(S(x) \cdot S(y) - c) \\ & + \lambda \sum_{\langle xy \rangle} (1 - S(x) \cdot S(y))\theta(c - S(x) \cdot S(y)) \end{aligned} \tag{4.1}$$

Here we use the representation of the elements of $\tilde{Y}$ as unit vectors S in R^4 obtained by the standard correspondence between $SU(2)$ and S_3

$$S \to S_o + i\vec{\sigma} \cdot \vec{S} \tag{4.2}$$

The constant c satisfies

$$\cos\frac{\pi}{3} < c < \cos\frac{\pi}{5} \tag{4.3}$$

i.e. it lies between the cosines of the smallest angle ($\frac{\pi}{5} = 36^o$) and the second smallest angle ($\frac{\pi}{3} = 69^o$) between spins in $\tilde{Y}$.

The standard action is obtained for $\lambda = 1$; but we also studied in detail $\lambda = 3$ and did some simulations at $\lambda = \infty$. Choosing $\lambda > 1$ obviously makes the configurations smoother by putting an extra penalty on jumps by more than 36° between neighboring spins.

According to our heuristic principle we expect the well known two transitions in the $Z(10)$ model, one near $\beta = 1$, the other one at

$$\beta_M(10) \geq \beta_c(2)\frac{2}{1-\cos\frac{\pi}{5}} \approx 4.6 \tag{4.4}$$

(cf. Section 3). Similarly we expect a freezing transition in the $\tilde{Y}$ models for any λ at a $\beta_M(\tilde{Y})$ satisfying

$$\beta_F(\tilde{Y}) \geq 4.6 \tag{4.5}$$

Because the identity in $\tilde{Y}$ has 12 neighbors, some of which are again neighbors of each other, we expect a higher effective entropy per link of the domain walls, due to the possibility of branching, than in the $Z(10)$ model and hence

$$\beta(\tilde{Y}) > \beta(Z(10)) \tag{4.6}$$

We found a peak in the specific heat for $Z(10)$ near $\beta = 4.6$ and for $\tilde{Y}(\lambda = 1$ and 3) near $\beta = 6$ in full agreement with these arguments.

The heresy arises with the prediction that the $\tilde{Y}$ models should have another transition at much smaller β. It is not so easy to give inequalities for this β_c because the entropies are a bit trickier to estimate. In [5] we derived a random loop representation of the $O(N)$ models, however, which leads to the prediction for the $O(N)$ models

$$\beta_c(N) \geq \frac{N}{2}\beta_c(2) \tag{4.7}$$

Since the $\tilde{Y}$ model with standard action $\lambda = 1$) should be a good approximation to the standard $O(4)$ model, we expect likewise

$$\beta_c(\tilde{Y}, \lambda = 1) \geq 2\beta_c(Z(10)) = 2\tilde{\beta}_M(10) \tag{4.8}$$

A slightly different way to estimate the transition temperatures would proceed as follows: Expand the link Gibbs factor $g(3)$ into Gegenbauer polynomials

$$g(\beta) = \sum_{k=0}^{\infty} C_k^N(S \cdot S^1)\, a_{k,N} \tag{4.9}$$

The Gibbs factor for the "dual" link defect is then

$$\frac{a_{1,N}}{a_{0,N}} = \frac{\int g_\beta(S \cdot S')\; S \cdot S'\; d\gamma(S) d\gamma(S^1)}{\int g_\beta(S \cdot S')\; d\gamma(S) d\gamma(S')} \tag{4.10}$$

where $d\gamma$ is the normalized measure on the unit sphere. An estimate for $\beta_c(N)$ in terms of $\beta_c(N)$ would then be given by

$$\frac{a_{1,N}(\beta_c(N))}{a_{0,N}(\beta_c(N))} \approx \frac{a_{1,2}(\beta_c(2))}{a_{0,2}(\beta_c(2))} \tag{4.11}$$

For $O(4)$ with standard action this gives

$$\beta_c(4) \approx 2\beta_c(2) \tag{4.12}$$

It is, however, not easy to see here if increasing N increases the entropy per link. There are competing effects: On the one hand there is an extra entropy factor N for each closed loop of defects, this tends to inhibit polymerization. On the other hand, with growing N there is an increased weight for self-intersections due to Clebsch-Gordon coefficients which facilitates polymerization.

A rough estimate for the modified action for $\lambda = 3$ (based on the activity of the high temperature polymers) suggests

$$\beta_c(\tilde{Y}, \lambda = 3) \approx \frac{1}{3}\beta_c(\tilde{Y}, \lambda = 1) \tag{4.13}$$

We found confirmation of all of these considerations in our numerical simulations (see [6] for details).

Since the prediction of a phase transition for the nonabelian models at small β is very contrary to conventional wisdom we tried to be very careful in establishing data that can be trusted and are not spoiled by critical slowing down or finite size effects. This limited somewhat our possibilities to go deep into the critical region.

Our main signal for the existence of a β_c comes from measuring the susceptibility

$$\chi = \sum_x \langle S(0)\ S(x)\rangle \tag{4.14}$$

on lattices with a linear dimension of at least 10 times the correlation length.

We measured the correlation length by determining the closest zero of

$$\hat{G}(k)^{-1} \tag{4.15}$$

to the real axis, where $\hat{G}(k)$ is the Fourier transform of the 2-point function

$$G(x) = \langle S(0)\ S(x)\rangle \tag{4.16}$$

We found that our data for $\lambda = 1, 3$ are well represented by power law fits

$$\chi_\lambda = c_\chi^{(\lambda)}(\beta_c^{(\lambda)} - \beta)^{-\gamma} \tag{4.17}$$

$$\xi_\lambda = c_\xi^{(\lambda)}(\beta_c^{(\lambda)} - \beta)^{-\nu} \tag{4.18}$$

with powers γ, ν that are the same for $\lambda = 1, 3$. Moreover we could establish that there is a linear relation

$$\beta^{(3)} = a\beta^{(1)} + b \tag{4.19}$$

that allows to represent the data for the two models $\lambda = 1, 3$ by a single power law fit

$$\chi = c_\chi^{(1)}(\beta_c^{(1)} - \beta^{(1)})^{-\gamma} \tag{4.20}$$

$$\xi = c_\xi^{(1)}(\beta_c^{(1)} - \beta^{(1)})^{-\nu} \tag{4.21}$$

Our best values for the parameters are:

$$\beta_c^{(1)} = 2.58 \pm .08 \tag{4.22a}$$

$$\beta_c^{(3)} = .76 \pm .02 \tag{4.22b}$$

$$\gamma = 3.29 \pm .30 \tag{4.22c}$$

$$\nu = 2.02 \pm .20 \tag{4.22d}$$

As announced, they are in accordance with the simple estimates presented before. In particular the dual Gibbs factor (4.10) has practically the same value for both models at their respective β_c: It is .511 for $\lambda = 3$ at $\beta^{(3)} = .76$ and .518 for $\lambda = 1$ at $\beta^{(1)} = 2.58$. It is also interesting that for $\beta \to 0, \beta\lambda \to \infty$ the Gibbs factor is .824 which is much larger than this critical value .5, so we should be beyond the transition in the massless phase. This is also confirmed clearly by our simulation [6].

(U. Heller [21] has published a numerical determination of ξ at $\beta = 2.2$ for the standard action $\tilde{Y}$ model that seems to be in conflict with our fit (4.21). We believe that his simulation is suffering from critical slowing down and his result is therefore not to be taken at face value; for a more detailed discussion of this issue see [22]. Even for somebody who does not believe in our critical point around $\beta = 2.6$ it would be rather strange to assume that the behavior of the $\tilde{Y}$ model near $\beta = 2$ is determined by a nonexisting critical point at $\beta = \infty$ and not by the freezing/melting point at $\beta = 6$ that does exist and is much closer.)

We also measured (in four ways) the critical exponent of the two-point function and found

$$\eta = .37 \pm .02 \tag{4.23}$$

Remarkably, we found also for the $Z(10)$ that a power law fit worked well, much better than an exponential fit as suggested by Kosterlitz and Thouless [23], and even more remarkably we found practically the same value of η as for the $\tilde{Y}$ model.

To check the scaling properties of the $\tilde{Y}$ models closer to the proposed critical point we did a detailed finite size scaling analysis. It turns out that

$$L^{\eta-2}\chi_L(\beta) \tag{4.24}$$

depends only on $z = L/\xi$ and is given by the same function $f(z)$ for $\lambda = 1$ and 3; furthermore, $\chi_L(\beta)/\chi_\infty(\beta)$ is given by another (related) universal function g of $z = L/\xi$. For more details see [6].

Measurements of the critical exponent η at $\beta > \beta_c$ gave values significantly smaller than .37, which was found at β_c. This supports the picture of a critical line (spin wave phase) with continuously changing η extending from β_c to β_M, where the discretely magnetized ("frozen") phase starts.

As a curiosity (so far) I want to mention that our results suggest the formula

$$\nu = \frac{N+2}{3}$$

for the $O(N)$ models starting with $N = 1$ (the Ising model).

5. Testing the Heuristic Principle (III): Nonabelian Gauge Models in $D = 4$

According to our heuristic principle lattice gauge theories (LGT) based on a compact Lie group should have a phase transition caused by the polymerization of the high temperature polymers in any dimension $D \geq 3$. I want to make some rough estimates of the location of that transition and compare it with numerical results from the literature as well as some results of simulations we are carrying out ourselves [7].

But first let us recall the conventional wisdom: It is believed that compact LGT in $D = 3$ has no transition. In $D = 4$ $SU(N)$ LGT is believed to have a first order transition for $N \geq 4$, whereas for $SU(2)$ and $SU(3)$ there is not supposed to be such a transition. The (first order) transitions for $N \geq 4$ are supposed to be not deconfining.

On the other hand it is a well known fact that so-called finite temperature LGT in $D = 4$ (that is LGT in which the lattice has a finite extent N_τ proportional to the inverse temperature in one direction) has a **deconfining transition** (see [24]). According to conventional wisdom this transition moves to zero coupling as $N_\tau \to \infty$ (i.e. as the temperature goes to zero).

The heretic thesis I want to propose here is that $4D$ $SU(N)$ LGT at zero temperature has a phase transition which is deconfining and for $N \geq 4$ coincides with the first order "bulk" transition observed there.

The Gibbs factor for a plaquette in $SU(N)$ or $U(N)$ LGT is

$$G_\beta(N) \equiv exp(\frac{\beta}{N} trU) \tag{5.1}$$

The "dual Gibbs factor" that should determine the location of the transition is

$$g_\beta(N) \equiv \frac{\int exp(\frac{\beta}{N} trU)\frac{1}{N} trU dU}{\int exp(\frac{\beta}{N} trU)} \tag{5.2}$$

For $U(1)$ we obtain

$$g_\beta(U(1)) = \frac{I_1(\beta)}{I_0(\beta)} \tag{5.3}$$

It is known [1, 3] that $U(1)$LGT (compact QED) in $4D$ has a deconfining transition; according to Evertz et al. [25] its location is

$$\beta_c \cong 1.01 \tag{5.4}$$

(and it is weakly first order).

As before we expect the transition for $SU(N)$ to occur at a $\beta_c(N)$ satisfying

$$g_{\beta_c(N)}(N) \approx \frac{I_1(1)}{I_0(1)} \approx .45 \tag{5.5}$$

We do not know very well the effects of branching and selfintersections of polymers; (5.5) is based on the activity of large unbranched polymers (simple surfaces made out of excited plaquettes). Therefore, we should not expect (5.5) to be extremely precise. But let us see how well it is satisfied anyhow.

For $SU(2)$ we obtain

$$g_\beta(SU(2)) = \frac{I_2(\beta)}{I_1(\beta)} \tag{5.6}$$

and from (5.5)

$$\beta_c(SU(2)) \approx 2.2 \tag{5.7}$$

For β/N large, $g_\beta(SU(N))$ can be shown to behave like [24, 26]

$$g_\beta(SU(N)) \approx e^{-\frac{1}{2\beta}(N^2-1)} \tag{5.8}$$

This suggests that β_c should go as N^2 for large N which would justify in turn the use of the approximation (5.8). (5.8) and (5.5) imply

$$\beta_c(SU(N)) \approx \ .66N^2 \tag{5.9}$$

In the literature we find the following values for the "bulk" transition:

$$\beta_c(SU(4)) \approx 10.24 \pm .32 \quad [27,28]$$
$$\beta_c(SU(5)) \approx 16.3 \ \pm .1 \quad [29]$$
$$\beta_c(SU(5)) \approx 16.6 \ \pm .3 \quad [27]$$
$$\beta_c(SU(6)) \approx 24 \quad \pm 1 \quad [30]$$

which gives values for β_c/N^2 of about .65 to .68. The agreement with the simple-minded estimate (5.0) is remarkable.

Next let us consider the so-called finite temperature deconfinement transition. It has been recognized in the literature that it is difficult to separate from the "bulk transition" considered so far [31, 32]. It is also recognized that this bulk transition is accompanied by a "sharp change in the value of the string tension" [32].

For this reason some people changed the rules and studied instead certain 1-point lattice models, the so-called twisted Eguchi-Kawai models, which in the limit $N \to \infty$ are supposed to be equal to the model on an infinite lattice. Clearly this equivalence, even if it is accepted for $N = \infty$, cannot be exact for finite N and in particular it also raises the question of the interchangeability of the $N \to \infty$ limit with the thermodynamic limit. There are published results on these one-point lattice models claiming to have seen that the deconfinement transition happens at larger β than the "bulk transition" for some values of N [31, 32, 33, 34]. The distance between the two transitions is not very large (for instance $\frac{\beta}{N^2} = .35$ and $\frac{\beta}{N^2} = .36$ in [33]). But the main point is that those results do not refer to the LGT as defined here, for which an infite lattice is essential.

Aida El-Khadra has simulated the standard $SU(10)$ lattice model on a 2×4^3 lattice [35]; she finds a deconfining transition at

$$\beta_c(SU(10)) = 65 \text{ to } 70 \tag{5.10}$$

Note that this value is again in agreement with (5.9), but now for the supposedly different deconfinement transition! Finally let us study the deconfinement transition for $SU(2)$ and $SU(3)$.

For $SU(2)$ there are the rather recent high precision data by Engels at al. [36] which give:

n_T	3	4	5
β_c	2.1768	2.2964	2.3726
$g_c^2 = \frac{4}{\beta_c}$	1.8376	1.7419	1.6859

These authors fit their data with an asymptotic scaling formula corresponding to the conventional wisdom, namely

$$N_T^{-1} = 41.63\; e^{-\frac{12\pi^2}{11g^2}} \left(\frac{24\pi^2}{11g_c^2}\right)^{\frac{51}{121}} \tag{5.11}$$

While the fit is certainly good, it still seems rather unnatural, when considered with a mind free of theoretical bias. If we plot N_T^{-1} against g^2 we obtain almost a straight line; in fact the three points give a slightly concave shape, while the fit (5.11) is convex. The only significance of (5.11) is that it represents the slope correctly. To look at the 3 points and claim that they indicate that $N_T \to \infty$ implies $g_c^2 \to 0$ seems rather farfetched. A direct linear extrapolation, on the other hand, suggests for $N_T \to \infty$

$$g_c^2(2) \approx \quad 1.46 \tag{5.12a}$$

$$\beta_c(2) \approx \quad 2.7 \tag{5.12b}$$

which is not too far from the estimate (5.7).

We simulated $SU(2)$ LGT ourselves on a 6^4 lattice [7] and found two things:

There seems to be a jump in a certain disorder variable around $\beta = 2.3$ and there is also a marked increase in the so-called Polyakov loop susceptibility there that might be an indication of deconfinement (note that [36] would suggest $\beta_c \approx 2.4$ for $N_T = 6$). This work is not yet finished. We are planning to run on larger lattices and see if the picture persists; we are also planning to test our disorder variable on an abelian model where we know that there is a deconfinement transition at $N_T = \infty$.

For $SU(3)$ deconfinement was studied in the past few years by Kennedy et al. [37], Gottlieb et al. [38] and Christ and Terrano [39]. If we again plot the data as N_T^{-1} against g_c^2, we find the following:

The data by Kennedy et al., who used $N_T \leq 10$ fall again pretty much on a straight line that extrapolates to

$$g_c^2(3) \approx \quad .94 \tag{5.13a}$$

$$\beta_c(3) \approx \quad 6.45 \tag{5.13b}$$

at $N_T = \infty$. Note again that this is not terribly far from (5.9). Gottlieb et al. [38] and Christ et al. [39] used lattices with N_T from 8 resp. 10 up to 14. Their results both show a much smaller slope than the straight line through the data of Kennedy et al. [37]. This seems to confirm asymptotic scaling and be difficult to reconcile with my claim of a finite $\beta_c \approx 6$ to 7 at $N_T = \infty$. I want to mention, however, that the results of [38] and [39] are mutually incompatible. Also, on such large lattices, one certainly has considerable slowing down of the algorithm. In particular one needs a long time to build up the necessary strong correlations, especially in time direction, needed to obtain a nonvanishing Polyakov loop expectation signalling deconfinement. I do not claim that this case is closed, but that further investigations should be carried out with an unbiased mind and attention to problems like critical slowing down. I must say that, leaving aside everything else, I find the non-monotomic approach to asymptotic scaling (the so-called "dip" in the β function) suggested by the results of [38, 39] rather hard to swallow.

It should also be mentioned that a first order transition in $SU(2)$ LGT was also observed when the action was modified by adding a term proportional to the trace of the adjoint action [40] of sufficient size. According to [40] the transition disappears when this term becomes small, but it naturally extrapolates to a point somewhere above $\beta = 2$ for the standard action, that is where we predict a transition.

Edgar [41] studied a different modification: Instead of coupling the gauge fields on plaquettes he used "windows" consisting of two adjacent plaquettes. He found a first order transition at a value of β that corresponds to something near 2 for the standard plaquette action.

Let me also recall the results on compact QED in $D = 3$ mentioned in Section 3 where we found a phase transition, apparently of first order, in the place suggested by our heuristic principle, which according to the conventional wisdom does not exist.

Finally, let me make some remarks concerning the order of the transitions: We expect them to be of first order in all LGT in $D \geq 3$. In compact QED_3 the transition can only be first order because according to the theorem of Göpfert and Mack there is a mass at weak coupling and the mass has to be a monotomically increasing function of the coupling. Making the dimension higher should make the transition "harder". The only thing that might be confusing is the fact that in $SU(2)$ LGT the finite temperature deconfining transition is supposed to be second order [36, 42]. But this second order nature refers to some very special observables, the Polyakov loops, so there is really no paradox if we identify this transition with the first order transition we have seen in [7].

6. Conclusions

We proposed a simple energy-entropy argument that predicts semiquantitatively phase transitions for a large number of spin and gauge models. These predictions seem to be confirmed in all cases where the existence of those transitions is uncontested.

The argument furthermore predicts transitions in models such as $O(N)$ models with $N \geq 3$ and $SU(2)$ and $SU(3)$ lattice gauge models that according to conventional wisdom do not have such transitions. We do find, however, numerical evidence for these transitions, partly by doing our own simulations, partly by looking at published data in an unbiased way.

The emerging picture, while in conflict with the prevailing view, seems nevertheless quite natural. It implies that the difference between abelian and nonabelian models has been overstated. $2D$ $O(N)$ models, according to our view, have a soft phase just like the $O(2)$ model; $4D$ $SU(N)$ LGT's have a first order transition to a deconfining weak coupling phase just like the $U(1)$ model.

REFERENCES

[1] A. Guth, *Phys. Rev.* **D 21** (1980) 2291.

[2] J. Fröhlich, T. Spencer,*Comm.Math.Phys.*, **81** (1981) 527.

[3] J. Fröhlich, T. Spencer, *Comm.Math.Phys.*, **83** (1982) 411.

[4] E.T. Tomboulis, *Phys. Rev. Lett.* **50** (1983) 885.

[5] A. Patrascioiu, E. Seiler, I.O. Stamatescu, From Ice to Deconfinement: A Unifying View of Certain Phase Transitions, MPI-PAE/PTh 75/87.

[6] E. Seiler, I.O. Stamatescu, A. Patrascioiu, V. Linke, Critical Bahavior, Scaling and Universality in Some Two-Dimensional Spin Models, MPI/PAE/PTh 76/87.

[7] A. Patrascioiu, I.O. Stamatescu, E. Seiler, work in progress.

[8] R. Peierls, *Proc. Camb. Phys. Soc.* **32** (1936) 447.

[9] M. Aizenmann, J. Bricmont, J. Lebowitz, Percolation of the Minority Spins in High Dimensional Ising Models, Rutgers University preprint 1987.

[10] E. Seiler, Gauge Theories as a Problem of Constructive Quantum Field Theory and Statistical Mechanics, Lecture Notes in Physics **159**, Springer-Verlag Berlin 1982.

[11] G. Bhanot, M. Creutz, *Phys. Rev.* **D 21** (1980) 2892.

[12] M. Suzuki, *Prog. Theor. Phys.* **37** (1967) 770.

[13] S. Elitzur, R.B. Pearson, J. Shigemitsu, *Phys. Rev.* **D 19** (1979) 3698.

[14] F.C. Alcaraz, R. Köberle, *J.Phys.A* **14** (1980) 1169.

[15] M.Aizenman, B. Simon,*Phys. Lett.* **76 A** (1980) 281.

[16] M. Creutz, L. Jacobs, C. Rebbi, *Phys. Rev.* **C 94** (1983) 101.

[17] A. Patrascioiu, E. Seiler, *Phys. Rev. Lett.* **60** (1988) 875.

[18] K. Binder in: Phase Transitions and Critical Phenomena, vol. 5B, C. Domb and M.S. Green (eds.), Academic Press, London 1976.

[19] A.M. Polyakov, *Phys. Lett.* **59 B** (1975) 79.

[20] E. Brézin, J. Zinn-Justin, *Phys. Rev.* **B 14** (1976) 3110.

[21] U. Heller, Monte Carlo Renormalization Group Investigation of the Two-dimensional O(4) Sigma Model, NSF-ITP-88-20.

[22] A. Patrascioiu, E. Seiler, I.O. Stamatescu, Critical Slowing Down, Monte Carlo Renormalization Group and All That, University of Arizona preprint to appear.

[23] J.M. Kosterlitz, D.J. Thouless, *J. Phys.* **C6** (1973)1181.

[24] C. Borgs, E. Seiler, *Nucl. Phys.* **B 215** (1983) 125; *Comm. Math. Phys.* **91** (1983) 329.

[25] H.G. Evertz, J. Jersák, T. Neuhaus, P.M. Zerwas, *Nucl. Phys.* **B 251** (1985) 279.

[26] M. Creutz, *Phys. Rev.* **D 15** (1977) 1128.

[27] M. Creutz, *Phys. Rev. Lett.* **46** (1981) 46.

[28] K.J.M. Moriarty, *Phys. Lett.* **106 B** (1981) 130.

[29] H. Bohr, K.J.M. Moriarty, *Phys. Lett.* **104 B** (1981) 217.

[30] M. Creutz, K.J.M. Moriarty, *Phys. Rev.* **D 25** (1982) 1724.

[31] S. Das, J. Kogut, *Phys. Lett.* **141 B** (1984) 105.

[32] S. Das, J. Kogut, *Nucl. Phys.* **B** (1985) 141.

[33] K. Fabricius, O. Haan, *Nucl. Phys.* **B 260** (1985) 285.

[34] K. Fabricius, O. Haan, F.R. Klinkhamer, *Phys. Rev.* **D 30** (1984) 2227.

[35] A. El-Khadra, Diplomarbeit F.U. Berlin 1987, private communication by I.O. Stamatescu.

[36] J. Engels, J. Jersak, K. Kanaya, E. Laermann, C.B. Lang, T. Neuhaus, H. Satz, *Nucl. Phys.* **B 280** (1987) 577.

[37] A.D. Kennedy, J. Kuti, S. Meyer, B.J. Pendleton, *Phys. Rev. Lett.* **54** (1985) 87.

[38] S.A. Gottlieb, J. Kuti, D. Toussaint, A.D. Kennedy, *Phys. Rev. Lett.* **55** (1985) 1958.

[39] N. Christ, E. Terrano, *Phys. Rev. Lett.* **56** (1986) 111.

[40] G. Bhanot, R. Dashen, *Phys. Lett.* **113 B** (1982) 299.

[41] R.C. Edgar, *Nucl. Phys.* **B 200** (1982) 345.

[42] B. Svetitsky, L. Yaffe, *Phys. Rev.* **D 26** (1982) 963; *Nucl. Phys.* **B 210** (1982) 423.

EQUILIBRIUM STATES FOR QUANTUM SYSTEMS AND APPLICATIONS

A. Verbeure
Instituut voor Theoretische Fysica
Universiteit Leuven
Celestijnenlaan 200 D
B-3030 Leuven (Belgium)

ABSTRACT

These lecture notes contain :

I. Introduction
II. Equilibrium conditions for classical lattice systems
III. Equilibrium conditions for quantum systems
IV. Equilibrium states for permutation invariant models
V. The spin-boson model
VI. Absence of spontaneous symmetry breaking

I. INTRODUCTION

In these lectures I want to explain a number of results about equilibrium statistical mechanics. We will concentrate mainly on equivalent characterizations of the equilibrium conditions and on some specific applications.

First of all I want to state that this is not an overview of all existing formulations because the field is by now too extensive to be treated in a few hours. I feel forced to make a selection, which might be not a canonical selection in the sense that it will reflect my personal flavour and ... knowledge. These notes will be based on work done in collaboration with students, co-workers and colleagues. I will not mention all their names, nevertheless I should not refrain from mentioning Mark Fannes, whose collaboration was most appreciated to obtain the results on which these lectures are based.

As said before I will make a selection, the basis of which will be the following underlying idea, namely that in order to have an appropriate idea of equilibrium of a system one has to look for the way the system behaves under local perturbations, in other words one has to look for the specific stability of the system under such perturbations.

Although I will be interested here in equilibrium states of quantum systems, I will start with an introduction of the classical equilibrium condition, because classical physics speaks more to our intuitive mind and the classical features should be included in quantum physics. Next I formulate some equilibrium conditions for quantum systems which express the same stability properties as the classical ones. The rest of these notes is devoted to the exposition of a number of applications. As far as these are concerned we give a solution of a spin-boson model, the treatment of mean field models and a general theory for proving the absence of spontaneous symmetry break down.

II. CLASSICAL LATTICE SYSTEMS

Consider a lattice S and for each lattice site $j \in S$, consider a phase space K_j, a copy of a compact Hausdorff space K (for Ising spins $K = \{0,1\}$). For each finite subset X of S one has the phase space $K_X = \prod_{j\in X} K_j$ and for the infinite system the phase space is $\Omega = \prod_{j\in S} K_j$.

Denote by C(X) the set of real valued continuous functions on K_X and by $C(\Omega)$ these on Ω; $C(\Omega)$ is also the sup-norm closure of $\bigcup_{X\subset S} C(X)$.

Denote by Q the group of transformations of $C(\Omega)$ induced by the invertible transformations of the configuration space Ω, changing the configurations in some finite volume only. In fact for the following we can restrict ourselves to any set of transformations separating the points of Ω.

A <u>relative Hamiltonian</u> is a map h from Q into $C(\Omega)$: $\tau \to h(\tau)$ satisfying the cocycle relation :

$$h(\tau_1\tau_2) = \tau_1 h(\tau_2) + h(\tau_1)$$

Remark that if a sequence of local Hamiltonians $\{H_\Lambda\}_{\Lambda\to S}$ is given, then

$$h(\tau) = \lim_{\Lambda \to S} \tau H_\Lambda - H_\Lambda$$

is a relative Hamiltonian if the limit $\Lambda \to S$ can be given a mathematical sense.

Conversely, if a relative Hamiltonian h is given, then one can find local Hamiltonians such that they yield the relative Hamiltonian. Take $x, y \in \Omega$ and $\Lambda \subset S$ such that

$$(\tau\, x)_i = x_i \quad \text{for} \quad i \in \Lambda^c \text{ (complement of } \Lambda)$$

$$= y_i \quad \text{for} \quad i \in \Lambda$$

then $\tau \in Q$ and define

$$H_\Lambda(y) = h(\tau)(x)$$

One checks that the H_Λ are well defined (i.e. independent of the choice of the τ's with the above property).

Hence a classical system can be defined by means of a relative Hamiltonian or equivalently by means of a sequence of local Hamiltonians.

In order to define equilibrium states we use the notion of relative Hamiltonian, it represents the energy difference of a configuration with its transformed one under the map τ. A classical physical system is therefore defined by giving the pair $(C(\Omega), h)$ where $C(\Omega)$ is the set (algebra) of observables and h its relative Hamiltonian.

Such a system might have an equilibrium state at a certain inverse temperature $\beta > 0$. A state is a probability measure on Ω, and

<u>Definition II.1</u>

A state ω_β of Ω satisfies the DLR-conditions [1,2] at $\beta > 0$ for h if for all $f \in C(\Omega)$ and $\tau \in Q$ holds :

$$\omega_\beta(\tau^{-1}(f)) = \omega_\beta(f\, e^{-\beta h(\tau)}) \tag{1}$$

Remark that one gets the original formulation of DLR if one takes for

f the characteristic functions of local configurations and one derives that the probability of finding a certain configuration in a finite volume, given particular boundary conditions, is proportional to the Gibbs factor. In the above definition one does not need the concept of conditional probability. In measure theoretic terminology formula (1) expresses the absolute continuity of the measure $\omega \cdot \tau^{-1}$ with respect to ω, with Radon-Nikodym derivative equal to $\exp -\beta h(\tau)$, or

$$\frac{\partial\, \omega_\beta \cdot \tau^{-1}}{\partial\, \omega_\beta} = e^{-\beta h(\tau)} \quad ; \quad \tau \in Q$$

We have the following characterization.

<u>Theorem II.2</u>

For a classical lattice system $(C(\Omega),h)$ we have the following equivalent conditions

(i) a state ω is a DLR-state (II.1)

(ii) a state ω satisfies the Energy-Entropy-Balance (EEB) inequality :

$\forall f \in C(\Omega)$, $f \geqslant 0$, $f \neq 0$ and $\tau \in Q$:

$$\beta\omega(f\ h(\tau)) \geqslant \omega(f) \ln \frac{\omega(f)}{\omega(\tau^{-1} f)} \tag{2}$$

<u>Proof</u> : (i) $\rightarrow$ (ii)

First we show that ω is faithful. Take $f \geqslant 0$, $f \neq 0$ with supp $f = \Lambda$. Take any other local function $g \geqslant 0$ such that supp $g \subset \Lambda$. Then one can always find a finite number of τ's say $\{\tau_i | i = 1,\ldots n\}$ and $\lambda_i > 0$ such that

$$g \leqslant \sum_i \lambda_i\ \tau_i^{-1} f$$

Then

$$0 \leqslant \omega(g) \leqslant \sum_i \lambda_i\ \omega(\tau_i^{-1} f)$$

$$= \sum_i \lambda_i \, \omega(f \, e^{-\beta h(\tau_i)})$$

$$\leq \sum_i \lambda_i \, \| e^{-\beta h(\tau_i)} \| \, \omega(f)$$

Hence $\omega(f) = 0$ implies $\omega = 0$ which is impossible, therefore ω is faithful.

Choose now again $f \geq 0$, $f \neq 0$ and $\tau \in Q$ then by Jensen's inequality :

$$\beta\omega(f \; h(\tau)) = -\omega(f) \ln \exp \cdot \left(- \frac{\omega(f \, \beta h(\tau))}{\omega(f)} \right)$$

$$\geq -\omega(f) \ln \frac{\omega(f \; \exp -\beta h(\tau))}{\omega(f)}$$

$$= -\omega(f) \ln \frac{\omega(\tau^{-1} f)}{\omega(f)}$$

The inverse (ii) → (i) is proved by taking for f a sequence of functions approximating the δ-functions with support the individual configurations (see [3]).

■

Remark :

Without going too much into details we mention that also for continuous classical systems the equilibrium conditions can be expressed along the lines of above.

Suppose that we have a continuous classical system defined on a phase space

$$K = \{x = (x_i) \,|\, i = 1,\ldots,n,\ldots \; ; \; x_i = (q_i,p_i) \; ,$$

$$q_i,p_i \in \mathbb{R}^\nu \; ; \; x \cap V \times \mathbb{R}^\nu \text{ finite for all finite volumes } V\}$$

K can be defined as a topological space (a Polish space).

Define again Q as the local transformations of K and denote now by $A = C(K)$ the set of observables. If the system is given by the local Hamiltonians H_V, then

$$h(\tau) = \lim_V (\tau H_\Lambda - H_\Lambda)$$

if the limit exists in some sense. Otherwise one starts from a given relative Hamiltonian h. A Poisson bracket is defined to be a map from $A \times A$ into A : $f,g \in A$, $x \in K$

$$[f,g](x) = \sum_i \frac{\partial f(x)}{\partial q_i}\frac{\partial g(x)}{\partial p_i} - \frac{\partial f(x)}{\partial p_i}\frac{\partial g(x)}{\partial q_i}$$

The E.E.B. inequalities are meaningful; an equilibrium state ω is a measure satisfying for all $\tau \in Q$ and $f \geqslant 0$, $f \neq 0$, $f \in A$:

$$\beta\omega(f h(\tau)) \geqslant \omega(f) \lim \frac{\omega(f)}{\omega(\tau^{-1} f)} \tag{3}$$

We show that the so-called classical KMS-condition of Gallavotti-Verboven [4] follows from the inequality.

Take for τ the one-parameter group of maps

$$\tau_s = \exp s[g,.] \quad , \quad s \in \mathbb{R}$$

with g a local observable; substitute in (3) $\tau = \tau_s$ and develop both sides with respect to s :

$$\lim_\Lambda \beta\omega(f(s[g,H_\Lambda] + O(s^2))) \geqslant s\omega([g,f]) + O(s^2)$$

for all $s \in \mathbb{R}$; hence

$$\lim_\Lambda \beta\omega(f[g,H_\Lambda]) = \omega([g,f]) .$$

III. QUANTUM SYSTEMS

For quantum systems there are also many alternative formulations of equilibrium states [see e.g. 5] . We limit ourselves here to a few of them, in particular we start from the best known equilibrium condition, namely the KMS-condition. Then we proceed to a couple of formulations expressing the behaviour under local perturbations. In these formulations we stress also the role played by the relative Hamiltonian, as we did for classical systems.

For quantum systems, one usually starts from a $C^\star$-dynamical system (A,α) of a $C^\star$-algebra A of observables and $\alpha = (\alpha_t)_{t\in\mathbb{R}}$ a strongly continuous one-parameter group of *-automorphisms of A, representing the time evolution.

Now we define a norm-dense α-invariant *-subalgebra A_α of A as follows. For any complex function f such that $f \in C^\infty(\mathbb{R})$ and its Fourier transform

$$\hat{f}(\lambda) = \int dt\, f(t)\, e^{i\lambda t}$$

has compact support one defines the Bochner integral

$$A(f) = \int dt\, f(t)\, \alpha_t(A)$$

as an element of the algebra A. Denote by A_α the linear *-subalgebra of A generated by the set of these elements i.e. by

$$\{A(f) \mid f \in C^\infty \;;\; \text{supp } \hat{f} = \text{compact} \;;\; A \in A\}$$

It is readily checked that A_α is a norm-dense α-invariant *-subalgebra of A and using :

$$\alpha_{i\beta}\, A(f) = \int dt\, f(t - i\beta)\alpha_t A$$

one can formulate :

Definition III.1

A state ω of a $C^\star$-dynamical system $(\mathcal{A},\alpha)$ is called a KMS- or equilibrium state at $\beta > 0$ if for all $A, B \in \mathcal{A}_\alpha$ holds

$$\omega(A\ \alpha_{i\beta}\ B) = \omega(BA). \tag{1}$$

Here we formulated the KMS-conditions for $C^\star$-dynamical systems but mutatis mutandis, they can also be formulated for $W^\star$-dynamical systems. For simplicity of notation we identify in the following the observables $A \in \mathcal{A}$ with their representatives $\pi_\omega(A)$ under the GNS-representations of the states ω. It readily follows that if ω is a KMS-state, then $\omega \cdot \alpha_t = \omega$ for all $t \in \mathbb{R}$ i.e. ω is time invariant and α_t has a unitary representation

$$\alpha_t(A) = e^{itH}\ A\ e^{-itH} \quad , \quad A \in \mathcal{A}$$

The self-adjoint operator H is called the Hamiltonian. Let

$$H = \int \lambda\ dE(\lambda)$$

be its spectral resolution.

It is clear that the condition (1) is equivalent to

$$\omega(A^\star \alpha_{i\beta} A) = \omega(A\ A^\star) \quad ; \quad A \in \mathcal{A}_\alpha \tag{2}$$

by polarization. After substitution $A \to A(f)$ in (2) one gets for all $f \in C^\infty$:

$$\int e^{-\beta\lambda}\ |\ \hat{f}(\lambda)\ |^2\ d\mu_A(\lambda) = \int |\hat{f}(\lambda)|^2\ d\nu_A(\lambda)$$

where $d\mu_A$ and $d\nu_A$ are the spectral measures defined by

$$d\mu_A(\lambda) = \omega(A^\star dE(\lambda)A)$$

$$d\nu_A(\lambda) = \omega(A\, dE(-\lambda)A^\star)$$

proving the following theorem.

Theorem III.2

A state ω of $(\mathcal{A},\alpha)$ is a β-KMS-state iff for all $A \in \mathcal{A}$ holds :

$$\frac{d\mu_A(\lambda)}{d\nu_A(\lambda)} = e^{\beta\lambda} \quad ; \quad \lambda \in \mathbb{R}.$$

∎

This formulation of the equilibrium condition shows already some similarity with the DLR-conditions for classical systems as given in II.A, in the sense that the Boltzmann factor is expressed as a Radon-Nikodym derivative of a spectral measure with respect to a perturbed one. However the type of perturbation is physically not so transparant as in the classical one.

Now we proceed to a formulation in terms of relative Hamiltonians. By assumption, $(\alpha_t)_{t\in\mathbb{R}}$ is a strongly continuous one-parameter group of $\star$-automorphisms, hence

$$\alpha_t = \exp t\,\delta$$

where δ is the generator of α with domain $\mathcal{D}(\delta)$. In a time-invariant state

$$\delta = i[H,.]$$

Let τ be an inner automorphism i.e.

$$\tau(A) = U^\star A\, U \quad , \quad A \in \mathcal{A}$$

where $U^{\star}U=UU^{\star}\in\mathcal{A}$; τ is called an <u>α-local automorphism</u> if $U\in\mathcal{D}(\delta)$. Denote now by Q the set of α-local automorphisms.

Remark that for spin systems i.e. $\mathcal{A}=\bigotimes_{i\in Z^{\nu}} M_n$ with M_n the set of $n\times n$ matrices, any automorphism leaving invariant everything except in a local volume, is inner; its implementing unitary belongs to $\mathcal{D}(\delta)$ if the dynamics is defined by short-range interactions.

For any $\tau\in Q$, then

$$\alpha_t^{\tau}=\tau\,\alpha_t\,\tau^{-1}\quad ;\quad t\in\mathbb{R}$$

is a strongly continuous one-parameter group of $\star$-automorphisms and

$$\alpha_t^{\tau}=\exp t\,\delta_{\tau}$$

with : $\mathcal{D}(\delta_{\tau})=\mathcal{D}(\delta)$

$$\delta_{\tau}(A)=\delta(A)+i[\,h(\tau),A]$$

$$h(\tau)=-i\,U^{\star}\delta(U)$$

Moreover it is easy to check that the unitary cocycle Γ_t^{τ} in $\mathcal{A}$, relating α_t and α_t^{τ} through

$$\alpha_t^{\tau}(A)=\Gamma_t^{\tau}\,\alpha_t(A)\,(\Gamma_t^{\tau})^{\star}$$

and defined by

$$\frac{d\,\Gamma_t^{\tau}}{dt}=i\,\Gamma_t^{\tau}\,\alpha_t(h)\quad ;\quad \Gamma_0^{\tau}=1$$

is given by

$$\Gamma_t^{\tau}=U^{\star}\,\alpha_t(U)$$

Let ω be a KMS-state, then $\omega \cdot \alpha_t = \omega$ and

$$\alpha_t(A) = e^{itH} A e^{-itH}$$

and

$$\omega(A) = (\Omega, A\Omega)$$

$$H\Omega = \Omega \quad ; \quad \delta(A) = i[H,A]$$

$$A''\Omega \subset \mathcal{D}(\Delta) \quad ; \quad \Delta = \exp - \frac{\beta H}{2} .$$

$$J \Delta^{1/2} A \Omega = A^{\star} \Omega$$

$$J\Omega = \Omega$$

also

$$\Omega \in \mathcal{D}\left(\exp - \frac{\beta}{2} (H + h(\tau))\right)$$

Clearly :

$$\delta_\tau(A) = \delta(A) + i[h(\tau), A]$$

becomes

$$\delta_\tau(A) = i[H,A] + i[U^{\star}[H,U], A]$$

$$= i[U^{\star} H U, A]$$

and

$$h(\tau) = U^{\star} H U - H$$

is the <u>relative Hamiltonian</u>, it is the increment of energy due to the α-local automorphism.

Now we compute :

$$\omega(A) = (\Omega, A\Omega)$$

$$= (J\Omega, A\ J\Omega)$$

$$= (J\ AJ\ UU^{\star}\Omega, \Omega)$$

$$= (J\ AJ\ U^{\star}\Omega, U^{\star}\Omega) \quad \text{(Remark that } J\ AJ \in A'\text{)}$$

$$= (J\ A\ J\ J\ \ \Delta U\ \Omega, J\ \Delta\ U\ \Omega)$$

$$= (\Delta U\ \Omega, A\ \Delta U\ \Omega)$$

$$= (U^{\star}\Delta U\ \Omega, U^{\star}A\ UU^{\star}\Delta U\ \Omega)$$

$$= (e^{-\frac{\beta}{2}(H+h(\tau))}\Omega, \tau(A)\ e^{-\frac{\beta}{2}(H+h(\tau))}\Omega)$$

This proves part of

<u>Theorem III.3</u>

A state ω of (A,α) is a β-KMS-state iff for all $A \in A$ and all $\tau \in Q$ holds

$$\omega(\tau^{-1}(A)) = (e^{-\frac{\beta}{2}(H+h(\tau))}\Omega, A\ e^{-\frac{\beta}{2}(H+h(\tau))}\Omega) \qquad (3)$$

for ω a time-invariant state.

Proof : One implication is proved above; the converse is left to the reader as an exercise (see [6]) ∎

This characterization of an equilibrium state is clearly given in terms of expressing the perturbed equilibrium state $\omega \cdot \tau^{-1}$ versus the equilibrium state ω itself. As one can read from the condition (3), the relative Hamiltonian h is the relevant quantity to establish the relation.

From a more mathematical point of view, if U is an α-analytic element then

$$U^{\star} \Delta U \Delta$$

is a well defined element of A and it is possible to show that

$$U^{\star} \Delta U \Delta \Omega$$

belongs to the natural positive cone associated to ω. Then using Araki's definition [7], one concludes that (3) expresses that

$$U^{\star} \Delta U \Delta$$

is the Radon-Nikodym derivative of $\omega \cdot \tau^{-1}$ with respect to ω.

Using specific algebras and particular dynamics, and local transformations τ of a specific nature, it should be possible to use this characterization of equilibrium states in the study of states with particular properties like spontaneous symmetry breaking.

Suppose that we specify the $C^{\star}$-algebra A to be the CCR or CAR algebra over some reasonable test-function space in terms of creation and annihilation operators $a^{\pm}(f)$. Then we can introduce the gauge transformations

$$\gamma_{\theta}(a^{\pm}(f)) = e^{\pm i\theta} a^{\pm}(f)$$

Using the characterization (3) it is possible to describe separately canonical and grand canonical equilibrium states by their behaviour under local transformations τ, depending on whether one chooses these transformations to be implemented by, respectively gauge invariant unitaries or by general unitaries in the field algebra.

Finally there is yet an other characterization of equilibrium states which we want to mention and which is the quantum mechanical version of Theorem II.2.

Theorem III.4

A state ω of $(\mathcal{A},\alpha)$ is a α-KMS-state iff for all $A \in \mathcal{D}(\delta)$ one has the E.E.B. inequality

$$-\beta\omega(A^{\star} i\delta(A)) \geqslant \omega(A^{\star}A) \ln \frac{\omega(A^{\star}A)}{\omega(AA^{\star})} \qquad (4)$$

or iff for all $A \in \mathcal{D}(\delta)$ and $\tau \in Q$ one has the E.E.B. inequality :

$$\beta\omega(A^{\star}(-i\delta(A) + h(\tau)A))$$

$$\geqslant \omega(A^{\star}A) \ln \frac{\omega(A^{\star}A)}{\omega(\tau^{-1}(AA^{\star}))} \qquad (5)$$

Proof : The derivation of (4) from the KMS-condition is as in the proof of Theorem II.2. The proof of the converse is left as an exercise for the reader.
That (4) and (5) are equivalent is a matter of substituting

$$A \to UA \quad ; \quad U^{\star}U = UU^{\star} = 1$$

in (4) and using the notations

$$\tau = U^{\star} \cdot U$$

$$h(\tau) = U^{\star}HU - H$$

(see also :[8,9]) ■

Above we studied separately classical and quantum systems and we tried to develop as much as possible similar presentations. It is clear that this is also a good form to develop the theory of equilibrium conditions for systems having a classical and quantum subsystem interacting with each other. We are not entering into these problems but refer to [10] for more details. Instead, we turn to applications in the next sections.

IV. EQUILIBRIUM STATES FOR PERMUTATION INVARIANT (MEAN FIELD) MODELS

Here we consider a discrete quantum spin system, described by the following $C^{\star}$-algebra of observables

$$A_{\mathbb{N}} = \bigcup_{\Lambda \subset \mathbb{N}} A_{\Lambda}$$

i.e. $A_{\mathbb{N}}$ is the inductive limit of the finite algebras A_{Λ}, $\Lambda \subset \mathbb{N}$, with

$$A_{\Lambda} = \bigotimes_{i \subset \Lambda} M_n$$

M_n is the algebra of $n \times n$ complex matrices. Denote by P the group of local permutations of $\mathbb{N}$ i.e. if $p \in P$ then p is a permutation of the set $\mathbb{N}$ leaving unchanged all elements of $\mathbb{N}$ except of a finite number. Denote also by the same symbol the induced automorhism of A i.e. if

$$p = \begin{pmatrix} 1 & 2 & 3 \dots \\ \\ p_1 & p_2 & p_3 \dots \end{pmatrix}$$

then for the $X_i \in M_n$ and $X = X_1 \otimes X_2 \otimes \dots \in A$

$$p(X) = X_{p_1} \otimes X_{p_2} \otimes \dots$$

A state ω of $A_{\mathbb{N}}$ is called <u>symmetric</u> if $\omega \cdot p = \omega$ for all $p \in P$.

Let ρ be a state of M_n, denote

$$\omega_\rho = \bigotimes_i \rho_i$$

where ρ_i is a copy of ρ, the corresponding state of $A_{\mathbb{N}}$. The ω_ρ are called the <u>symmetric product states</u>. Størmer [11] proved the following theorem.

<u>Theorem IV.1</u>

The symmetric states of $A_{\mathbb{N}}$ form a simplex whose extreme points are the symmetric product states.

■

It follows immediately from this theorem that any symmetric state ω of $A_{\mathbb{N}}$ can be decomposed in a way into symmetric product states i.e. for any such state ω there exists a unique probability measure μ on $\mathcal{B}$, the state space of M_n such that

$$\omega = \int_{\mathcal{B}} d\mu(\rho)\ \omega_\rho \tag{1}$$

Recently, Theorem IV.1, has been extended to the situation of the symmetric states of a composite algebra of the type $A \otimes A_{\mathbb{N}}$ (see [12]).

Instead of turning into this generalization we keep our attention to an application of Theorem IV.1. We use this theorem in order to give a characterization of the equilibrium states of mean-field models. We will prove that the equilibrium states of these models are of the type given by (1) such that the support of the measure μ is on those symmetric product states satisfying the so-called gap equation. We give here a revised proof of what appeared in [13].

The mean-field systems which we have in mind are described by the following local Hamiltonians :

$$H_\Lambda = \sum_{i\in\Lambda} A_i + \frac{1}{2\ \#\ \Lambda} \sum_{\substack{i,j\in\Lambda \\ i\neq j}} B_{ij} \tag{2}$$

where $A_i \in M_n = \mathcal{A}_{\{i\}}$

$B_{ij} \in \mathcal{A}_{\{i,j\}}$

are copies of $A \in M_n$ and $B \in M_n \otimes M_n$ and where B is permutation invariant (i.e. B commutes with the permutation $\phi \otimes \psi \to \psi \otimes \phi$; $\phi,\psi \in C^n$).

The local Gibbs states at $\beta > 0$ are given in terms of the local Hamiltonians H_Λ (2) :

$$\omega_{\beta,\Lambda}(\ \cdot\ ) = \frac{\text{tr}\ e^{-\beta H_\Lambda}\ \cdot}{\text{tr}\ e^{-\beta H_\Lambda}} .$$

We are asking for the characterization of all possible limiting Gibbs states as $\Lambda \to \mathbb{N}$ in the sense that Λ eventually absorbs any finite subset of $\mathbb{N}$. From section III, Theorem III.4 the state $\omega_{\beta,\Lambda}$ is uniquely determined by the E.E.B.-inequalities :

$$\beta\omega_{\beta,\Lambda}(X^\star[H_\Lambda,X]) \geqslant \omega_{\beta,\Lambda}(X^\star X)\ \ln \frac{\omega_{\beta,\Lambda}(X^\star X)}{\omega_{\beta,\Lambda}(XX^\star)}$$

for all $X \in \mathcal{A}_\Lambda$.

Consider now any limiting point ω_β of the $\omega_{\beta,\Lambda}(\Lambda \to \mathbb{N})$ which exists by $W^\star$-compactness. As all $\omega_{\beta,\Lambda}$ are "locally" symmetric, the limit point ω_β will be symmetric. Hence there exists a unique probability measure $\mu_\beta(d\rho)$ on the state space $\mathcal{B}$ of M_n such that (Theorem IV.1 and (1)) :

$$\omega_\beta = \int_{\mathcal{B}} \mu_\beta\ (d\rho)\ \omega_\rho \tag{3}$$

First we prove, what is called in the physics literature, the existence of effective Hamiltonians.

Lemma IV.2

For any $X \in \mathcal{A}_\Lambda$, we have

$$\beta \int \mu_\beta(d\rho)\, \omega_\rho(X^\star [H_\Lambda^\rho, X]) \geq \omega_\beta(X^\star X) \ln \frac{\omega_\beta(X^\star X)}{\omega_\beta(XX^\star)} \tag{4}$$

where

$$H_\Lambda^\rho = \sum_{i\in\Lambda} H_i^\rho \quad ; \quad H_i^\rho \in \mathcal{A}_{\{i\}} \quad ; \quad H_i^\rho \text{ is a copy of } H^\rho \in M_n$$

and

$$H^\rho = A + B^\rho ,$$

B^ρ is the partial trace of B :

$$(\sigma \otimes \rho)(B) = \sigma(B^\rho) \quad \text{for all} \quad \sigma \in \mathcal{B}$$

Proof : Let $\{\Lambda_\alpha\}_\alpha$ be a net of finite volumes tending to $\mathbb{N}$ such that

$$\omega_\beta = W^\star - \lim_\alpha \omega_{\beta,\Lambda_\alpha}$$

For all $\Lambda_\alpha \supset \Lambda$, $X \in \mathcal{A}_\Lambda$, using the symmetry of $\omega_{\beta,\Lambda_\alpha}$ one gets :

$$\omega_{\beta,\Lambda_\alpha}(X^\star [H_{\Lambda_\alpha}, X])$$

$$= \omega_{\beta,\Lambda_\alpha}(X^\star \sum_{i\in\Lambda} A_i, X)$$

$$+ \frac{1}{2\,\#\,\Lambda_\alpha}\, \omega_{\beta,\Lambda_\alpha}(X^\star [\sum_{\substack{i,j\in\Lambda \\ i\neq j}} B_{ij}, X])$$

$$+ \frac{\# \Lambda_\alpha - \# \Lambda}{\# \Lambda_\alpha} \omega_{\beta,\Lambda_\alpha}(X^\star[\sum_{\substack{i\in\Lambda \\ k\in\Lambda_\alpha \\ k\notin\Lambda}} B_{i,k} , X])$$

Taking the limit α and using formula (3) one gets :

$$\lim_\alpha \omega_{\beta,\Lambda_\alpha}(X^\star[H_{\Lambda_\alpha},X]) = \int \mu_\beta(d\rho)\ \omega_\rho(X^\star[H^\rho_\Lambda,X])$$

■

Next we prove that the support of the measure μ_β is on the symmetric product states, which are invariant for their respective effective time evolutions.

Lemma IV.3

For μ_β-almost all ρ one has $[\rho,H^\rho] = 0$.

Proof : Use the result of Lemma IV.2 and substitute :

$$X = 1 + \lambda Y \quad ; \quad \lambda \in \mathbb{R} \quad ; \quad Y^\star = Y \in A_\Lambda$$

then for all $\lambda \in \mathbb{R}$

$$\beta \int \mu_\beta(d\rho)\ \omega_\rho((1 + \lambda Y)[H^\rho_\Lambda,\lambda Y]) \geqslant 0$$

Hence for all $X \in A_\Lambda$:

$$\int \mu_\beta\ (d\rho)\ \omega_\rho([H^\rho_\Lambda,X]) = 0\ .$$

This expresses the time invariance of the state ω_β. Choose now : $Y,Z_1,\ldots,Z_k \in M_n$ and take

$$X = Y \otimes Z_1 \otimes \ldots \otimes Z_k$$

then one gets

$$\int \mu_\beta(d\rho)\{\rho([H^\rho,Y])\rho(Z_1)\ \dots\ \rho(Z_k) +$$

$$\rho(Y) \sum_{i=1}^{k} \rho(Z_1)\ \dots\ \rho([H^\rho,Z_i])\ \dots\ \rho(Z_k)\} = 0$$

Substituting $Y \to Y + \lambda 1$, $\lambda \in \mathbb{R}$, then one checks immediately that

$$\int \mu_\beta(d\rho)\ \rho([H^\rho,Y])\ \rho(Z_1)\ \dots\ \rho(Z_k) = 0$$

for all $Y,Z_i \in M_n$.

As the set of functions

$$\tilde{Z} : \rho \to \tilde{Z}(\rho) = \omega_\rho(Z) \quad , \quad Z \in \bigcup_\lambda A_\Lambda$$

are continuous, contain the constant function (for Z = 1) and separate the state space of M_n, the Lemma follows from the theorem of Stone-Weierstrass.

■

Finally we are able to prove the main theorem of this section proving that the support of the measure μ_β consists of density matrices of M_n satisfying the gap equation.

Theorem IV.4

For μ_β-almost all ρ one has

$$\rho = \frac{\exp -\beta H^\rho}{\mathrm{tr}\ \exp -\beta H^\rho}$$

Proof : Take an arbitrary local observable Z, say $Z \in A_\Lambda$.
Take for N = 1,2,... copies $Z_1,Z_2,..Z_N$ of Z such that

$$Z_i \in A_{\Lambda_i} \quad ; \quad \Lambda_i \cap \Lambda_j = \phi \quad ; \quad \text{if} \quad i \neq j$$

and $1 \notin \Lambda_i$ for all i.

Take

$$X^N = \sum_{k=1}^{N} Y \otimes 1_{\Lambda_1} \otimes \ldots \otimes Z_k \otimes \ldots \otimes 1_{\Lambda_N}$$

with $Y \in A_{\{1\}}$. Then using IV.3

$$\int \mu_\beta(d\rho)\ \omega_\rho(X^{N\star}[H^\rho_{\{1\}\cup\Lambda_1\cup\ldots\cup\Lambda_N}, X^N])$$

$$= (N^2 - N) \int \mu_\beta(d\rho)\ \rho(Y^\star[H^\rho, Y])\,|\rho(Z)|^2 + O(N)$$

with the simplified notation $\rho(Z) = \omega_\rho(Z)$.

$$\omega_\beta(X^{N\star} X^N) = (N^2 - N) \int \mu_\beta(d\rho)\ \rho(Y^\star Y)\,|\rho(Z)|^2 + O(N)$$

$$\omega_\beta(X^N X^{N\star}) = (N^2 - N) \int \mu_\beta(d\rho)\ \rho(YY^\star)\,|\rho(Z)|^2 + O(N)$$

Substitute these expressions in the inequality (4) and perform the limit $N \to \infty$. One gets :

$$\beta \int \mu_\beta(d\rho)\ \rho(Y^\star[H^\rho, Y])\,|\rho(Z)|^2 \geqslant$$

$$\int \mu_\beta(d\rho)\ \rho(Y^\star Y)\,|\rho(Z)|^2 \ln \frac{\int \mu_\beta(d\rho)\ \rho(Y^\star Y)\,|\rho(Z)|^2}{\int \mu_\beta(d\rho)\ \rho(YY^\star)\,|\rho(Z)|^2}$$

for all $Y \in M_n$ and $Z \in \bigcup_\Lambda A_\Lambda$.

If we can show that the set $\tilde{D}^+$ of functions

$$|\tilde{Z}(\rho)|^2 = |\rho(Z)|^2 = |\omega_\rho(Z)|^2 \; ; \quad Z \in \bigcup_\Lambda A_\Lambda$$

is dense in the continuous positive functions $C^+(D_n)$ on D_n, the inequality of above can only hold if for μ_β almost everywhere

$$\beta\rho(Y^\star[H^\rho, Y]) \geq \rho(Y^\star Y) \ln \frac{\rho(Y^\star Y)}{\rho(YY^\star)} \; ; \quad Y \in M_n$$

But by III.4 this is equivalent to

$$\rho = \frac{\exp -\beta H^\rho}{\mathrm{tr}\exp -\beta H^\rho}$$

Finally the density of $\tilde{D}^+$ in $C^+(D_n)$ follows again from the fact that the set of functions $\tilde{D}$:

$$\tilde{X} : \rho \in D_n \to \tilde{X}(\rho) = \omega_\rho(X) \; ; \quad X \in \bigcup_\Lambda \Lambda$$

is an algebra under pointwise multiplication, containing the constants and separating the points of D_n. This is easily seen by approximating the square root of the positive continuous functions by an element of $\tilde{D}$. This finishes the proof of the theorem. ∎

The result of Theorem IV.4 is based on Störmer's Theorem IV.1 and the correlation inequality III.4. Störmer's theorem deals with permutation invariant states. A typical example is the strong coupling BCS-model

$$H_\Lambda = \varepsilon \sum_{i\in\Lambda} \sigma_i^3 + \frac{\lambda}{\#\Lambda} \sum_{i,j\in\Lambda} \sigma_i^+ \sigma_j^- \tag{5}$$

where σ^+, σ^- and σ^3 are the Pauli matrices.

However, a model of the type

$$H_\Lambda = \sum_{i\in\Lambda} \varepsilon_i \sigma_i \sigma_i^3 + \frac{\lambda}{\#\Lambda} \sum_{i,j\in\Lambda} \sigma_i^+ \sigma_j^- \; ; \; \varepsilon_i \neq \varepsilon_j \quad \text{if} \quad i \neq j \qquad (6)$$

can not be handled by means of Theorem IV.4 because this Hamiltonian is not permutation invariant, and the equilibrium states are not yet completely characterized. Recently the classical Hamiltonians of this type i.e.

$$H_\Lambda = \sum_{i\in\Lambda} \varepsilon_i \sigma_i^3 + \frac{\lambda}{\#\Lambda} \sum_{i,j\in\Lambda} \sigma_i^3 \sigma_j^3 \qquad (7)$$

where treated. Fannes [14] generalized in a non trivial way Störmer's theorem as follows. Denote $\phi_i = \varepsilon_i \sigma_i^3$, then

$$p\,H_\Lambda - H_\Lambda = \sum_{i\in\Lambda} (p\,\phi_i - \phi_i) \equiv \Phi_\Lambda(p)$$

Suppose that the $\lim_\Lambda \Phi_\Lambda(p) = \Phi(p)$ exists, he introduces a new notion of symmetry as follows. A probability measure μ on $K = \{\pm 1\}^{\mathbb{N}}$ satisfies the Φ-symmetry if for all $f \in C(K)$ and $p \in P$ one has

$$\mu(p^{-1} f) = \mu(f\; e^{-\Phi(p)})$$

Then he proves :

<u>Theorem IV.5</u> [14]

If ϕ is a uniformly bounded potential (i.e. $\sup_i \|\phi_i\| < \infty$), then the set of Φ-symmetric states of $C(K)$ is a simplex whose extreme points are the Φ-symmetric product states.

It is clear that one can characterize the equilibrium states of Hamiltonians of the type (7), in the sense of Theorem IV.4 using this result and the classical correlation function (II.2). An analogous result as Theorem IV.5 in the non-commutative case is still an open problem.

V. THE SPIN-BOSON MODEL [15,16,17]

In this section we illustrate a method to solve the KMS-condition (III.1) in the case of a model which is so far not explicitly soluble, given by the following Hamiltonian :

$$H = \mu\, \sigma_x + \int dk\ \varepsilon(k) a^+(k) a(k) + \sigma_z \int dk\ \lambda(k)(a^+(k) + a(k)) \tag{1}$$

where the real functions λ and ε satisfy

$$\int dk\ \lambda(k)^2 < \infty \quad ; \quad \int dk\ \frac{\lambda(k)^2}{\varepsilon(k)} < \infty ,$$

and σ_x, σ_z are the Pauli matrices, $a^{(+)}$ are the Bose creation and annihilation operators.

The algebra of observables of this system is of the form $M_2 \otimes A$, where M_2 is the spin $\frac{1}{2}$ algebra of the complex 2×2 matrices and A is the CCR-algebra of the Bose field. We postpone the particular choice we have to make for A. Anyway the elements of $M_2 \otimes A$ are represented by

$$X = \begin{pmatrix} X_{11} & X_{12} \\ X_{21} & X_{22} \end{pmatrix} \tag{2}$$

where the $X_{ij} \in A$. A state of $M_2 \otimes A$ defines the linear functionals ω_{ij} (i,j = 1,2) of A

$$\omega = \begin{pmatrix} \omega_{11} & \omega_{12} \\ \omega_{21} & \omega_{22} \end{pmatrix} \tag{3}$$

and

$$\omega(X) = \sum_{ij} \omega_{ji}(X_{ij})$$

The representation (3) defines a state if and only if

$$\omega_{11}(1) + \omega_{22}(1) = 1$$

$$\overline{\omega_{12}(x)} = \omega_{21}(x^\star)$$

$$|\omega_{12}(x^\star y)|^2 \leq \omega_{11}(x^\star x)\omega_{22}(y^\star y)$$

$$\omega_{11}(x^\star x) \geq 0 \quad ; \quad \omega_{22}(y^\star y) \geq 0$$

for all $x, y \in A$.

We will solve the problem by perturbation theory. Therefore we consider first the system

$$H_0 = \int dk\ \varepsilon(k) a^+(k) a(k) + \sigma_z \int dk\ \lambda(k)(a^+(k) + a(k)) \qquad (4)$$

which we call the unperturbed system.

Consider for A a von Neumann algebra containing the Weyl CCR-algebra built on a Hilbert space $\mathcal{H}$ and hence generated by the Weyl operators

$$W(f) = \exp i(a^+(f) + a(f))$$

$f \in \mathcal{H}$, acting on Fock space.
Now compute on $M_2 \otimes A$:

$$\alpha_t^0 = e^{itH_0} \cdot e^{-itH_0}$$

This yields :

$$\alpha_t^0(\sigma_\pm) = \sigma_\pm W(\pm \frac{2i\lambda}{\varepsilon}(1 - e^{it\varepsilon}))$$

with $\sigma_\pm = \frac{1}{2}(\sigma_x \pm i\sigma_y)$

$$\alpha_t^0(\sigma_z) = \sigma_z$$

$$\alpha_t^0(W(f)) = W(e^{it\varepsilon}f) \exp i\, \sigma_z \operatorname{Im}(\frac{2i\lambda}{\varepsilon}, (1 - e^{it\varepsilon})f)$$

Take now for A the smallest von Neumann algebra such that α_t^0 extends to a continuous one-parameter group of *-automorphisms of $M = M_2 \otimes A$. This way we defined a $W^\star$-dynamical system

$$(M, \alpha_t^0).$$

Now it is well known that the full dynamics α_t can be defined on M by means of the Dyson-expansion :

$$\alpha_t(X) = \alpha_t^0(X) + \sum_{n=1}^{\infty} (i\mu)^n \int_0^t ds_1 \int_0^{s_1} ds_2 \cdots \int_0^{s_{n-1}} ds_n$$

$$[\alpha_{s_n}^0(\sigma_x), [\ \ldots [\alpha_{s_1}^0(\sigma_x), \alpha_t^0(X)] \ldots] \qquad (5)$$

We get then the $W^\star$-dynamical system (M, α_t).

Now we use the following theorem about the stability of KMS-states, describing the thermal wave function as a perturbation series of the unperturbed one.

Theorem V.1 [5]

Let (M,α_t^0) be a $W^\star$-dynamical system, Ω_0 a vector of the Hilbert space $\mathcal{H}$ which is cyclic for M and such that the state

$$\omega_\beta^0(\cdot) = (\Omega_0, \cdot\Omega_0)$$

of (M,α^0) is β-KMS. If α_t is the perturbed evolution (5), then

(i) there exists an element $\Omega \in \mathcal{H}$, given by

$$\Omega = \Omega_0 + \sum_{n=1}^{\infty} (-\mu)^n \int_0^{\beta/2} ds_1 \ldots \int_0^{s_{n-1}} ds_n\, \alpha_{is_n}^0(\sigma_x) \ldots \alpha_{is_1}^0(\sigma_x)\Omega$$

(ii) $\Omega \neq 0$ and

$$\omega_\beta(\cdot) = \frac{(\Omega, \cdot\Omega)}{\|\Omega\|^2}$$

is a state of (M,α) which is β-KMS.

(iii) there exists a bijection between the normal β-KMS states of (M,α^0) and of (M,α).

■

In particular it follows from this theorem that ω_β is the unique β-KMS-state of (M,α) if ω_β^0 is the unique β-KMS-state of the unperturbed dynamical system (M,α^0). The message is clear, solve the KMS-condition for the unperturbed system (M,α^0). We are able to prove the following theorem.

Theorem V.2

For all $\beta > 0$, there exists a unique β-KMS-state ω_β^0 of the $W^\star$-dynamical system (M,α^0) given by :

$$\omega_\beta^0 = \frac{1}{2}\begin{pmatrix} \omega_+ & 0 \\ 0 & \omega_- \end{pmatrix}$$

$$\omega_\pm(W(f)) = \exp\{\pm\, i\, \mathrm{Im}\, (\frac{2i\lambda}{\varepsilon}, f) - \frac{1}{2}(f, \mathrm{cth}(\frac{\beta\varepsilon}{2}) f)\}$$

Proof : The proof is divided into different steps. By (3) any solution of the KMS-condition is of the form

$$\omega_\beta^0 = \begin{pmatrix} \omega_{11} & \omega_{12} \\ \omega_{21} & \omega_{22} \end{pmatrix}$$

(a) first we prove that $\omega_{12} = \omega_{21} = 0$. From the equation

$$\omega_\beta^0(A\ \alpha_{i\beta}^0\ B) = \omega_\beta^0(BA) \ ; \quad A, B \in M$$

with $A = \sigma_x\, W(f)$, $B = \sigma_z$ one has

$$-2i\ \omega_\beta^0(\sigma_y\ W(f)) = \omega_\beta^0(\sigma_x\ W(f)\sigma_z)$$

$$= \omega_\beta^0(\sigma_x\ W(f)\ \alpha_{i\beta}^0\ \sigma_z) = 2i\ \omega_\beta^0(\sigma_y\ W(f))$$

Hence $\omega_\beta^0(\sigma_y\ W(f)) = 0$;

analogously $\omega_\beta^0(\sigma_x\ W(f)) = 0$.

Therefore

$$\omega_{12}(W(f)) = \omega_{21}(W(f)) = 0 \quad \text{for all } f \text{ or } \omega_{12} = \omega_{21} = 0$$

and any solution is of the form

$$\omega_\beta^0 = \begin{pmatrix} \omega_1 & 0 \\ 0 & \omega_2 \end{pmatrix}$$

(b) Now we check that the $\omega_{1,2}$ are β-KMS-positive functionals of the $W^\star$-dynamical systems $(A,\alpha^\pm)$, where A is the CCR-algebra and

$$\alpha_t^\pm(W(f)) = W(e^{it\varepsilon}f)\exp \pm i \, \mathrm{Im}(\frac{2i\lambda}{\varepsilon},(1 - e^{it\varepsilon})f)$$

These are quasi-free evolutions and it is well-known [15] that for these dynamics the KMS-condition has a unique solution. We conclude immediately that

$$\omega_1 = \eta \, \omega_+$$

$$\omega_2 = (1 - \eta)\omega_-$$

where $\omega_\pm$ are given as in the theorem, and $\eta \in [0,1]$.
We derived that all solutions are of the form

$$\omega_\beta^0 \equiv \omega_\eta = \begin{pmatrix} \eta\omega_+ & 0 \\ 0 & (1 - \eta)\omega_- \end{pmatrix}$$

(c) Finally we prove the unicity by proving that $\eta = 1/2$. We derive this again from the KMS-condition :

$$\eta = \omega_\eta(\sigma_+ \, \alpha_{i\beta}^0 \, \sigma_-) = \omega_\eta(\sigma_-\sigma_+) = 1 - \eta$$

■

This finishes the proof of the existence and unicity of the KMS-conditions for the spin-boson model. The solution is given in terms of a perturbation series for the thermal wave functions. We remark that the problem of the ground state solutions is still open.

VI. ABSENCE OF SPONTANEOUS SYMMETRY BREAKING

In this we develop a general argument which can be used to prove the absence of spontaneous symmetry breaking for quantum mechanical systems with an internal symmetry. What is the problem ? Suppose we have a $W^{\star}$-dynamical system (M,α) given by a von Neumann algebra M on a Hilbert space $\mathcal{H}$ and a continuous one-parameter group of $\star$-automorphisms $(\alpha_t)_{t\in\mathbb{R}}$. Without loss of generality for what follows we may assume that the time evolution is implemented by unitaries

$$\alpha_t = e^{itH} \cdot e^{-itH}$$

where $H = H^{\star}$ is called the Hamiltonian, a self-adjoint operator on the Hilbert space $\mathcal{H}$. Again we are interested in the state ω_β of (M,α) which satisfies the β-KMS-condition :

$$\omega_\beta(A\ \alpha_{i\beta}\ B) = \omega_\beta(BA) \tag{1}$$

for all $A,B \in M_\alpha$ = a dense α-invariant subalgebra of M.

Suppose now that τ is a symmetry of the system (M,α) i.e. τ is a $\star$-automorphism of M such that $\tau^{-1}\ \alpha_t\ \tau = \alpha_t$ $(t \in \mathbb{R})$. It is immediately clear from this, that $\omega_\beta \cdot \tau$ is also a β-KMS-state of (M,α) if ω_β satisfies (1).

If there exists a state ω_β satisfying (1) such that $\omega_\beta \cdot \tau \neq \omega_\beta$ then there exist solutions of (1) which are not τ-invariant. One says that the symmetry τ is spontaneously broken.

Now we are developping a general argument in order to disprove the occurence of spontaneous symmetry breaking based on the E.E.B.-correlation inequality. First we mention the mathematical theorems on which the argument will be based.

Theorem VI.1 [5, Theorem 5.3.30]

Denote by K_β the set of β-KMS-states of a dynamical system (M,α), then

(i) K_β is a convex set

(ii) K_β is $W^\star$-compact

(iii) K_β is a simplex

(iv) $\omega \in K_\beta$ is an extremal point iff ω is a factor state.

■

It follows from this theorem that each β-KMS-state ω has a unique decomposition into extremal factor states. As the KMS-condition is an equation linear in the state, it is sufficient to study only the extremal ones. Moreover, one has also the following result :

Theorem VI.2 [5, Proposition 5.3.29]

Let $\omega \in K_\beta$ and suppose that ρ is an other state of M such that $\rho(X^\star X) \leqslant \lambda\omega(X^\star X)$ for all $X \in M$, then the following are equivalent :

(i) $\rho \in K_\beta$

(ii) there exists a positive operator T in the center $M \cap M'$ (M' = commutant of M) such that : $\forall\, X \in M$;

$$\rho(X) = \omega(T^{1/2}\; X\; T^{1/2})$$

Moreover, T is unique if (i) or (ii) is satisfied.

■

Now we are able to show

Theorem VI.3

Suppose τ is a symmetry of the system (M,α) and $\omega \in K_\beta$, then $\omega \cdot \tau = \omega$ if there exists a constant $\lambda \in \mathbb{R}^+$ such that

$$\omega \cdot \tau \leqslant \lambda\; \omega\; .$$

Proof : By Theorem VI.1 it is sufficient to prove the theorem for the extremal components of $\omega \in K_\beta$.

If ω is extremal the result follows from Theorem VI.2 and the fact that ω is a factor state, then

$$T \in M \cap M' = \mathbb{C}\, 1$$

or

$$T = \lambda\, 1 \quad ; \quad \lambda \in \mathbb{R}^+ \quad \text{and} \quad \omega \cdot \tau = \lambda\, \omega$$

but by normalization of $\omega \cdot \tau$ one has $\lambda = 1$.

■

Suppose that we have the realistic situation of a $W^\star$-dynamical system (M,α), and a state ω satisfying the E.E.B. inequality

$$\beta\omega(X^\star[H,X]) \geqslant \omega(X^\star X) \ln \frac{\omega(X^\star X)}{\omega(XX^\star)} \tag{2}$$

for all $X \in \mathcal{D}(H)$ and

$$\omega(\,\cdot\,) = (\Omega, \,\cdot\, \Omega) \text{ , } \Omega \text{ is cyclic for } M.$$

Normally one has

$$\omega(X^\star[H,X]) = \lim_\Lambda \omega(X^\star[H_\Lambda,X])$$

where $\{H_\Lambda\}_\Lambda$ is a sequence of given local Hamiltonians.

Let τ be a *-automorphism of M satisfying

(i) τ is approximately inner : i.e. $\exists(U_n)_{n\geqslant 1}$ a sequence of unitaries of M, such that for all $X \in M$:

$$\lim_n \| \tau(X) - U_n^\star X\, U_n \| = 0 \tag{3}$$

(ii) τ is a symmetry of the system : for all $t \in \mathbb{R}$

$$[\alpha_t,\tau] = 0 \text{ and for all } n \in \mathbb{N} :$$

$$U_n \in \mathcal{D}(H)$$

and

$$\sup_n \| [H, U_n] \| = K < \infty \tag{4}$$

this means that α_t commutes almost with the local approximations U_n of τ. This is a rather strong condition which can be weakened as will be clear from the proof.

Theorem VI.4

Let τ be a symmetry as above then if $\omega \in K_\beta$ there exists a constant C such that for all $X \in \mathcal{M}$:

$$\omega(\tau(X^\star X)) \leqslant C\ \omega(X^\star X)$$

and by VI.3, there is no spontaneous breaking of the τ-symmetry.

Proof : As in section III consider the dense α-invariant *-subalgebra $\mathcal{M}_\alpha$ of $\mathcal{M}$, generated by the elements

$$A(f)\ ,\ f \in C^\infty(\mathbb{R})$$

For any $\varepsilon > 0$ one finds a decomposition of the identity by a sequence $(h_n)_{n\geqslant 1}$ of functions in C^∞ such that

$$\hat{h}_n \geqslant 0$$

$$\sum_{n\geqslant 1} \hat{h}_n^2 = 1$$

$$\text{supp } \hat{h}_n \text{ contained in an interval of length less than } \varepsilon.$$

One computes :

$$\omega(AA^\star) = \int d\nu_A(-\lambda) = \sum_n \int \hat{h}_n(\lambda)^2 d\nu_A(-\lambda)$$

$$= \sum_n \omega(A(h_n)A(h_n)^*) \qquad (*)$$

Substitute in the E.E.B. inequality (2)

$$X \to U_m \, X(h_n)$$

then one gets

$$\omega(X(h_n)^* X(h_n)) \ln \frac{\omega(X(h_n)^* X(h_n))}{\omega(U_m \, X(h_n) \, X(h_n)^* U_n^*)}$$

$$\leq \beta\omega(X(h_n)^* U_m^* [H, U_m \, X(h_n)])$$

$$= \beta\omega(X(h_n)^* U_m^* [H, U_m] X(h_n))$$

$$+ \beta\omega(X(h_n)^* [H, X(h_n)])$$

After adding and substracting a term :

$$\omega(X(h_n)^* X(h_n)) \ln \frac{\omega(X(h_n) \, X(h_n)^*)}{\omega(U_m \, X(h_n) \, X(h_n)^* U_m^*)}$$

$$- \beta\omega(X(h_n)^* U_m^* [H, U_m] X(h_n))$$

$$\leq \beta\omega(X(h_n)^* [H, X(h_n)])$$

$$- \omega(X(h_n)^* X(h_n)) \ln \frac{\omega(X(h_n)^* X(h_n))}{\omega(X(h_n) \, X(h_n)^*)} \qquad (**)$$

One proves afterwards that the right hand side of this inequality is majorized by

$$\varepsilon\ \omega(X(h_n)^* X(h_n)) \qquad (***)$$

Using (4) :

$$|\omega(X(h_n)^* U_m [H,U_m] X(h_n))| \leq K\ \omega(X(h_n)^* X(h_n))$$

one gets from (**) :

$$\omega(X(h_n)X(h_n)^*) \leq e^{(K+\varepsilon)} \omega(U_m\ X(h_n)\ X(h_n)^* U_m^*)$$

And after summation over n, using (*) and (3)

$$\omega(XX^*) \leq e^{(K+\varepsilon)} \omega(\tau^{-1}(XX^*))$$

Substituting $X \to \tau X$ proves the relative boundedness of $\omega \cdot \tau$ with respect to ω.

Finally we prove (***). This follows from the following argument. Suppose that $\text{supp}\ \mu_A \subset I = [\lambda_1,\lambda_2] \subset \mathbb{R}$, where μ_A is the spectral measure

$$d\mu_A(\lambda) = \omega(A^* dE(\lambda)A)$$

defined in section III, then

$$0 \leq \beta\omega(A^*[H,A]) - \omega(A^*A) \ln \frac{\omega(A^*A)}{\omega(AA^*)}$$

$$= \int \beta\lambda\ d\mu_A(\lambda) - \int d\mu_A(\lambda) \ln \frac{\int d\mu_A(\lambda)}{\int e^{-\lambda\beta} d\mu_A(\lambda)}$$

$$\leq (\lambda_2 - \lambda_1) \int d\mu_A(\lambda) = (\lambda_2 - \lambda_1)\omega(A^*A)$$

where we used Theorem III.2 and the fact that $\lambda_2 \leqslant \inf \operatorname{supp} \mu_A$

$$\lambda_2 \geqslant \sup \operatorname{supp} \mu_A$$

■

The applications of this technique which we described above to disprove the occurence spontaneous symmetry breaking are numerous. One can follow this line of proof for continuous symmetry groups [see e.g. 19,20,21,22] but this technique has the advantage on the Bogoliubov inequality, that it is also applicable for discrete symmetry groups [23,24] .

REFERENCES

1. Dobrushin R.L., Theory Prob. Appl. 13, 197 (1968).
2. Lanford O.E. and Ruelle D., Commun. Math. Phys. 13, 194 (1969).
3. Fannes M., Vanheuverzwijn P. and Verbeure A., J. Stat. Phys. 29, 547 (1982).
4. Gallavotti G. and Verboven E., Il Nuovo Cimento B 28, 274 (1975).
5. Bratteli O. and Robinson D.W., Operator Algebras and Quantum Statistical Mechanics Vol. I and II (Berlin - Springer 1979, 1981).
6. Vanheuverzwijn P., Lett. Math. Phys. 7, 373 (1983).
7. Araki H., Pac. J. Math. 50, 309 (1974).
8. Fannes M. and Verbeure A., Commun. Math. Phys. 57, 165 (1977).
9. Fannes M. and Verbeure A., J. Math. Phys. 19, 558 (1978).
10. Fannes M. and Verbeure A., J. Phys. A : Math. Gen. 20, (1987).
11. Störmer E., J. Funct. Anal. 3, 48 (1969).
12. Fannes M., Lewis J.T. and Verbeure A., Lett. Math. Phys. (1988).
13. Fannes M. Spohn H. and Verbeure A., J. Math. Phys. 21, 355 (1980).
14. Fannes M., "Non-translation invariant product states of classical lattice systems", to appear.
15. Fannes M., Nachtergaele B. and Verbeure A., Europhysics Lett. 4, 963 (1987).

16. Fannes M., Nachtergaele B. and Verbeure A., Commun. Math. Phys. (1987).

17. Fannes M., Nachtergaele B. and Verbeure A., J. Phys. A : Math. Gen. 21, (1988).

18. Rocca F., Sirugue M. and Testard D., Commun. Math. Phys. 19, 119 (1970).

19. Mermin N.D. and Wagner H., Phys. Rev. Lett. 17, 1133 (1966).

20. Hohenberg P.C., Phys. Rev. 158, 383 (1967).

21. Bonato C.A., Perez J.F. and Klein A., J. Stat. Phys. 29, 159 (1982).

22. Fröhlich J. and Pfister C.E., Commun. Math. Phys. 81, 277 (1981).

23. Fannes M., Vanheuverzwijn P. and Verbeure A., Phys. Lett. 94A, 418 (1983).

24. Fannes M., Vanheuverzwijn P. and Verbeure A., J. Math. Phys. 25, 76 (1984).

THE RENORMALIZATION GROUP ON HIERARCHICAL LATTICES

P.M. Bleher

The Keldysh Institute of Applied Mathematics
of the USSR Academy of Sciences
Moscow A-47, Miusskaya sq., 4
USSR

1. INTRODUCTION.

On the Migdal-Kadanoff approximate scheme.

The renormalization group (RG) was implemented into the statistical physics in early seventies (see [1]). It gives a general description of critical phenomena and phase transitions. The aim of my lecture is to present some basic RG ideas on the example of a simple classical spin model, the Ising model on hierarchical lattices. This model can be viewed as a caricature of real models. As a matter of fact it arises in the Migdal-Kadanoff approximate RG scheme. The history of the model is the following.

At the beginnig of the seventies Berezinskii [2] proposed a very nice idea of a successive integration of spin variables which leads to a system of recurrent equations for restricted partition functions. To explain it let us consider square S_n with side*) 2^n on square lattice $\mathbb{Z}^2$. Let Γ_n be the boundary of S_n and $\sigma(\Gamma_n) = \{\sigma(i),\ i\in\Gamma_n\}$

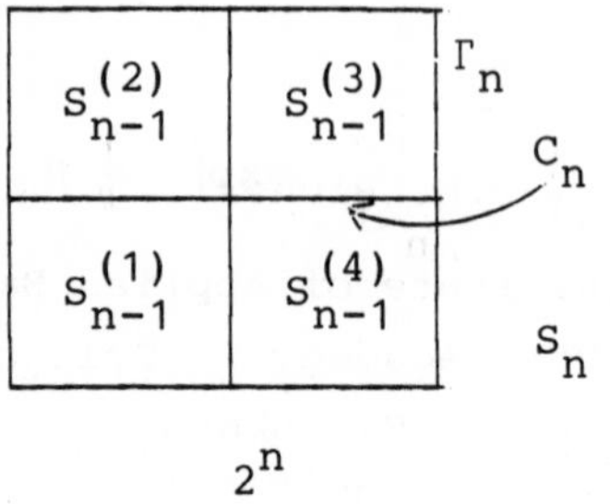

Fig. 1

be the set of boundary spins. Square S_n consists of 4 squares $S_{n-1}^{(i)}$ with side 2^{n-1} (see Fig.1). Denote their boundaries $\Gamma_{n-1}^{(i)}$ and let $\sigma(\Gamma_{n-1}^{(i)}) = \{\sigma(j), j\in\Gamma_{n-1}^{(i)}\}$. Consider the restricted partition functions

$$Z_n(T,\sigma(\Gamma_n)) = \sum_{\{\sigma(i),\, i\in S_n\setminus\Gamma_n\}} \exp\left(-\frac{1}{T} H_{S_n}(\sigma)\right) \tag{1.1}$$

of the Ising model on S_n. Here T is the temperature and

$$H_{S_n}(\sigma) = -J \sum_{\substack{|i-j|=1,\\ i,j\in S_n}} \sigma(i)\sigma(j) - h \sum_{i\in S_n} \sigma(i) \tag{1.2}$$

is the Hamiltonian of the Ising model, $\sigma(i) = \pm 1$, J is the coupling constant, h is the external magnetic field. Note that in (1.1) the values of boundary spins $\sigma(i)$, $i\in\Gamma_n$, are

*) We mean here that the length of the side of S_n is equal 2^n so the side contains 2^n+1 integer points.

fixed. $Z_n(T,\sigma(\Gamma_n))$ can be calculated in the following way. Let $\bar{\Gamma}_n = \bigcup_{i=1}^{4} \Gamma_{n-1}^{(i)}$. $\bar{\Gamma}_n$ consists of Γ_n and the "cross" C_n inside of S_n. Let

$$Z_n(T,\sigma(\bar{\Gamma}_n)) = \sum_{\{\sigma(i),\, i\in S_n\setminus\bar{\Gamma}_n\}} \exp\left(-\frac{1}{T} H_{S_n}(\sigma)\right)$$

be a restricted partition function with fixed values $\sigma(i)$, $i\in\bar{\Gamma}_n$. As only neighbour spins interact in (1.2), $Z_n(T,\sigma(\bar{\Gamma}_n))$ is factorized,

$$Z_n(T,\sigma(\bar{\Gamma}_n)) = \prod_{i=1}^{4} Z_{n-1}(T,\sigma(\Gamma_{n-1}^{(i)})) ,$$

so

$$Z_n(T,\sigma(\Gamma_n)) = \sum_{\{\sigma(i),\, i\in C_n\}} \prod_{i=1}^{4} Z_{n-1}(T,\sigma(\Gamma_{n-1}^{(i)})). \qquad (1.3)$$

These are just the Berezinskii recurrent equations for restricted partition functions $Z_n(T,\sigma(\Gamma_n))$.

Starting with the recurrent equations (1.3) Migdal developed in [3] as approximate scheme which leads to simple recurrent equations for the restricted partition functions. Now we shall describe it briefly. Consider again square S_n decomposition into squares $S_{n-1}^{(i)}$, $i=1,\ldots,4$. Each of squares $S_{n-1}^{(i)}$ can be decomposed in his own turn into four squares $S_{n-2}^{(ij)}$, $j=1,\ldots,4$, with side 2^{n-2} and so on. Thus we have partitions $\xi_k = \{S_{n-k}^{(\alpha)}, \alpha=i_1 i_2 \ldots i_k, i_1=1,\ldots,4\}$ of square S_n such that any element $S_{n-k}^{(\alpha)}$ of partition ξ_k consists of four elements $S_{n-k-1}^{(\alpha j)}$, $j=1,\ldots,4$, of partition ξ_{k+1} .

Assume now three things:

(i) Spins $\sigma(i)$ "live" on the elementary faces (edges) of square S_n (and not on the vertices): $i \in L(S_n)$, where $L(S_n)$ is the set of elementary faces in S_n,

(ii) For any element $S_{n-k}^{(\alpha)}$ of partitions ξ_k the values of spin $\sigma(i)$ are the <u>same</u> along any side of square $S_{n-k}^{(\alpha)}$.

The point (ii) is the main assumption in the Migdal approximation. To formulate it in a more formal way we say that two elementary faces $i,j \in L(S_n)$ are <u>equivalent</u> if i,j belong to the same side of an element (square) $S_{n-k}^{(\alpha)}$. Then (ii) means $\sigma(i) = \sigma(j)$ for any equivalent i,j. To finish the description of the Migdal approximation we say that two elementary faces are <u>opposite</u> iff they are parallel and belong to the same elementary square of S_n. Assume

(iii) The spins on the opposite faces interact in the same way as in the Ising model, i.e. they give the input $-J\sigma(i)\sigma(j)$ to the Hamiltonian.

The points (i)-(iii) describe the Migdal approximate model. Its Hamiltonian is

$$H_n(\sigma) = - \sum_{i,j \in L(S_n)} U(i,j)\sigma(i)\sigma(j) - h \sum_{i \in L(S_n)} \sigma(i) \, , \tag{1.4}$$

where

$$U(i,j) = \begin{cases} \infty & \text{if } i \text{ and } j \text{ are equivalent,} \\ J & \text{if } i \text{ and } j \text{ are opposite,} \\ 0 & \text{otherwise} \end{cases}$$

For the Migdal approximate model the Berezinskii recurrent equations (1.3) become

$$Z_n(1,2,3,4) = \sum_{5,6,7,8} Z_{n-1}(1,5,8,4)\, Z_{n-1}(1,2,6,5)$$

$$Z_{n-1}(6,2,3,7)\, Z_{n-1}(8,7,3,2) \qquad (1.5)$$

where we write i instead of $\sigma(i)$ for brevity and $\sigma(1),\ldots,$ $\sigma(4)$ are the spin values on the sides of Γ_n and $\sigma(5),\ldots,$ $\sigma(8)$ are the same on the cross C_n (see Fig.2.).

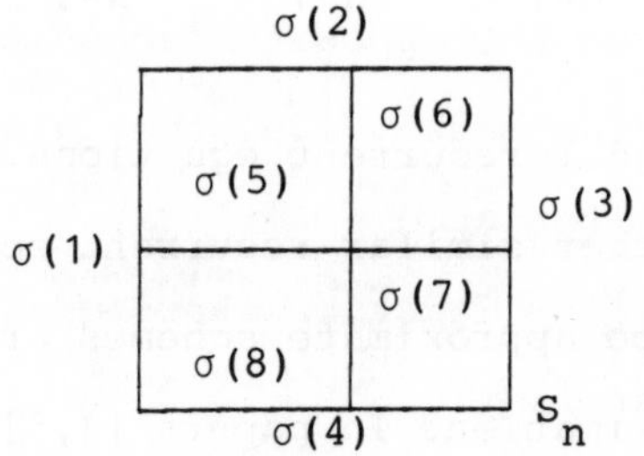

Fig.2. Spins in Migdal's approximation.

The equations (1.5) are factorized. Let $I_k(S_n)$, k=1,2,

be the set of faces orthogonal to the k-th axis. Then

$$H_n(\sigma) = H_{n1}(\sigma^{(1)}) + H_{n2}(\sigma^{(2)}), \tag{1.6}$$

where $\sigma^{(k)} = \{\sigma(j), j \in L_k(S_n)\}$, k=1,2, and

$$H_{nk}(\sigma^{(k)}) = - \sum_{i,j \in L_k(S_n)} U(i,j)\sigma(i)\sigma(j) - h \sum_{i \in L_k(S_n)} \sigma(i), \tag{1.7}$$

which means that spins on non-parallel faces do not interact so

$$Z_n(1,2,3,4) = Z_n(1,3)Z_n(2,4)$$

and (1.5) becomes

$$Z_n(1,3) = \sum_{6,8} Z_{n-1}(1,8)Z_{n-1}(1,6)Z_{n-1}(6,3)Z_{n-1}(8,3). \tag{1.8}$$

These are just the Migdal recurrent equations. Later Kadanoff described some other similar recurrent schemes (see [3]) and therefore such type approximate schemes are called the Migdal-Kadanoff approximation. In papers [4,5] it was noted that the Migdal equations (1.8) are exact for the model with the Hamiltonian (1.4).

The spin system with the Hamiltonian (1.7) permits a simple interpretation (see [6]). As the spins on equivalent elementary faces take the same value it is natural to unite them into one spin. So let us decompose set $L_1(S_n)$ into the subsets of equivalent faces, $L_1(S_n) = \bigcup_{v \in V_n} M_v$. Consider V_v as a set of vertices of graph Γ^n with the condition that any two vertices $v,w \in V_n$ are connected by an edge if there exist opposite faces $i,j \in L_1(S_n)$ such as $i \in M_v$, $j \in M_w$. The resulting graph Γ^n is just what is called a hierarchical lattice. Below we shall give a general definition of the hierarchical lattices. A spin system with the Hamiltonian (7) is reduced to one on hierarchical lattice Γ^n with the Hamiltonian

$$H_n(\sigma) = - J \sum_{\substack{<v,w> \\ v,w \in V_n}} \sigma(v)\sigma(w) - h \sum_{v \in V_n} m(v)\sigma(v)$$

where $m(v)$ is "multiplicity" $|M_v|$ of spin $\sigma(v)$, the number of equivalent faces in a given class M_v. The last formula can be rewritten also as

$$H_n(\sigma) = - J \sum_{\substack{<v,w> \\ v,w \in V_n}} \sigma(v)\sigma(w) - \frac{h}{2} \sum_{\substack{<v,w> \\ v,w \in V_n}} [\sigma(v) + \sigma(w)]$$

This is a starting formula for subsequent considerations (for technical reasons we shall use below the factor $\frac{J}{2}$ instead of J).

The setup of remainder of the paper is the following. We define hierarchical lattices in section 2 and the Ising model on such lattices in section 3. Furthermore we derive

RG equations for partition functions and discuss some properties of these equations. In section 4 we calculate free energy, spontaneous magnetization and some other characteristics of the model under consideration. In section 5 we discuss the limit Gibbs states of the model and in sections 6, 7 we present some other results concerning spin systems on hierarchical lattices.

The main results of this work were obtained by the author in collaboration with E. Žalys (see [7,8]).

2. HIERARCHICAL LATTICES

Let G be a set of connected oriented (i.e. each edge has its initial and final vertices) graphs $\Gamma=(V,L,i)$ with two marked vertices $\alpha,\tau\in V$ which we shall call sometimes external ones. For $\Gamma,\Gamma'\in G$ define a product $\Gamma'' = \Gamma'\Gamma\in G$.

Definition. Let $\Gamma,\Gamma'\in G$ and $\alpha,\tau\in V$, $\alpha',\tau'\in V'$ be marked vertices of Γ and Γ' . Insert instead of each edge $l\in L$ of Γ graph Γ' identifying α' with the initial vertex of l and τ' with its final vertex. The resulting graph Γ'' is called the product of Γ' and Γ, $\Gamma'' = \Gamma'\Gamma$. Its marked vertices α'',τ'' are defined as the images of α and τ with respect to the described operation.

Example

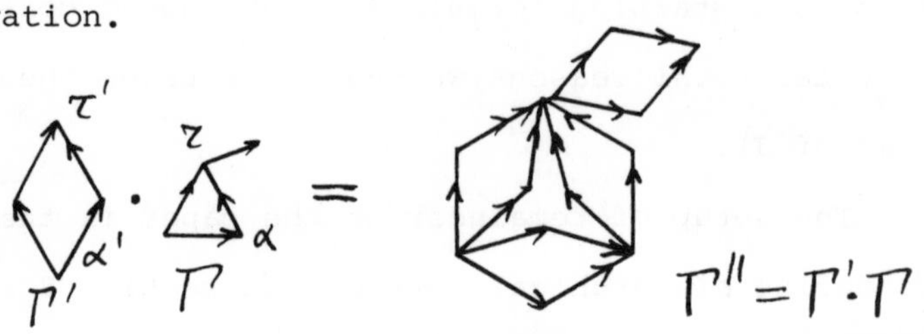

Fig. 3. Multiplication of graphs.

The operation of the product $\Gamma'\cdot\Gamma$ is not commutative but associative, i.e.

$$(\Gamma^{(1)}\cdot\Gamma^{(2)})\cdot\Gamma^{(3)} = \Gamma^{(1)}\cdot(\Gamma^{(2)}\cdot\Gamma^{(3)}). \tag{2.1}$$

The graph $\Gamma_0=(V_0,L_0,i)$, $V_0=\{\alpha,\tau\}$, $L_0=\{1\}$ such that α is the initial vertex of 1 and τ is its final vertex (see Fig.4) plays the role of the unit, i.e.

$$\Gamma_0\cdot\Gamma = \Gamma\cdot\Gamma_0 = \Gamma.$$

<u>Definition</u>. Let $\Gamma\in G$. The graph $\Gamma^n=\underbrace{\Gamma\cdot\ldots\cdot\Gamma}_{n}$ is called a finite hierarchical lattice of the n-th level with the generating graph Γ.

<u>Examples</u> (for simplicity we do not indicate any orientation of graphs)

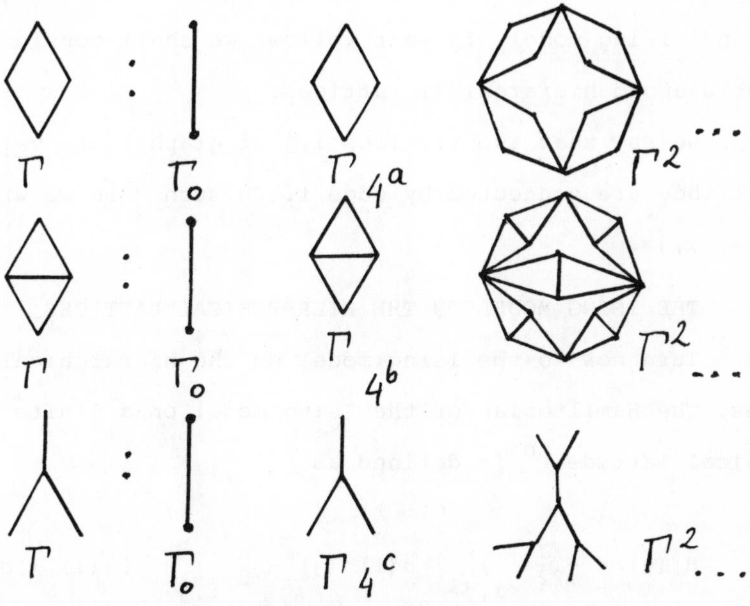

Fig.4. Hierarchical lattices.

Definition. Let $\Gamma \in G$ be a graph with two external vertices α, τ and 2^{d-1} internal ones $v_1, \ldots, v_{2^{d-1}}$ which are connected by edges with α and τ (see Fig.5). Then Γ^n is called a finite d-dimensional diamond hierarchical lattice of n-th level.

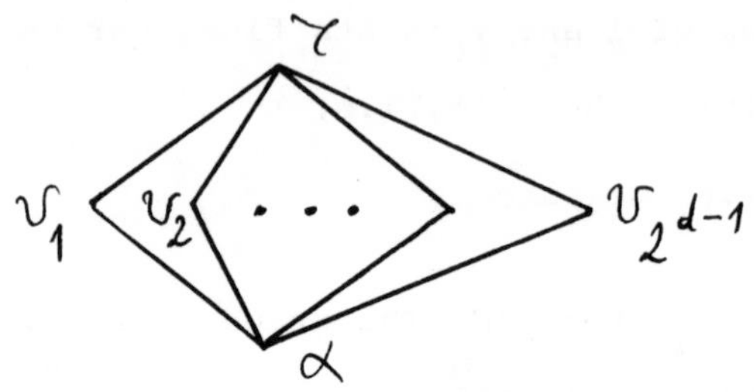

Fig. 5. Generating graph of the diamond hierarchical lattice.

Note that just the d-dimensional diamond hierarchical lattices appear in the Migdal approximation of the d-dimensional Ising model. In what follows we shall consider mainly the diamond hierarchical lattices.

We say that two vertices i,j of graph Γ are neighbours iff they are connected by edge l. In such case we write $l = \langle i,j \rangle$.

3. THE ISING MODEL ON THE HIERARCHICAL LATTICES

Turn now to the Ising model on the hierarchical lattices. The Hamiltonian of the Ising model on a finite hierarchical lattice Γ^n is defined as

$$H_n(\sigma) = -\frac{J}{2} \sum_{\substack{\langle i,j \rangle \\ i,j \in \Gamma^n}} \sigma(i)\sigma(j) - \frac{h}{2} \sum_{\substack{\langle i,j \rangle \\ i,j \in \Gamma^n}} [\sigma(i) + \sigma(j)]. \tag{3.1}$$

Somewhat unusual form of the linear term with respect to $\sigma(i)$ in the last formula is explained by the "multiplicity" of spins in the Migdal approximation. Spins $\sigma(i)$ take values ± 1. The partition functions is

$$Z_n = \sum_{\sigma} \exp(-\frac{1}{T} H_n(\sigma)), \tag{3.2}$$

the finite Gibbs measure is

$$\mu_n(\sigma) = Z_n^{-1} \exp(-\frac{1}{T} H_n(\sigma)). \tag{3.3}$$

Define restricted (conditioned) partition functions

$$\left.\begin{aligned} P_n &= \sum_{\sigma:\sigma(\alpha)=\sigma(\tau)=+1} \exp(-\frac{1}{T} H_n(\sigma)), \\ Q_n &= \sum_{\sigma:\sigma(\alpha)=+1,\sigma(\tau)=-1} \dots, \quad R_n = \sum_{\sigma:\sigma(\alpha)=\sigma(\tau)=-1} \dots, \end{aligned}\right\} \tag{3.4}$$

where α, τ are the external vertices of Γ^n. Then

$$Z_n = P_n + 2Q_n + R_n \tag{3.5}$$

and recurrent equations for P_n, Q_n, R_n hold. Let us derive them. Let Γ^n be the diamond hierarchical lattice. As $\Gamma^{n+1} = \Gamma^n \cdot \Gamma$ the graph Γ^{n+1} can be viewed as the graph Γ in which each edge is replaced by graph Γ^n (see Fig.6). Calculate restricted partition functions P_{n+1}, Q_{n+1}, R_{n+1} in two steps.

Let

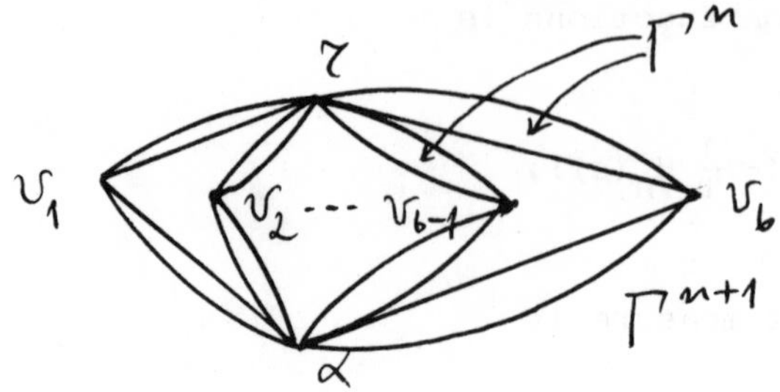

Fig. 6. Two step summation.

$\{\alpha,\tau,v_1,\ldots,v_b\} \subset \Gamma^{n+1}$ $b=2^{d-1}$, be the images of graph Γ vertices with respect to operation $\Gamma^n\cdot\Gamma$ (see Fig.6). First we fix the values $\sigma(\alpha)$, $\sigma(\tau)$, $\sigma(v_1),\ldots,\sigma(v_b)$ and take the sum over all the other spins. Since only neighbouring spins interact the restricted partition function $Z_{n+1}(\sigma(\alpha),\sigma(\tau),$ $\sigma(v_1),\ldots,\sigma(v_b))$ is factorized,

$$Z_{n+1}(\sigma(\alpha),\sigma(\tau),\sigma(v_1),\ldots,\sigma(v_b)) = \prod_{j=1}^{b}\Big\{Z_n(\sigma(\alpha),\sigma(v_j)) \, Z_n(\sigma(v_j),\sigma(\tau))\Big\}.$$

Next we take the sum over $\sigma(v_1),\ldots,\sigma(v_b)$ and get

$$Z_{n+1}(\sigma(\alpha),\sigma(\tau)) =$$

$$= \sum_{\sigma(v_1),\ldots,\sigma(v_b)=\pm 1} \prod_{j=1}^{b}\Big\{Z_n(\sigma(\alpha),\sigma(v_j))\,Z_n(\sigma(v_j),\sigma(\tau))\Big\}.$$

This is a recurrent equation for restricted partition functions. By this equation,

$$\left.\begin{aligned} P_{n+1} &= \sum_{j=1}^{b} \binom{b}{j} P_n^{2j}\, Q_n^{2b-2j} = (P_n^2 + Q_n^2)^b , \\ Q_{n+1} &= Q_n^b (P_n + R_n)^b, \quad R_{n+1} = (R_n^2 + Q_n^2)^b . \end{aligned}\right\} \tag{3.6}$$

Besides, since Γ_0 is the graph consisting of a single edge,

$$\left.\begin{aligned} P_0 &= \exp\left(\frac{J}{2T} + \frac{h}{T}\right), \quad Q_0 = \exp\left(-\frac{J}{2T}\right), \\ R_0 &= \exp\left(\frac{J}{2T} - \frac{h}{T}\right). \end{aligned}\right\} \tag{3.7}$$

The equations (3.6), (3.7) determine recurrently the restricted partition functions P_n, Q_n, R_n. Now the original definition of the model can be forgotten and only the equations (3.5)-(3.7) can be considered.

It is convenient to introduce relative variables

$$z_n = \left(\frac{R_n}{P_n}\right)^{1/b} , \qquad t_n = \left(\frac{Q_n^2}{P_n R_n}\right)^{1/b} . \tag{3.8}$$

By (3.6), (3.7)

$$z_{n+1} = \left(\frac{R_{n+1}}{P_{n+1}}\right)^{1/b} = \frac{R_n^2 + Q_n^2}{P_n^2 + Q_n^2} = \frac{(R_n/P_n) + (Q_n^2/P_n R_n)}{(P_n/R_n) + (Q_n^2/P_n R_n)} ,$$

$$z_{n+1} = \frac{z_n^b + t_n^b}{z_n^{-b} + t_n^b} \quad , \quad t_{n+1} = \frac{z_n^b + z_n^{-b} + 2}{z_n^b + z_n^{-b} + t_n^b + t_n^{-b}} , \tag{3.9}$$

$$z_0 = \exp\left(- \frac{2h}{bT} \right) \quad , \quad t_0 = \exp\left(- \frac{2J}{bT} \right). \tag{3.10}$$

Define the map

$$R : (z,t) \to \left(\frac{z^b + t^b}{z^{-b} + t^b} , \frac{z^b + z^{-b} + 2}{z^b + z^{-b} + t^b + t^{-b}} \right). \tag{3.11}$$

Then $(z_{n+1}, t_{n+1}) = R(z_n, t_n)$. R is called the renormalization group transformation for the model under consideration. According to (3.10) it is natural to write z_n, t_n in the form

$$z_n = \exp\left(- \frac{2h_n}{bT_n} \right) \quad , \quad t_n = \exp\left(- \frac{2J}{bT_n} \right). \tag{3.12}$$

Values h_n, T_n are called the renormalized external field and the renormalized temperature respectively. A very important feature of the diamond hierarchical lattices is worth noticing; namely that for them the renormalization group transformation is reduced to the one of the external field and the temperature. For more realistic lattices such as $\mathbb{Z}^d$ the RG transformation has a very complicated structure and acts in the huge space of multiparticle Hamiltonians (see e.g. [9,10]).

4. CALCULATION OF THERMODYNAMICAL FUNCTIONS

Calculate now the free energy in terms of z_n, t_n. We have by (3.5), (3.8)

$$Z_n = P_n + 2Q_n + R_n = P_n\left(1 + 2\frac{Q_n}{P_n} + \frac{R_n}{P_n}\right) =$$

$$= P_n\left(1 + 2\, z_n^{b/2}\, t_n^{b/2} + z_n^b\right),$$

$$P_n = \left(P_{n-1}^2 + Q_{n-1}^2\right)^b = P_{n-1}^{2b}\left(1 + z_{n-1}^b\, t_{n-1}^b\right)^b = \ldots =$$

$$= P_0^{(2b)^n}\left(1 + z_{n-1}^b\, t_{n-1}^b\right)^b \cdot \ldots \cdot \left(1 + z_0^b\, t_0^b\right)^{b(2b)^{n-1}} .$$

So the free energy

$$F = -T\lim_{n\to\infty}\frac{\ln Z_n}{(2b)^n} = -T\ln P_0 - \frac{T}{2}\sum_{n=0}^{\infty}\frac{1}{(2b)^n}\ln(1 + z_n^b\, t_n^b)$$

$$- T\lim_{n\to\infty}\frac{1}{(2b)^n}\ln\left(1 + 2\, z_n^{b/2}\, t_n^{b/2} + z_n^b\right).$$

Assume that $0<T<\infty$ and $h \geqq 0$, then $0<t_0<1$ and $0<z_0\leqq 1$. By (3.9) we get easily $0<t_1<1$, $0<z_1\leqq 1$, ..., $0<t_n<1$, $0<z_n\leqq 1$,.... . So the series $\sum_{n=0}^{\infty}\frac{1}{(2b)^n}\ln(1 + z_n^b t_n^b)$ converges uniformly and $\lim_{n\to\infty}\frac{1}{(2b)^n}\ln(1 + 2z_n^{b/2}\, t_n^{b/2} + z_n^b) = 0$. Besides by (3.7) $-T\ln P_0 = -\frac{J}{2} - h$.

Thus

$$F = -\frac{J}{2} - h - \frac{T}{2}\sum_{n=0}^{\infty} \frac{1}{(2b)^n} \ln\left(1 + z_n^b t_n^b\right) . \tag{4.1}$$

When h=0 we have (3.9) $1=z_0=z_1= \dots$ and

$$t_{n+1} = \frac{4}{2 + t_n^b + t_n^{-b}} \equiv f(t_n) . \tag{4.2}$$

Then we shall assume that $b=2^{d-1}>1$. Note that $f(0)=0, f(1)=1$ $f'(0) = f'(1) = 0$.

Therefore $t^c=f(t^c)$ for some $0<t^c<1$ (see Fig.7). Thus

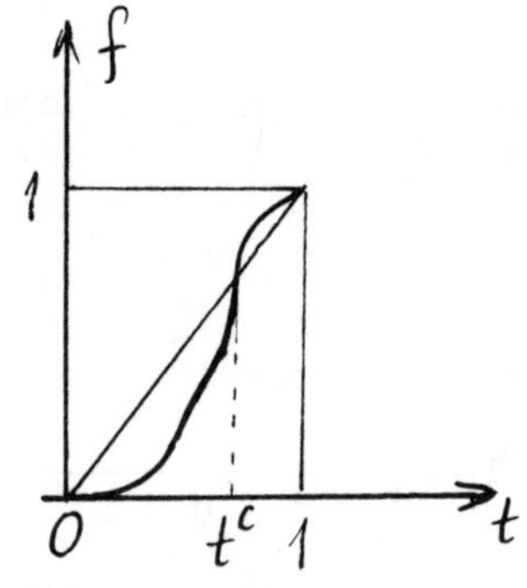

Fig.7. Function f(t).

t^c is a critical point and $t^c=\exp\left(-\frac{2J}{bT_c}\right)$ determines the critical temperature of the Ising model on the diamond hierarchical lattice.

Let us prove the uniqueness of t^c. Let $t^c=\exp(-\tau^c)$, where $0<\tau^c<\infty$. Then $\tau^c=g(\tau^c)$, where $g(\tau)=-\ln\left\{\frac{2}{1+\frac{(\exp(b\tau)+\exp(-b\tau))}{2}}\right\}=$

$$=\ln\left(\frac{1+\cosh(b\tau)}{2}\right) = 2\ln\cosh(\frac{b\tau}{2}) .$$

As $g''(\tau) = \frac{b^2}{2} \cosh^{-2}(\frac{b\tau}{2}) > 0$, the function $g(\tau)-\tau$ is convex so it has two simple zeroes at the most. As $g(0)-0=0$, τ^c is another simple zero, which was stated. We have also $f'(t^c)>1$.

It be proved easily for any sequence $\{t_n\}$ determined by the recurrent equation (4.2) that

$$\lim_{n\to\infty} t_n = \begin{cases} 0 , & \text{if} \quad 0 \leqq t_0 < t^c \\ t^c , & \text{if} \quad t_0 = t^c \\ 1 , & \text{if} \quad t^c < t_0 \leqq 1. \end{cases} \tag{4.3}$$

As t=0,1 are superstable fixed points, the convergence has a super-exponential character.

Let us calculate now the spontaneous magnetization $M(T) = \lim_{h\to 0+} (-\frac{\partial F}{\partial h})$. By (4.1)

$$M = 1 + \frac{T}{2} \sum_{n=0}^{\infty} \frac{1}{(2b)^n} \cdot \frac{\partial}{\partial h} \ln(1 + z_n^b\, t_n^b)\Big|_{h=0+} . \tag{4.4}$$

Let us find $\frac{\partial z_n}{\partial h}\Big|_{h=0+}$ and $\frac{\partial t_n}{\partial h}\Big|_{h=0+}$. We have $z_n\big|_{h=0+}=1-$ and

$$\frac{\partial t_n}{\partial h}\Big|_{h=0+} = \frac{\partial t_n}{\partial t_{n-1}}\Big|_{z_{n-1}=1-} \cdot \frac{\partial t_{n-1}}{\partial h}\Big|_{h=0+} + \frac{\partial t_n}{\partial z_{n-1}}\Big|_{z_{n-1}=1-} \cdot \frac{\partial z_{n-1}}{\partial h}\Big|_{h=0+} .$$

The function $\rho(z) \equiv z^b + z^{-b}$ has a minimum at $z=1$ so $\rho'(1)=0$.

It shows that by (3.9) $\left.\frac{\partial t_n}{\partial z_{n-1}}\right|_{z_{n-1}=1-} = 0$, so $\left.\frac{\partial t_n}{\partial h}\right|_{h=0+} =$ $= \lambda_n \left.\frac{\partial t_{n-1}}{\partial h}\right|_{h=0+}$, where $\lambda_n = \left.\frac{\partial t_n}{\partial t_{n-1}}\right|_{z_{n-1}=1-}$. Iterating this equality we get $\left.\frac{\partial t_n}{\partial h}\right|_{h=0+} = \lambda_n \lambda_{n-1} \cdots \lambda_1 \left.\frac{\partial t_0}{\partial h}\right|_{h=0+}$ but t_0 does not depend on h so

$$\left.\frac{\partial t_n}{\partial h}\right|_{h=0+} = 0 \; . \tag{4.5}$$

Similarly $\left.\frac{\partial z_n}{\partial h}\right|_{h=0+} = \left.\frac{\partial z_n}{\partial z_{n-1}}\right|_{z_{n-1}=1-} \cdot \left.\frac{\partial z_{n-1}}{\partial h}\right|_{h=0+}$. Differentiating of (3.9) gives $\left.\frac{\partial z_n}{\partial z_{n-1}}\right|_{z_{n-1}=1-} = \frac{2b}{1 + t_{n-1}^b}$, so $\left.\frac{\partial z_n}{\partial h}\right|_{h=0+} = \frac{2b}{1 + t_{n-1}^b} \cdot \left.\frac{z_{n-1}}{h}\right|_{h=0+}$. Iterating this equality we get

$$\left.\frac{\partial z_n}{\partial h}\right|_{h=0+} = \left[\prod_{j=0}^{n-1} \left(\frac{2b}{1 + t_j^b}\right)\right]\left(-\frac{2}{bT}\right) \quad , \tag{4.6}$$

as $\left.\frac{\partial z_0}{\partial h}\right|_{h=0+} = -\frac{2}{bT}$. By (4.4)-(4.6)

$$M = 1 + \frac{T}{2} \sum_{n=0}^{\infty} \frac{1}{(2b)^n} \frac{t_n^b}{1 + t_n^b} \cdot b \left.\frac{\partial z_n}{\partial h}\right|_{h=0+} =$$

$$= 1 - \sum_{n=0}^{\infty} \frac{1}{(2b)^n} \cdot \frac{t_n^b}{1 + t_n^b} \prod_{j=0}^{n-1} \left(\frac{2b}{1 + t_j^b}\right) = 1 - \sum_{n=0}^{\infty} t_n^b \prod_{j=0}^{n} \frac{1}{1 + t_j^b} =$$

$$= 1 - \frac{t_0^b}{1+t_0^b} - \frac{t_1^b}{(1+t_0^b)(1+t_1^b)} - \ldots = \frac{1}{1+t_0^b} - \frac{t_1^b}{(1+t_0^b)(1+t_1^b)} -$$

$$- \ldots = \frac{1}{(1+t_0^b)(1+t_1^b)} - \ldots = \frac{1}{(1+t_0^b)(1+t_1^b)} \quad \ldots \quad .$$

Thus we have the final formula

$$M(T) = \prod_{n=0}^{\infty} \left(\frac{1}{1+t_n^b} \right) , \tag{4.7}$$

where $t_{n+1}=f(t_n)$, $t_0=\exp(-\frac{2J}{bT})$. It is worth noticing that for a full proof of the last formula one has to prove also a uniform differentiability of the series (4.1) at h=0+ (see details in [8]).

Let us analyze briefly formula (4.7). If $T>T_c$ then $t_0>t^c$ and by (4.3) $\lim_{n\to\infty} t_n = 1$, so $M(T)=0$. If $T=T_c$ then $t_0=t^c$ and $t_n=t^c$, $n>0$, so $M(T_c)=0$. Finally if $T<T_c$ then $t_0<t^c$ and $t_n \to 0$ in a super-exponential manner so $M(T)>0$ (see Fig.8).

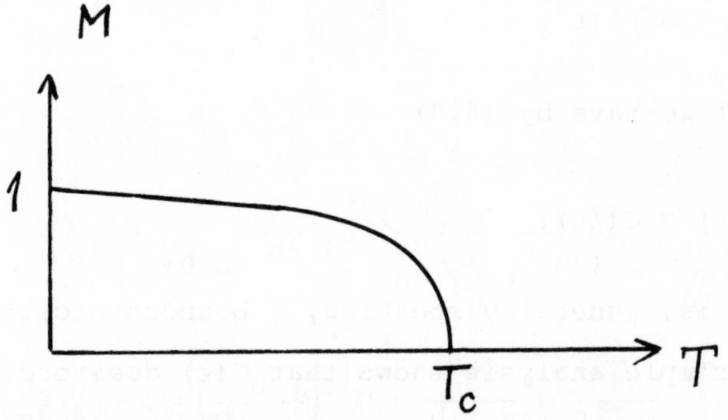

Fig.8. Spontaneous magnetization.

Furthermore $M(T) \sim C|T-T_c|^{\beta}$ as $T \to T_c^-$. Let us find β . Consider M as a function of $t_0 = \exp(-\frac{2J}{bT})$. Then by (4.7)

$$M(t_0) = \frac{1}{1 + t_0^b} M(t_1) \; . \tag{4.8}$$

This is a renormalization group equation for the spontaneous magnetization. Let $t_0 = t^c - \varepsilon$, $\varepsilon \to 0+$. Then $t_1 = f(t_0) \cong t^c - \lambda\varepsilon$, where $\lambda = f'(t^c) > 1$. So when $\varepsilon \to 0+$,

$$M(t^c - \varepsilon) \cong \frac{1}{1 + (t^c)^b} M(t^c - \lambda\varepsilon) . \tag{4.9}$$

This scaling relation for M shows that

$$M(t^c - \varepsilon) = C(\varepsilon)\varepsilon^{\beta} , \tag{4.10}$$

where $\frac{1}{1 + (t^c)^b} \lambda^{\beta} = 1$, or

$$\beta = \frac{\log(1 + (t^c)^b)}{\log f'(t^c)} \; . \tag{4.11}$$

For $C(\varepsilon)$ we have by (4.9)

$$C(\varepsilon) \cong C(\lambda\varepsilon),$$

so $C(\varepsilon)$ is, generally speaking, a bounded oscillating factor. A simple analysis shows that $C(\varepsilon)$ does oscillate as $\varepsilon \to 0+$. The appearence of such oscillating factors is typical for various critical asymptotics for spin models on hierarchical lattices (see [11]).

As was shown in [12] the magnetic susceptibility $\chi = \frac{\partial M}{\partial h}$ diverges at h=0 when $T>T_c$. Its leading singularity was found in [8]. Namely if $T>T_c$, $h \to 0+$ then

$$\chi(T,h) = \begin{cases} \frac{A^2(T)}{2T} \log_2 \frac{1}{|h|} + P\left(\frac{2A(T)}{T}|h|\right) + \tau(T,h), & \text{when } b=2, \\ \frac{A^2(T)}{2T}\left(\frac{2A(T)}{T}|h|\right)^{-\gamma} P\left(\frac{2A(T)}{T}|h|\right) + \tau(T,h), & \text{when } b>2 \end{cases} \qquad (4.12)$$

where $\gamma = 1 - \frac{\log 2}{\log b}$, $A(T) = \prod_{n=0}^{\infty} \left(\frac{2}{1+t_n^b}\right) > 0$, $t_{n+1} = f(t_n)$, $t_0 = \exp(-\frac{2J}{bT})$, $P(t)$ is analytic for $t>0$ and $P(bt) = P(t)$ for any $t>0$, so $P(t)$ is oscillating when $t \to 0+$. Moreover $P(t)>0$ for any $t>0$ when $b>2$. The term $\tau(T,h)$ has a finite limit as $h \to 0+$.

Sketch of the proof of formulae (4.12). By (4.1)

$$\chi = \frac{T}{2} \sum_{n=0}^{\infty} \frac{1}{(2b)^n} \frac{\partial^2}{\partial h^2} \ln\left(1 + z_n^b t_n^b\right).$$

Let $z_n = \exp(-\phi_n)$. An analysis shows that the divergence of χ at h=0+ is determined by the quantity

$$\chi_0 = \frac{Tb^2}{2} \sum_{n=0}^{\infty} \frac{1}{(2b)^n} \frac{1}{1+\cosh(b\phi_n)} \left(\frac{\partial \phi_n}{\partial h}\right)^2 . \qquad (4.13)$$

Moreover $\frac{\partial\phi_n}{\partial h} \cong A(T)b^n\left(-\frac{2}{bT}\right)$, $\phi_n \cong A(T)b^n\left(-\frac{2}{bT}\right)h$, so that $\frac{\partial\phi_n}{\partial h}$, ϕ_n in (4.13) can be replaced by these expressions. It gives

$$\chi_0 \cong \frac{A^2(T)}{2T} \sum_{n=0}^{\infty} \left(\frac{b}{2}\right)^n \frac{1}{1+\cosh\left(\frac{2A(T)}{T}b^n h\right)} \tag{4.14}$$

In the last sum there is a competition between increasing factor $\left(\frac{b}{2}\right)^n$ and decreasing factor $\frac{1}{1+\cosh\left(\frac{2A(T)}{T}b^n h\right)}$. Find asymptotics of this sum as $h \to 0+$. Let $b>2$. Consider the sum

$$\omega(h) = \sum_{n=-\infty}^{\infty} \left(\frac{b}{2}\right)^n \frac{1}{1+\cosh\left(\frac{2A(T)}{T}b^n h\right)} .$$

It differs from the sum in (4.14) only by a finite term $\sum_{n=-\infty}^{-1} \left(\frac{b}{2}\right)^n \left(1+\cosh\left(\frac{2A(T)}{T}b^n h\right)\right)^{-1}$. Moreover

$$\omega(bh) = \frac{2}{b} \sum_{n=-\infty}^{\infty} \left(\frac{b}{2}\right)^{n+1} \left(1+\cosh\frac{2A(T)}{T}b^{n+1}h\right)\Big)^{-1} = \frac{2}{b}\,\omega(h)$$

so

$$\omega(h) = h^{-\gamma}\,C(h)$$

where $b^\gamma = \frac{b}{2}$, $\gamma = 1 - \frac{\log 2}{\log b}$, and

$$C(bh) = C(h)$$

so C(h) is an oscillating analytic and positive at h>0 function. This explains (4.12) for b>2. For b=2 we define

$$\omega(h) = \sum_{n=-\infty}^{-1} \left[\left(1 + \cosh\left(\frac{2A(T)}{T} 2^n h\right)\right)^{-1} - 1\right] +$$

$$+ \sum_{n=0}^{\infty} \left(1 + \cosh\left(\frac{2A(T)}{T} 2^n h\right)\right)^{-1}$$

and use similar considerations

The magnetic susceptibility divergence at $T>T_c$ means that the half-line $\{T \geqq T_c, h=0\}$ is a set of points of phase transition in h of order higher than one. At $\{T<T_c, h=0\}$ we have usual phase transition of the first order (see Fig.9). It is

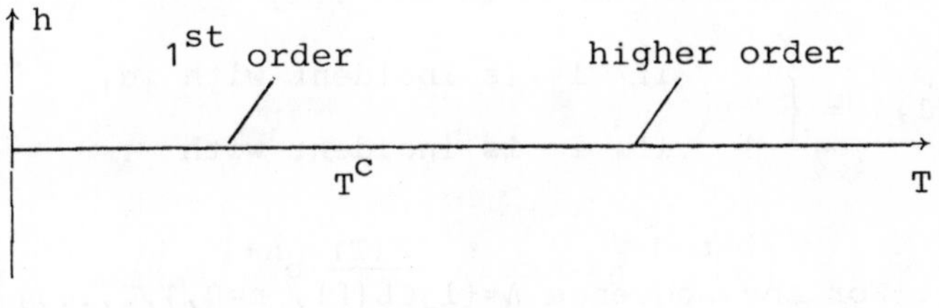

Fig. 9. Phase diagram

worth to note that for $T<T_c$ the magnetic susceptibility is finite and even infinitely differentiable at h=0+ (see [8]).

5. LIMIT GIBBS STATES

To consider limit Gibbs states we have to define first an infinite hierarchical lattice. If $\Gamma'' = \Gamma' \cdot \Gamma$ then by definition for each edge $l \in L(\Gamma)$ we have an imbedding $i_l : \Gamma' \to \Gamma''$

which is induced by the insertion of Γ' instead of l in Γ. An infinite hierarchical lattice is defined as an inductive limit of finite hierarchical lattices.

<u>Definition.</u> Let an infinite sequence $\Lambda=\{l_n \in L(\Gamma),\ n=0,1,2,\ldots\}$ of edges of $\Gamma \in G$ be given. Consider imbeddings $i_{l_n}: \Gamma^n \to \Gamma^{n+1} = \Gamma^n \cdot \Gamma$. Then the inductive limit $\varinjlim_{n\to\infty} \Gamma^n = \Gamma^\infty = \Gamma^\infty(\Lambda)$ with respect to these imbeddings is called an infinite hierarchical lattice with generating graph Γ associated with Λ.

As was noted in [6] infinite lattices $\Gamma^\infty(\Lambda)$ can be non-isomorphic as graphs for different Λ. Let Γ be the generating graph of the diamond hierarchical lattice (see Fig.3). Define the function

$$s(l) = \begin{cases} 1 & \text{if } l \text{ is incident with } \alpha, \\ -1 & \text{if } l \text{ is incident with } \tau \end{cases}$$

on $L(\Gamma)$. For any sequence $\Lambda=\{l_n \in L(\Gamma),\ n=0,1,2,\ldots\}$ define $s(\Lambda)=\{s(l_n)=\pm 1,\ n=0,1,2,\ldots\}$. It can be seen that if $s(\Lambda)$ and $s(\Lambda')$ coincide for all but finite numbers of $s(l_n)$ then $\Gamma^\infty(\Lambda)$ is isomorphic to $\Gamma^\infty(\Lambda')$. Moreover $\Gamma^\infty(\Lambda)$ is isomorphic to $\Gamma^\infty(\Lambda')$ if $s(\Lambda')=-s(\Lambda)\equiv\{-s(l_n),\ n=0,1,2,\ldots\}$. In [7] it is proved that if $s(\Lambda)$ differs from both $s(\Lambda')$ and $-s(\Lambda')$ for infinite number of n's then $\Gamma^\infty(\Lambda)$ and $\Gamma^\infty(\Lambda')$ are non-isomorphic.

To define limit Gibbs states on $\Gamma^\infty(\Lambda)$ consider the so-called finite Gibbs measure on Γ^n with boundary conditions.

For that purpose let a spin configuration $\{\sigma'(i),\ i\in\Gamma^{\infty}(\Lambda)\setminus\Gamma^{n}\}$ outside of Γ^n be given. We consider Γ^n as being embedded in $\Gamma^{\infty}(\Lambda)$ with the help of $\ldots\ i(l_{n+1})i(l_n)$. The configuration $\{\sigma'(i),\ i\in\Gamma^{\infty}(\Lambda)\setminus\Gamma^{n}\}$ is a boundary condition.

Definition. Finite Gibbs measure on Γ^n with boundary condition $\sigma'=\{\sigma'(i),\ i\in\Gamma^{\infty}(\Lambda)\setminus\Gamma^{n}\}$ is

$$\mu_n(\sigma;T,\sigma') = Z_n^{-1}(T,\sigma')\exp\left(-\frac{1}{T}H_n(\sigma,\sigma')\right) \tag{5.1}$$

where

$$H_n(\sigma,\sigma') = H_n(\sigma) - \frac{J}{2}\sum_{\substack{<i,j>\\ i\in\Gamma^n,\, j\in\Gamma^{\infty}(\Lambda)\setminus\Gamma^n}} \sigma(i)\sigma'(j), \tag{5.2}$$

$$Z_n(T,\sigma') = \sum_{\sigma}\exp\left(-\frac{1}{T}H_n(\sigma,\sigma')\right). \tag{5.3}$$

Limit Gibbs state is $\mu_{\infty}(\sigma;T,\{\sigma'_n\}) = \lim_{n\to\infty}\mu_n(\sigma;T,\sigma'_n)$ if the limit exists for some sequence of boundary conditions σ'_n.

It is worth noticing that as a matter of fact only external vertices $\alpha=\alpha_n$ and $\tau=\tau_n$ of Γ^n appear in (5.2) as $i\in\Gamma^n$. Note also that it is assumed tacitly in (5.2) that the number of neighbours $j\in\Gamma^{\infty}(\Lambda)\setminus\Gamma^n$ of α_n and τ_n is finite. It can be invalid in some degenerate cases. Namely we call a sequence $\Lambda=\{l_n\in\Gamma,\ n=0,1,2,\ldots\}$ and respectively a lattice $\Gamma^{\infty}(\Lambda)$ degenerate iff $s(l_n)$ is constant beginning from some n_0. It can be seen that the number of neighbours $j\in\Gamma^{\infty}(\Lambda)$ of the vertices α_n, τ_n is finite for all n iff Λ is non-degenerate.

Main problems connected with limit Gibbs states are to prove the existence of $\lim_{n\to\infty} \mu_n(\sigma;T,\{\sigma'_n\})$ for some sequence of boundary conditions (at least for some subsequence of n's), the existence problem, and to describe all the limit Gibbs states, the uniqueness problem. These two problems were considered in [7] where the following result was proved.

Theorem. Let $\Gamma^\infty(\Lambda)$ be an infinite diamond non-degenerate hierarchical lattice. So if $h\neq 0$ or $h=0$ and $T\geq T_c$ then a limit Gibbs state exists and is unique. If $h=0$ and $0<T<T_c$ then limit Gibbs states $\mu_\infty(\sigma;T,\{\pm 1\})$ with $\{+1\}$ and $\{-1\}$ boundary conditions exist. Any other limit Gibbs state $\mu_\infty(\sigma;T,\{\sigma'_n\})$ is a convex combination of $\mu_\infty(\sigma;T,\{\pm 1\})$, i.e. $\mu_\infty(\sigma;T,\sigma'_n) = =p\mu_\infty(\sigma;T,\{+1\}) + (1-p)\mu_\infty(\sigma;T,\{-1\})$ for some $0<p<1$.

The proof of this theorem is based on some ideas of papers [13],[14]. Namely, consider a sequence of boundary spins $\{\sigma(\alpha_n),\sigma(\tau_n),\ n=0,1,2,\ldots\}$ of finite lattices $\Gamma^n\subset\Gamma^\infty(\Lambda)$. It is verified that this sequence of random variables forms a half-infinite non-homogeneous Markov chain whose transition probabilities are expressed via the quantities z_n, t_n defined in (3.8). Using properties of these quantities it can be proved that a Markov chain with the given transition probabilities exists and is unique if $h\neq 0$ or $h=0$ and $T\geq T_c$, and if $h=0$ and $T<T_c$ there exist exactly two different ergodic Markov chains with these transition probabilities. This reasoning proves the theorem.

6. SOME ANALYTICAL PROPERTIES

Equations (3.9), (3.10), (4.1) are well applicable for seeking of zeroes of the partition function and of complex singularities of the free energy. Let first $h=0$. Then we have a rational map $R: t \to f(t) \equiv \dfrac{4}{2 + t^b + t^{-b}}$ and the free energy

$$F = -\frac{J}{2} - \frac{T}{2} \sum_{n=0}^{\infty} \frac{1}{(2b)^n} \ln(1 + t_n^n),$$

where $t_{n+1} = f(t_n)$, $t_0 = \exp(-\frac{2J}{bT})$. The complex singularities of F as a function of t_0 lie on the Julia set of the function $f(t)$. A detailed study of the Julia set, fine computer pictures of this set and calculations of the free energy singularities on the Julia set are represented in papers [15,12].

Another question is connected with complex singularities of F in h for real values of T. By the famous Lee-Yang theorem the zeroes of the partition function Z_n lie on the imaginary axis h, or on the unit circle $|z_0|=1$, $z_0=\exp(-\frac{2h}{bT})$. Consider a finite cylinder $C=\{|z|=1\}\times\{0\leq t<1\}$. It can be verified easily that the renormalization group transformation (3.11) maps C into C, $R: C \to C$. By formulae of section 4

$$Z_n = \left(1 + 2z_n^{b/2}\, t_n^{b/2} + z_n^b\right) P_0^{(2b)^n} \prod_{j=0}^{n-1}\left(1 + z_j^b t_j^b\right)^{b\cdot(2b)^{n-1}} .$$

It is clear that $P_0=\exp(J/2T + h/T) \neq 0$. If $0\leq T<\infty$ then $0\leq t_0<1$,

$|z_0|=1$, i.e. $(z_0,t_0)\in C$, so $(z_j,t_j)\in C$, $j\geq 0$, so $|z_j^b t_j^b| = t_j^b < 1$, so $1 + z_j^b t_j^b \neq 0$ for any j. Therefore the zeroes of Z_n coincide with those of $1 + 2z_n^{b/2} t_n^{b/2} + z_n^b = 2z_n^{b/2}\left[t_n^{b/2} + (z_n^{b/2} + z_n^{-b/2})/2\right]$. Thus the zeroes of Z_n are determined from the equation

$$t_n^{b/2} + \left(z_n^{b/2} + z_n^{-b/2}\right)/2 = 0.$$

This equation can be rewritten in a recurrent form. Denote Σ_n the set of zeroes of Z_n in the cylinder C. Then

$$\Sigma_{n+1} = R^{-1}(\Sigma_n),$$

$$\Sigma_0 = \left\{(z,t)\in C \,|\, t^{b/2} + \left(z^{b/2} + z^{-b/2}\right)/2 = 0\right\},$$

where R^{-1} means the full preimage. Using some simple properties of the map R it can be shown that $\bigcup_{n\geq 0}\Sigma_n$ is everywhere dense in C (see [8]). That means that at any $T\geq 0$ the zeroes of the partition functions Z_n in h form an everywhere dense set on the imaginary axis. It is worth to note here that for the Ising model on $\mathbf{Z}^d$ at high temperatures there exist lacunae on the imaginary axis h free of zeroes.

7. OTHER SPIN MODELS ON HIERARCHICAL LATTICES

The simplicity of the renormalization group equations makes hierarchical lattices (HL) very attractive for the study of phase transitions and critical phenomena in various spin systems. We list briefly some results of such type. Phase diagram of the Potts model on HL was studied in papers [16-20]. In [5] it was rigorously proved the existence of long-range order at low temperatures and its absence at high temperatures in XY-model on the diamond HL if the dimension d>2. In [21] numerical evidences were given for absence of long-range order at any temperature in XY-model if d=2. Recently Ito proved it rigorously not only for XY-model but also for any n-model (see [22,23]). Moreover Ito's result gives exponential decay of correlations for the two-dimensional n-model on the diamond HL. Hence there is no analog of the Kosterlitz-Thouless phase transition on the diamond HL.

In a series of papers [24-26] (see [26] for further references) a spin glass model on HL was studied. In particular it was proved (see [24]) that the Edwards-Anderson order parameter has a phase transition in such a model.

Percolation on HL was studied in [27].

REFERENCES

1. Wilson, K.G. Renormalization group and critical phenomena, I,II, Phys.Rev.B4,3174-3194(1971).
2. Berezinskii, V.L., unpublished.
3. Migdal, A.A., Recursion equations in gauge field theories, ZhETF, 69,N9, 810-822,N10, 1457-1465(1975).
Kadanoff, L.P.,Notes on Migdal's recursion formulas. Ann. Phys.100,1-2, 359-394(1976).
4. Berker, A.N., Ostlund, S., Renormalization group calculations of finite systems.J.Phys.C12,4961-4975(1979).

5. Bleher, P.M., Zalys, E. Existence of long-range order in the Migdal recursion equations. Commun.Math.Phys.67,1, 17-42(1979)

6. Griffiths, R.B., Kaufman, M., Spin systems on hierarchical lattices. Introduction and thermodynamic limit. Phys. Rev.B26,5022-5039(1982).

7. Bleher, P.M., Zalys, E., Limit Gibbs states for the Ising model on hierarchical lattices. Lithuanian Math. Sborn.27(1988).

8. Bleher, P.M., Zalys, E., Asymptotics of the susceptibility for the Ising model on hierarchical lattices (in preparation).

9. Wilson, K.G., Kogut, J., The renormalization group and the ε-expansion. Phys.Rep.12C,2,75-199(1974).

10. Ma, S.K., Modern theory of critical phenomena, W.A. Benjamin. Inc., London et al., 1976.

11. Derrida, B., Ytzykson, C., Luck, J.M., Oscillatory critical amplitudes in hierarchical models. Commun.Math. Phys.94,1, 115-127(1984).

12. Kaufman, M., Griffiths, R.B., Infinite susceptibility at high temperatures in the Migdal-Kadanoff scheme, J.Phys. A.Math.Gen.15,L239-L242(1982).

13. Bleher, P.M.,Construction of non-Gaussian self-similar random fields with hierarchical structure. Commun.Math. Phys., 84,557-578(1982).

14. Bleher, P.M., Major, P., Critical phenomena and universal exponents in statistical physics. On Dyson's hierarchical model. Ann.Prob., 15,2,431-477(1987).

15. Derrida, B., De Seze, L,, Ytzykson, C., Fractal structure of zeroes in hierarchical models. J.Stat.Phys.30, 3,559-570(1983).

16. Kaufman, M., Griffiths, R.B., Yeomans, J.M., Fisher, M. E., Three-component model and tricritical points. A renormalization group study. Phys.Rev.B23,3448(1981).

17. Griffiths, R.B., Kaufman, M., Spin systems on hierarchical lattices. Some examples of soluble models. Phys.Rev. B30,1, 244-249(1984).

18. Kaufman, M., Griffiths, R.B., First-order transitions in defect structures at a second-order critical point for the Potts model on hierarchical lattices. Phys.Rev.B26, 5282-5284(1982.

19. Kaufman, M., Kardar, M., Pseudodimensional variation and tricriticality of Potts models, Phys.Rev.B30,1609-1611 (1984).

20. Kaufman, M., Duality and Potts critical amplitudes on a class of hierarchical lattices. Phys.Rev.B30,413-414 (1984).

21. Bleher, P.M., Numerical simulation of approximate equations of Migdal's renorm-group, preprint Inst.Appl.Math. USSR, Ac.Sci.48,1976.

22. Ito, K.R., Absence of the Kosterlitz-Thouless fixed points in the Migdal-Kadanoff recursion formulas. Phys. Rev.Lett.55,558(1985).

23. Ito, K.R., Mass generations in two-dimensional hierarchical Heisenberg model of Migdal-Kadanoff type. Commun. Math.Phys.110,237-246(1987).

24. Collet, P., Eckmann, J.P., A spin glass model with random couplings. Commun.Math.Phys.93,379-407(1984).

25. Collet, P., Eckmann, J.P., Glaser, Y., Martin, A., Study of the iterations of a mapping associated to a spin glass model. Commun.Math.Phys.,94,353-370(1984).

26. Collet, P., Systems with random couplings on diamond lattices. In: Stat.Phys.and Dynam.Syst.Rigor. Results., Progress in Physics, V,10, Eds. J.Fritz, A.Jaffe, D. Szasz, Birkhauser, Boston e.a., 1985, 105-126.

27. Bovin, V.P., Was'kin, V.V., Shneiberg, I.Ya., Recursive models in percolation theory, Journ.Theor.and Math.Phys. 54,175-181(1983).

SPECTRAL PROPERTIES OF THE KIRKWOOD-SALSBURG OPERATOR AND THE UNIQUENESS OF THE GIBBS STATES

R. Gielerak

BiBoS Research Center, University of Bielefeld, FRG

and

Institute of Theoretical Physics, Wroclaw University

Wroclaw, ul.Cybulskiego 36, Poland

ABSTRACT

A new method of analysis of the Kirkwood-Salsburg identities is presented. This method enlarges the domain of uniqueness and analiticity of the limiting Gibbs state.

0. INTRODUCTION

A large amount of information on the correlation functions of statistical mechanical systems in equilibrium can be derived from families of coupled integral equations satisfied by these functions [1]. The Kirkwood-Salsburg equations and various generalizations there of have been most useful in the derivation of rigorous results. In particular, they have been used to establish the existence and uniqueness of the thermodynamic limit and some analiticity properties of dilute for high-temperature systems. In the present contribution we describe the extensions of these methods and results to cover the whole resolvent set of the corresponding Kirkwood-Salsburg operator. Previously only contraction map principle were used to extract some information about correlation functions.

In the next section we invent a new method for the analysis of the corresponding Kirkwood-Salsburg identities. By simple topological arguments we obtain results about the uniqueness of the limiting Gibbs states which include the old ones and which are certainly stronger than those previously obtained. Additionally we obtain a larger analicity domain than those given by methods based on the contraction map principle.

As an illustration of our general method we consider classical continuous gas in which the interaction is given by superstable and (strongly) lower regular pair potential. The general method explained in the next section is certainly applicable to larger classes of systems like quantum continuous gases [2] , spin systems [3], random spin systems and lattice gauge theories [4].

1. EXPOSITION OF THE METHOD

Let R_Λ be some statistical mechanical system enclosed in some bounded region $\Lambda \subset R^d$. The system might be continuous or discrete (in which case R^d is replaced by some lattice), quantum or classical. In many situations the Gibbs states describing the system in the region Λ are described completely by a sequence of certain moments that are called correlation functions (or functionals in the quantum case). We denote them by $\rho(\omega)$, where ω denotes the "typical" boundary condition. Thre is a particular boundary condition ω which we will call the free boundary condition, ω=free. The following conditions are frequently fulfilled in many applications.

01). There exists a Banach space $\mathcal{B}$ such that for any $|\Lambda|<\infty$ and any "typical" boundary condition $\rho_\Lambda(\omega)\in\mathcal{B}$.

02). Thre exists a bounded, linear operator $K_\Lambda(\omega)\in L(\mathcal{B})$ and a vector $\alpha_\Lambda(\omega)\in\mathcal{B}$ such that the following identity holds:

$$\rho_\Lambda(\omega) = K_\Lambda(\omega)\rho_\Lambda(\omega) + \alpha_\Lambda(\omega).$$

Assume that there exists another Banach space $*\mathcal{B}$, the dual space of which is equal to $\mathcal{B}$, i.e.

$$(*\mathcal{B})* = \mathcal{B}.$$

The following assumptions are crucial for the applications of our method.

A1). There exist operators $*K_\Lambda(\omega)$, $*K_\infty(\text{free}) \in L(*\mathcal{B})$ such that:

i) $(*K_\Lambda(\omega))* = K_\Lambda(\omega)$

ii) let (Λ_n) be an arbitrary monotonic sequence of bounded regions $\Lambda_n \subset R^d$ which tends to R^d by inclusion.

Then for any "typical" ω :

$$*K_\Lambda(\omega) \xrightarrow[\text{in } *\mathcal{B}]{\text{strongly}} *K_\infty(\text{free})$$

iii) There exists $K_\infty(\text{free}) \in L(\mathcal{B})$ such that

$$(*K_\infty(\text{free}))* = K_\infty(\text{free})$$

A2). There exists a vector $\alpha_\infty(\text{free}) \in \mathcal{B}$ such that for any "typical" ω:

$$\alpha_\Lambda(\omega) \xrightarrow{\text{weak-*}} \alpha_\infty(\text{free})$$

For any "typical" boundary condition ω

A3) the set $\{\rho_\Lambda(\omega)\}_{\Lambda \in \mathcal{F}(R^d)} \subset \mathcal{B}$, where $\mathcal{F}(R^d)$ denotes the collection of bounded regions in R^d is precompact in the weak-* topology of $\mathcal{B}$.

Then we have:

Statement 1.

Every accumulation point $\rho_\infty(\omega)$ of the set $\{\rho_\Lambda(\omega)\}_{\Lambda\in F(R^d)}$ fulfills the relation

$$\rho_\infty(\omega) = \alpha_\infty(\text{free}) + K_\infty(\text{free})\rho_\infty(\omega). \tag{1-1}$$

Proof:

Let $\rho_\infty(\omega)$ be any of the accumulation points of the set $\{\rho_\Lambda(\omega)\}_{\Lambda\in F(R^d)}$. Then there exists a net $(\Lambda_\beta)\subset F(R^d)$, monotonously tending to R^d, such that $\rho_\Lambda(\omega) \to \rho_\infty(\omega)$ in the weak-* topology. From assumptions A1) and A2) then it follows that

$$\bigwedge_{\psi\in{}^*B} \lim_\beta \langle\psi,\rho_{\Lambda_\beta}(\omega)\rangle = \langle\psi,\rho_\infty(\omega)\rangle =$$

$$= \lim_\beta[\langle\psi,\alpha_{\Lambda_\beta}(\omega)\rangle + \langle{}^*K_{\Lambda_\beta}(\omega)\psi,\rho_{\Lambda_\beta}(\omega)\rangle]$$

$$= [\langle\psi,\alpha_\infty(\text{free})\rangle + \langle{}^*K_\infty(\text{free})\psi,\rho_\infty(\omega)\rangle]$$

$$= \langle\psi,\alpha_\infty(\text{free})\rangle + \langle\psi,K_\infty(\text{free})\rho_\infty(\omega)\rangle$$

∎

Statement 2.

Assume that $1 \notin$ spectrum $(K_\infty(\text{free}))$. Then the set of the accumulation points of $\{\rho_\Lambda(\omega)\}_\Lambda$ consists exactly of one element $\rho_\infty(\text{free})$ which is defined by:

$$\rho_\infty(\text{free}) = (1 - K_\infty(\text{free}))^{-1}\alpha_\infty(\text{free})\in B$$

and moreover

$$\rho_\Lambda(\omega) \xrightarrow{\text{weak-*}} \rho_\infty(\text{free}) \ .$$

Proof:

From Statement 1 it follows that any accumulation point $\rho_\infty(\omega)$ of the set $\{\rho_\Lambda(\omega)\}$ fulfills the relation (1-1). From the hypothesis $1 \notin$ spectrum $(K_\infty(\text{free}))$ it follows that there exists a unique solution $\rho_\infty(\text{free}) \in \mathcal{B}$ of (1-1) ∎

Statement 3.

Let the operator $K_\infty(\text{free})$ depend on some parameters $z \in \mathcal{O} \subset C^n$ and let $\mathcal{O}^*$ be a maximal open subset of $\mathcal{O}$ on which $1 \notin$ spectrum $(K_\infty(\text{free}))$. Assuming that the map

$$\mathcal{O} \ni z \quad \to \quad K_\infty(\text{free}) \in L(\mathcal{B})$$

is strongly analitic in z we conclude that the map

$$\mathcal{O}^* \ni z \to \quad \rho_\infty(\text{free}) \in \mathcal{B}$$

is strongly analitic on $\mathcal{O}^*$ ∎

In all the results we could find in the existing literature the emphasis has been placed on checking the condition $\| K_\infty(\text{free}) \| < 1$ from which it follows that $1 \notin$ spectrum $(K_\infty(\text{free}))$ ∎

2. APPLICATION. CONTINUOUS, CLASSICAL GASES.

Let us consider a gas of classical particles enclosed in some bounded region $\Lambda \subset R^d$. We assume that particles interacts throughout two-body forces which are described by a two-body potential V about which we assume

V1) V is superstable

V2) V is lower (strongly) regular.

Configurations (or sometimes subsets $\{x_1,\dots,x_n\}$) $(x_1,\dots,x_n \subset \Lambda^{\otimes n}$ will be denoted (shortly) by ξ_n. The energy of a given configuration ξ_n is defined by

$$E(\xi_n) = \sum_{1\leq i<j\leq n} V(x_i - x_j) \tag{2-1}$$

and energy of interactions between two disjoint configurations $\xi_n=(x_1,\dots,x_n)$ and $\zeta_m=(y_1,\dots,y_m)$ is given by

$$E(\xi_n|\zeta_m) = \sum_{i=1}^{n}\sum_{j=1}^{m} V(x_i - x_j) \tag{2-2}$$

Let $\Omega^T(R^d)$ be the space of all locally finite configurations ω which are tempered. The restriction of a given $\omega\in\Omega^T(R^d)$ to a set $\Lambda\subset R^d$ is denoted as $\omega(\Lambda)$. The conditioned by $\omega\in$ $\in\Omega^T(R^d)$, finite volume Λ, grand canonical ensemble Gibbs measure $\mu_\Lambda^\omega(z,\beta)$ at a given value of the chemical activity z and the (inverse) temperature β is completely described by the sequence of the conditioned, finite volume correlation functions $\rho_\Lambda^\omega(z,\beta)=\{\rho_\Lambda^\omega(z,\beta|\zeta_n)\}_{n=1,2,\dots}$ which are given by the following formulae:

$$\rho_\Lambda^\omega(z,\beta|\xi_n) = (Z_\Lambda^\omega(z,\beta))^{-1}\sum_{m=0}^{\infty}\frac{z^{m+n}}{m!}\int d\zeta_m e^{-\beta E(\xi_n\vee\zeta_m)}\cdot$$

$$\cdot\, e^{-\beta E(\xi_n\vee\xi_m|\omega(\Lambda^c))}, \tag{2-3}$$

$$Z_\Lambda^\omega(z,\beta) = \sum_{n=0}^{\infty}\frac{z^n}{n!}\int d\xi_n\, e^{-\beta E(\xi_n)}\, e^{-\beta E(\xi_n|\omega(\Lambda^c))}. \tag{2-4}$$

Let $\mathbb{B}_\zeta$ be the Banach space of all sequences $\mathbb{f}=\{f_n(\zeta_n)\}_{n=1,2,\dots,n}$, where each f_n is a measurable function on R^{dn}, for which the following norm

$$||\mathbb{f}||_\zeta = \sup_n \zeta^{-n} \operatorname*{ess\,sup}_{\xi_n \in R^{dn}} |f_n(\xi_n)| \text{ is finite.}$$

The space $\mathbb{B}_\zeta$ is the dual space of the Banach space *B composed from the sequences $\boldsymbol{\phi}=\{\phi_n\}_{n=1,2,...}$, where each ϕ_n is a L_1-integrable function on R^{dn}, for which the following norm

$$*||\boldsymbol{\phi}_n||_\zeta = \sum_{n=1}^{\infty} \zeta^n \int d\xi_n |\phi_n(\xi_n)|$$

is finite. The number ζ will be chosen to be $\zeta^{-1} = \int dx|e^{-\beta v(x)} - 1|$. Let $F(R^d)$ be a class of all bounded, regular subsets of R^d.

Lemma 2.1.

Let V be superstable and strongly lower regular and let $\omega \in \Omega^T(R^d)$. Then for any $\Lambda \in F(R^d)$:

(1) $\rho_\Lambda^\omega(z,\beta) \in \mathbb{B}_\zeta$

(2) The set $\{\rho_\Lambda^\omega\}_{\Lambda \in F(R^d)}$ is weakly * precompact in $\mathbb{B}_\zeta$.

Let us define:

$\alpha_\Lambda(\omega) \equiv \{\chi_\Lambda \exp-\beta E(x_1|\omega(\Lambda^c),0,0,...,0,...\} \in \mathbb{B}_\zeta$ and

$\alpha_\infty \equiv \{z,0,...,0,...\} \in \mathbb{B}_\zeta$

Lemma 2.2.

For any ω $\Omega^T(R^d)$

$$\alpha_\Lambda(\omega) \xrightarrow[\Lambda \uparrow R^d]{\text{weak-*}} \alpha_\infty \qquad \blacksquare$$

In the space $*\mathbb{B}_\zeta$ let us define the following linear operators:

$$(^*k_\infty\phi)_n(\xi_n) = \sum_{k=0}^{n} \int_{R^d} dy K(y|\zeta_k)\phi_{1+n-k}(yv(\zeta_n - \zeta_k)) \ , \tag{2-5}$$

$$(^*\Pi_\Lambda\phi)_n(\xi_n) = (\prod_{i=1}^{n} \chi_\Lambda(x_i))\phi_n(\xi_n) \ , \tag{2-6}$$

$$(e^{-\beta E_\Lambda^\omega}\phi)_n(\xi_n) = \exp - \beta E(x_1|\omega(\Lambda^c)\vee\xi_n')\phi_n(\xi_n) \tag{2-7}$$

where $K(\cdot|\cdot)$ are the familar Kirkwood-Salsburg kernels. Finally let J be an index juggling operator of Ruelle acting in the space $\mathbb{B}_\xi$ and let $^*J$ be its dual in the space $^*\mathbb{B}_\xi$ (see [5] for an explicit form of $^*J$). Then we define the following linear operators acting continuously in the space $^*\mathbb{B}_\xi$:

$$^*\mathbb{K}_\Lambda^\omega\phi = {}^*\Pi_\Lambda \circ {}^*k_\infty \circ {}^*J \circ e^{-\beta E_\Lambda^\omega}\Phi \tag{2-8}$$

and

$$^*\mathbb{K}_\infty\phi = {}^*k_\infty \circ {}^*J \circ e^{\cdot\beta E^{\omega=\phi}}\Phi \ . \tag{2-9}$$

Then we have

Lemma 2.3.

Let $^*\mathbb{K}_\Lambda^\omega$ and $\mathbb{K}_\infty$ be respectively conditioned and infinite volume (Ruelle-)Kirkwood-Salsburg operators. Then their duals $(\mathbb{K}_\Lambda^\omega)^*$ and $(\mathbb{K}_\infty)^*$ in the dual pair $(^*\mathbb{B}_\xi, \mathbb{B}_\xi)$ are given respectively by (2-8) and (2-9) .

Proposition 2-3.

Let V be superstable and strongly lower regular. Then

$$^*\mathbb{K}_\Lambda^\omega \xrightarrow[\Lambda\uparrow R^d]{} {}^*\mathbb{K}_\infty \text{ strongly on } {}^*\mathbb{B}_\zeta .$$

Let us denote by $\rho(J\,K_\infty)$ the resolvent set of (Ruelle)-Kirkwood-Salsburg operator.

On the basis of the previous discussion we can conclude that the following theorem is valid:

Theorem.

Let V be superstable and strongly lower regular two-body potential. Let $z^{-1}\in\rho(J\,K_\infty)$. Then there exists a unique, tempered grand-canonical equilibrium Gibbs measure $\mu_\infty(z,\beta)$ the correlation functions of which are given by

$$\rho_\infty(z,\beta) = (1 - z(JK_\infty))^{-1}\alpha_\infty$$

Moreover, this unique Gibbs measure $\mu_\infty(z,\beta)$ is analitic on the set $\rho(JK_\infty)$.

For an extension to many-body forces, see [8] and for an another recent applications of the method presented here [5,6,7].

REFERENCES

1. Ruelle, D. Statistical Mechanics Rigorous Results, Benjamin, New York, 1969.

2. Ginibre, J. Journ.Math.Phys.6,p.238,1965.

3. Gallavotti, G., Miracle-Sole, S., Comm.Math.Phys.7,p.274 1968.

4. Gielerak, R., (in preparation).

5. Gielerak, R., Existence of the transfer matrix formalism for a class of continuous systems. Journ.Stat.Phys.(in print).

6. Gielerak, R., Some Gibbsian aspects of the sine-Gordon transformation, Theor.Math.Phys.73/3/,p.323,1987.

7. Gielerak, R., Trigonometric perturbations of the Gaussian generalized random fields I, Ann.Inst.H'Poincare (in print)

8. Gielerak, R., On the Mayer-Lee-Yang hypothesis for a class of continuous systems. Journ.of Phys.A (in print).

A REMARK ON THE FORMATION OF CRYSTALS AT ZERO TEMPERATURE

S. Albeverio[1,2], R. Høegh-Krohn[1,3†], H. Holden[1,4], T. Kolsrud[5], M. Mebkhout[1,6].

[1]Centre de Physique Théorique, CNRS-Luminy Case 907, F-13288 Marseille, France.

[2]Institut für Mathematik, Ruhr-Universität Bochum, D-4630 Bochum 1, FRG. BiBoS Research Center, Universität Bielefeld, D-4800 Bielefeld, FRG

[3]Matematisk institutt, Universitetet i Oslo, N-0316 Oslo 3, Norway

[4]Institutt for matematiske fag, Universitetet i Trondheim, N-7034 Trondheim-NTH, Norway.

[5]Department of Mathematics, Royal Institute of Technology, S-10044 Stockholm, Sweden.

[6]Université d'Aix Marseille II, Faculté des Sciences de Luminy, F-13288 Marseille, France.

[†]Deceased on January 24, 1988.

ABSTRACT

We study the formation of crystals at zero temperature in classical statistical mechanics.

1. INTRODUCTION.

The questions of why and how crystals form at low temperature have been and still are among the major open problems in theoretical physics [1,2,3].

Assuming the system of N particles to be classical with Hamiltonian

$$H = - \sum_{j=1}^{N} \frac{p_j^2}{2m_j} + \sum_{\substack{i,j=1 \\ i<j}}^{N} V(x_i - x_j) \tag{1}$$

where x_j, $p_j \in \mathbb{R}^3$ denote the position and momentum respectively of the jth particle with mass m_j, one would like to show that this system has a ground state, in the sense of statistical mechanics with Gibbs factor $e^{-\beta H}$, H given by (1) and $\beta = 1/kT$ is proportional to the inverse temperature T, where the particles form a regular structure at low temperatures.

While this problem still is unsolved, more progress has been made at zero temperature where the problem reduces to minimizing the potential energy

$$U(x_1,\dots,x_N) = \sum_{\substack{i,j=1\\ i<j}}^{N} V(x_i-x_j) \tag{2}$$

Our understanding of this problem is fairly complete, both for finite systems ($N < \infty$) and in the thermodynamic limit ($N \to \infty$) in one space dimension. See the excellent survey paper[3] and references therein. In two dimensions our understanding is more restricted[3] and in higher dimensions only few results are known[3].

In this paper we give some results on the formation of crystal at zero temperature in three dimensions. In addition we discuss an explicit example in three dimensions where the potential is of very short range with hard core. Some of the proofs will appear elsewhere[5].

2. CLASSICAL STATISTICAL MECHANICS AT ZERO TEMPERATURE.

Consider N classical particles in a box $\Omega \subset \mathbb{R}^3$ interacting via a two-body interaction V. The Hamiltonian H of this system is then given by (1). The <u>partition function</u> Z reads

$$Z = Z(\Omega,N,\beta) = \int_{\Omega^N} \prod_{i=1}^{N} dx_i \int_{\mathbb{R}^{3N}} \prod_{i=1}^{N} dp_i \; e^{-\beta H} \quad . \tag{3}$$

By integrating out the momenta we find

$$Z = \left(\frac{2\pi m}{\beta}\right)^{\frac{3N}{2}} \int_{\Omega^N} \prod_{i=1}^{N} dx_i \exp\left[-\beta \sum_{\substack{i,j=1\\ i<j}}^{N} V(x_i-x_j)\right] \tag{4}$$

where we for implicity have assumed the particles to have identical mass ($m_i = m$ for $i = 1,\dots,N$).

At zero temperature, i.e. as $\beta \to \infty$, we see that the main contribution to Z comes at the <u>minimum</u> value of the total potential energy, viz.

$$U = U(x_1,\dots,x_N) = \sum_{\substack{i,j=1\\ i<j}}^{N} V(x_i-x_j) \tag{5}$$

and therefore <u>the crystal problem</u> in this context amounts to answering the question as to whether V has a minimum at an almost regular location $x_1,\dots,x_N$ of the N particles or not.

Stated in this form the crystal problem raises additional questions.

A finite system cannot of course be expected to form a completely regular structure due to e.g. surface effects. But it is difficult to express mathematically what is meant by "almost regular". A convenient way out of this problem is to consider the thermodynamic limit, i.e. to consider the limit $N \to \infty$, $\Omega \to \mathbb{R}^3$ keeping the density $\rho = \frac{N}{|\Omega|}$ nonzero and finite, and to show that the points $x_1,\dots,x_N$ approach a perfect lattice Λ in this limit.

The second question is the crucial one: Should there correspond a regular structure to every minimum of V? This cannot be true in general since one should also expect the possibility of encountering quasiregular or even disordered structures at minima of V. The third question is a natural consequence of the second - does the absolute minimum of V correspond to a regular structure?

Let now Λ be a Bravais lattice in $\mathbb{R}^3$, i.e.

$$\Lambda = \{n_1a_1+n_2a_2+n_3a_3 \in \mathbb{R}^3 | (n_1,n_2,n_3) \in \mathbb{Z}^3\} \tag{6}$$

where $a_1,a_2,a_3 \in \mathbb{R}^3$ form a basis in $\mathbb{R}^3$. By relabling and changing of coordinates we may write (2) as

$$U(\{y_\lambda\}_{\lambda\in\Lambda\cap\Omega}) = \frac{1}{2} \sum_{\substack{\lambda,\mu\in\Lambda\cap\Omega \\ \lambda\neq\mu}} V(\lambda-\mu+y_\lambda-y_\mu) \tag{7}$$

where y_λ now measures the deviation from the lattice point λ and we assume

$$V(x) = V(-x) . \tag{8}$$

(7) can be rewritten as

$$U(\{y_\lambda\}_{\lambda\in\Lambda\cap\Omega}) = \frac{1}{2} \sum_{\substack{\lambda,\mu\in\Lambda\cap\Omega \\ \lambda\neq\mu}} [V(\lambda-\mu+y_\lambda-y_\mu)-V(\lambda-\mu)] + \frac{1}{2} \sum_{\substack{\lambda,\mu\in\Lambda\cap\Omega \\ \lambda\neq\mu}} V(\lambda-\mu). \tag{9}$$

Now

$$U(\{y_\lambda\}_{\lambda\in\Lambda\cap\Omega}) = \frac{1}{2} \sum_{\substack{\lambda,\mu\in\Lambda\cap\Omega \\ \lambda\neq\mu}} [V(\lambda-\mu+y_\lambda-y_\mu)-V(\lambda-\mu)]$$
$$+\frac{1}{2}\eta \frac{|\Omega|}{|Q|} \sum_{\substack{\lambda\in\Lambda\cap\Omega \\ \lambda\neq 0}} V(\lambda) \tag{10}$$

where

$$\eta \equiv |Q| \sum_{\substack{\lambda,\mu\in\Lambda\cap\Omega \\ \lambda\neq\mu}} V(\lambda-\mu)\Big[|\Omega| \sum_{\substack{\lambda\in\Lambda\cap\Omega \\ \lambda\neq 0}} V(\lambda)\Big]^{-1} \tag{11}$$

and

$$Q = \{s_1a_1+s_2a_2+s_3a_3 \in \mathbb{R}^3 | s_j \in [-\tfrac{1}{2},\tfrac{1}{2}]\} \quad . \tag{12}$$

We have the simple result.

Lemma 1. Assume V to have compact support. Then

$$\lim_{\Omega\uparrow\mathbb{R}^3} \eta = 1 \quad . \tag{13}$$

■

Hence we can split the function U into two parts, i.e. by introducing the lattice function $U_{lat}(\Lambda)$ and the deviation function $U_{dev}(\{y_\lambda\}_{\lambda\in\Lambda\cap\Omega})$, viz.

$$U_{lat}(\Lambda) = \frac{1}{2} \sum_{\substack{\lambda\in\Lambda \\ \lambda\neq 0}} V(\lambda) \tag{14}$$

$$U_{dev}(\{y_\lambda\}_{\lambda\in\Lambda\cap\Omega}) = \frac{1}{2} \sum_{\substack{\lambda,\mu\in\Lambda\cap\Omega \\ \lambda\neq\mu}} [V(\lambda-\mu+y_\lambda-y_\mu)-V(\lambda-\mu)]$$

we have

$$U = U_{dev} + \frac{|\Omega|}{|Q|} U_{lat} + o(1) \tag{15}$$

as $\Omega \to \mathbb{R}^3$. If we assume that the particles form a lattice Λ in the thermodynamic limit, the second term $|\Omega|\ U_{lat}$ dominates. Hence the structure of the Bravais lattice is determined by minimazing U_{lat}.

The function U_{lat} is only a function of a_1, a_2, a_3, hence we write

$$U_{lat}(a_1,a_2,a_3) = U_{lat}(\Lambda) \ . \tag{16}$$

Proposition 2. The stationary points of U_{lat} are given by

$$\sum_{n\in\mathbb{Z}^3} n_i \nabla V(n_1a_1+n_2a_2+n_3a_3) = 0 \quad , \quad i = 1,2,3 \quad . \tag{17}$$

A stationary point is a local minimum if the 9×9 matrix

$$\left[\sum_{\substack{n\in\mathbb{Z}^3\\ n\neq 0}} n_i n_j \frac{\partial^2 V}{\partial x_k \partial x_\ell}(n_1a_1+n_2a_2+n_3a_3)\right] = 0 \ , \quad i = 1,2,3 \tag{18}$$

is positive definite.

■

If V is central, i.e.

$$V(x) = \Phi(|x|) \ , \tag{19}$$

(17) simplifies to

$$\sum_{\substack{n\in\mathbb{Z}^3\\ n\neq 0}} n_i n_j \frac{\Phi'(|\lambda|)}{|\lambda|} = 0 \quad , \quad i,j = 1,2,3. \tag{20}$$

We see that the crystalline structure is determined by a set of 9 equations with 9 unknowns involving only the two-body interaction V.

Consider now, as an example, the lattice $\Lambda = \alpha\mathbb{Z}^3$, where $\alpha > 0$ is some constant to be determined. (20) can then be reduced to one equation, viz.

$$\sum_{n\in\mathbb{Z}^3} |n|\Phi'(\alpha|n|) = 0 \tag{21}$$

by using the symmetry of the lattice. If $\Phi \in C^2([0,\infty))$, supp $\Phi \subset [0,\sqrt{2}]$

and $\Phi'(a_0) = 0$ for a unique $a_0 \in (0,\sqrt{2})$, $\Phi''(a_0) > 0$, then (21) is solved with $\alpha = a_0$.

We now turn to the analysis of U_{dev}. Observe first that the sum is convergent also in the limit $\Omega \to \mathbb{R}^3$ provided $\{y_\lambda\}_{\lambda\in\Lambda}$ decays at infinity, which amounts to, in a certain sense, fixing the lattice Λ at infinity, or in other words to a boundary condition at infinity. Of course this sum is trivially convergent with a potential of compact support.

<u>Proposition 3.</u> Assume $\Phi \in C^2((\epsilon,\infty))$, $\epsilon > 0$, to have compact support and let

$$V(x) = \Phi(|x|). \tag{22}$$

Then U_{dev} defined by

$$U_{dev}(\{y_\lambda\}_{\lambda\in\Lambda}) = \frac{1}{2} \sum_{\substack{\lambda,\mu\in\Lambda \\ \lambda\neq\mu}} [V(\lambda-\mu+y_\lambda-y_\mu)-V(\lambda-\mu)] \tag{23}$$

has a stationary point at $y_\lambda = 0$, $\lambda \in \Lambda$, which is a local minimum for U_{dev} if

$$(y,D^2_\Lambda y) \equiv \frac{1}{2} \sum_{\substack{\lambda,\mu\in\Lambda \\ \lambda\neq\mu}} (y_\lambda-y_\mu, D^2V(\lambda-\mu)(y_\lambda-y_\mu))$$

$$= \frac{1}{2} \sum_{\substack{\lambda,\mu\in\Lambda \\ \lambda\neq\mu}} \left\{\Phi''(|\lambda-\mu|) \frac{[(\lambda-\mu)\cdot(y_\lambda-y_\mu)]^2}{|\lambda-\mu|^2}\right. \tag{24}$$

$$\left. + \frac{\Phi'(|\lambda-\mu|)}{|\lambda-\mu|}\left[|y_\lambda-y_\mu|^2 - \frac{(\lambda-\mu)\cdot(y_\lambda-y_\mu)^2}{|\lambda-\mu|^2}\right]\right\} > 0$$

where D^2V denotes the Hessian of V, i.e.

$$D^2V(x) = \left[\frac{\partial^2 V}{\partial x_i \partial_j}(x)\right]^3_{i,j=1} \tag{25}$$

■

We will now investigate on alternative condition to (24) to ensure that the stationary point is a local minimum. For this we will essentially make a Fourier-transform of (24).

Let Γ denote the dual lattice of Λ, i.e.

$$\Gamma = \{n_1b_1+n_2b_2+n_3b_3 \in \mathbb{R}^3 \mid (n_1,n_2,n_3) \in \mathbb{Z}^3\},$$
$$b_i a_j = \delta_{ij}, \qquad i,j = 1,2,3 \tag{26}$$

The dual group, $\hat{\Lambda}$, reads

$$\hat{\Lambda} = \mathbb{R}^3/\Gamma = \{\alpha_1b_1+\alpha_2b_2+\alpha_3b_3 \in \mathbb{R}^3 \mid \alpha_i \in [-\tfrac{1}{2},\tfrac{1}{2})\} \tag{27}$$

Define now

$$A_\Lambda(\theta) = \sum_{\substack{\lambda\in\Lambda \\ \lambda\neq 0}} D^2V(\lambda)e^{i\theta\lambda}, \qquad \theta \in \hat{\Lambda}. \tag{28}$$

Then $A_\Lambda(\theta)$ is for any fixed $\theta \in \hat{\Lambda}$ a symmetric 3×3 matrix.

<u>Proposition 4.</u> Λ is a local minimum for U_{dev} on $\ell^2(\Lambda)$ if

$$A_\Lambda(0)-A_\Lambda(\theta) \geq 0\,, \qquad \theta \in \hat{\Lambda}\,, \tag{29}$$

i.e. if $A_\Lambda(0)-A_\Lambda(\theta)$ is a positive definite matrix for all $\theta \in \hat{\Lambda}$.

■

In order to apply Prop. 4 we have to control the following two quantities:

(i) That $A_\Lambda(\theta) = A_\Lambda(0)-c\theta \otimes \theta + \mathcal{O}(|\theta|^3)$, $c > 0$, as $\theta \to 0$.

(ii) That $A_\Lambda(\theta) = A_\Lambda(0)$ iff $\theta = 0$.

Define

$$P_\Lambda = \{V \in C_0^2(\mathbb{R}^3-\{0\}) \mid V(x) = \Phi(|x|),\ V \text{ satisfies (i) and (ii)}\} \tag{30}$$

<u>Theorem 5.</u> The set P_Λ is nonempty open set in the Whitney topology.

■

3. CRYSTALS IN THREE DIMENSIONS. AN EXAMPLE.

The example we want to discuss here is based on the following classical result[4]: Consider spheres of equal size. Then at most twelve spheres can simultaneously touch any given sphere.

This can essentially be accomplished in two different ways. Consider spheres arranged hexagonally in the plane. A second layer, again with hexagonally ordered spheres, can be put on top of the first layer either in the position marked B or C on Fig. 1. (Of course the choice is immaterial until a third layer is introduced.)

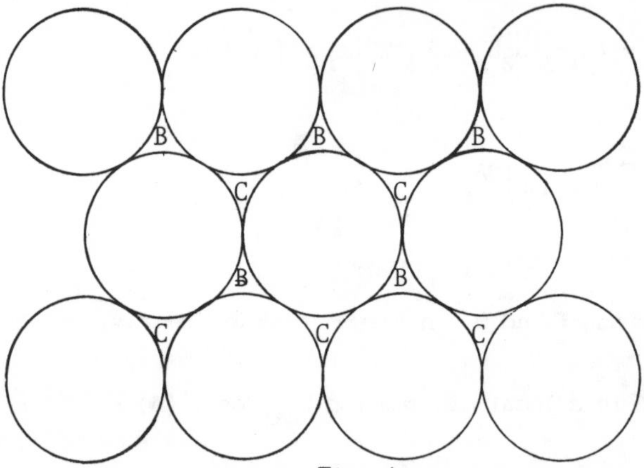

Fig. 1.

A third layer can be placed in the same position as the first one (denoted by A) or in the position not chosen for the second layer. This procedure gives either the sequence ABA or ABC, the sequences ACA or ACB being identical to the first two. By repeating this ad infinitum we obtain the periodic structures .... ABAB .... , which is called hexagonal close packing (hcp), or .... ABC ABC ..., which is called face centered cubic (fcc). However, one also has the possibility of combining the two as follows, ... ABC AB ABC AB ... or even in a nonperiodic combination of AB and ABC . In all cases each sphere has exactly twelve neighbours, all of which are touching the center sphere.

Define for given $\delta > 0$

$$C_\rho(n) = \{\{x_1,\dots,x_n\}\subset\mathbb{R}^3 \mid |x_i|>\rho,\ |x_i-x_j|>\rho,\ i\neq j,\ i,j=1,\dots,n\} \tag{31}$$

and let

$$\delta_\rho = \inf_{n>12}\ [\inf_{\{x_1,\dots,x_n\}\in C_\rho(n)} [\max\{|x_1|,\dots,|x_n|\}]]-2\rho \tag{32}$$

Thus δ_ρ measures the minimum separation of the outermost sphere when one tries to squeeze 13 or more spheres as close as possible to any given sphere. We have

$$\delta_\rho > 0. \tag{33}$$

We assume that the potential V is spherically symmetric, and satisfies (22) with

$$\Phi(r) = \begin{cases} \infty & , 0 \le r \le \rho_0 \\ \Phi(r), & \rho_0 < r < \rho_2 \\ 0 & , r > \rho_2 \end{cases} \tag{34}$$

where $\Phi(r) \to \infty$ as $r \to \rho_0+$, $\Phi \in C^2([\rho_0,\rho_2])$ and Φ has a unique absolute minimum at $\rho_1 \in (\rho_0,\rho_2)$ with $\Phi(\rho_1) < 0$. Assume that

$$\rho_0 < \rho_1 < \rho_2 < \rho_0+\delta_{\rho_0}. \tag{35}$$

To determine the lattice structure, given the potential V, one should according to the scheme outlined in section 2 minimaze U_{lat} given by (14) by varying the lattice parameters a_1,a_2,a_3. This will determine the lattice structure among simple Bravais lattices. Since the hexagonal lattice is not a Bravais lattice while the face centered is a Bravais lattice we have the following result.

<u>Theorem 6.</u> The Bravais-lattice minimum of (14) for the potential V satisfying (22) and (34) is obtained when a_1,a_2 and a_3 span a face centered cubic lattice of size ρ_1. The minimum equals $6\Phi(\rho_1)$.

■

Reformulating the statement somewhat we find

<u>Theorem 7.</u> Let

$$U(\{x_i\}_{i\in\mathbb{N}}) = \sum_{i=1}^{\infty} V(x_i), \tag{36}$$

where $x_i \in \mathbb{R}^3$, $i \in \mathbb{N}$, are arbitrary and V is given by (22) and (34). Then

$$\inf U(\{x_i\}) = 12\Phi(\rho_1). \tag{37}$$

Acknowledgements. Two days after this talk was delivered we received the tragic news that our close friend, inspirator and colleague Raphael Høegh-Krohn had suddenly passed away. An outstanding mathematician and a dear friend is no longer among us.

REFRENCES.

1. Anderson, P.W., Basic notions of condensed matter physics, Benjamin-Cummings 1984.

2. Simon, B., Fifteen problems in mathematical physics. In : Perspectives in mathematics. Anniversary of Oberwolfach 1984. Editors W. Jager, J. Moser, R. Remment, Birkhäuser, Basel, 1984, p. 442.

3. Radin, C., "Low temperature and the origin of crystalline symmetry", Intl. J. Mod. Phys. B1 (1987) 1157.

4. Schütte, K., van der Waerden, B.L., "Das problem der dreizehn Kugeln", Math. Annal. 125, 325-334 (1953).

5. Albeverio, S., Høegh-Krohn, R., Holden, H., Kolsrud, T. Mebkhout, M., in preparation.

EQUILIBRIUM CRYSTAL SHAPES - A MICROSCOPIC PROOF OF THE WULFF CONSTRUCTION

R.L.Dobrushin+), R.Kotecký++), S.B.Shlosman+)

+) Inst. for Problems of Information Transmission
Acad. of Sciences, 19 Yermolova St.,
101447 Moscow GSP-4, USSR

++) Dept. Mathematical Physics, Charles University
V Holešovičkách 2, 180 00 Praha 8, Czechoslovakia

1. INTRODUCTION

The study of equilibrium crystals has been vigorously pursued in recent years. Experiments on Helium crystals or micron-scale metal crystals confirm different theoretical concepts like e.g. roughening (faceting) transitions. (See ref. 1-3) for reviews covering the theory.) The thermodynamics of equilibrium crystal shapes is well understood. Given the anisotropy that is reflected in directional dependence of the interface free energy (surface tension), the shape of a crystal in equilibrium that yields a minimal overall surface tension can be computed by means of the so called Wulff construction. What is lacking however, and what was called for [1,2]*), is a direct

*)"Rigorous proofs even for simple cases would be welcome!"[1] "Rigorous results on the Wulff construction in statistical mechanics would be most timely"[2]

microscopic verification of the Wulff construction. Namely, a priori it is not clear that the surface tension computed for a large planar interface of a given direction may be used for the corresponding infinitesimal area on the surface of rounded crystal and to what extent the resulting shape of a large but finite crystal is well approximated by the Wulff shape.

In the present paper we address ourselves to this problem in a simple model situation. For concreteness we shall consider the Ising model on a square lattice and discuss the typical shape of a large droplet of "- phase" surrounded by "+ phase".

We shall set the problem in more details in the next section. In particular we mention the classical Minlos-Sinai result and introduce the Wulff construction. Our principal result states, roughly speaking, that at low temperatures "the shape of the big droplet with overwhelming probability closely follows the shape of an appropriately scaled ideal Wulff droplet".

The proof of the statement is rather complex. It involves a detailed study of extremal properties of the functional yielding the overall surface tension and a precise evaluation of direction dependent interface free energy and corresponding partition functions by means of local limit theorems. These ingredients are then used to evaluate the probabilities of large droplets of different shapes. Here we shall confine ourselves to the statement of the main results. The full account will be published elsewhere[4].

2. RESULTS

For the discussion of crystal shapes is especially well suited the small canonical ensemble. In the case of the Ising model in a finite volume we define the small canonical ensemble by considering only the configurations with the restriction

$$M_V(\sigma)=\sum_{t\in V}\sigma_t = m\cdot|V|$$

on the total magnetization $M_V(\sigma)$ (m is fixed, $|m|\leq 1$). The corresponding state $\langle\ \rangle_V^{\beta,m}$ yields for $m\in(-m^*(\beta),m^*(\beta))$ in the thermodynamic limit $V\nearrow\mathbb{Z}^2$ a mixture of the pure phases $\langle\ \rangle^{+,\beta}$ and $\langle\ \rangle^{-,\beta}$ that may be constructed in the grand canonical ensemble under + and - boundary conditions, respectively (equivalence of small and grand canonical ensembles in the thermodynamical limit).

Thus locally the state $\langle\ \rangle_V^{\beta,m}$ resembles either + or - phase. However, it is interesting to see how the two phases are separated - i.e. how a typical configuration looks like globally. The first rigorous result in this respect was the Minlos-Sinai "droplet" theorem[5]. It states that considering the small canonical ensemble with + boundary conditions and

$$m=(1-\alpha)m^*(\beta)+\alpha(-m^*(\beta))=(1-2\alpha)m^*(\beta),$$

where $m^*(\beta)=\langle\sigma_0\rangle^{+,\beta}$ is the spontaneous magnetization, a typical configuration σ_V contains, for β large, a single "big" contour Γ (with the length $|\Gamma|>c\ln|V|$). Moreover, this contour encloses the volume $\mathrm{Int}\,\Gamma$ of the area about $\alpha|V|$, it is of almost square shape, and the magnetization inside is roughly $-m^*(\beta)$, while outside it is $+m^*(\beta)$:

i) $\big|\,|\mathrm{Int}\,\Gamma|-\alpha|V|\,\big|<\varkappa(\beta)|V|^{3/4}$

ii) $\left| |\Gamma| - 4\sqrt{\alpha |V|} \right| < \delta(\beta) |V|^{1/2}$

iii) $\left| M^{(\sigma)}_{\mathrm{Int}\Gamma} + m^*(\beta)\alpha |V| \right| \leq \varkappa(\beta) |V|^{3/4}$ and

$\left| M^{(\sigma)}_{\mathrm{Ext}\Gamma} - m^*(\beta)(1-\alpha) |V| \right| \leq \varkappa(\beta) |V|^{3/4}$

with $\varkappa(\beta)$ and $\delta(\beta)$ vanishing exponentially fast with $\beta \to \infty$.

A much more precise information about the shape of the droplet Γ is suggested by the Wulff construction[6,1-3]. It consists in the following. Let n be a unit vector, $n \in S$, and $\mathbb{1}_n$ a plane orthogonal to n. The interface free energy $\tau(\beta, n)$ with respect to this plane is defined by considering the volume V_{KL} with the boundary conditions $\pm \ell_n$ as shown in Fig. 1

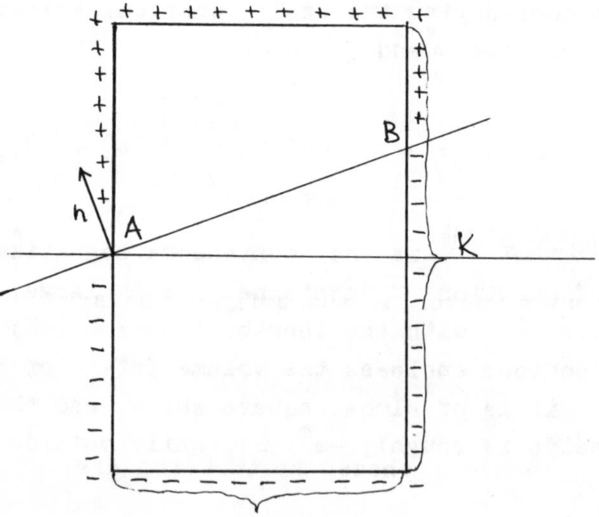

and putting

$$\tau(\beta,n) = \lim_{L\to\infty} \lim_{K\to\infty} -\frac{1}{\beta|A-B|} \ln \frac{Z_{KL}(\beta,\pm \ell_n)}{Z_{KL}(\beta,+)} . \quad (2)$$

Here $Z_{KL}(\beta,\pm \ell_n)$ is the partition function under the considered boundary conditions, while $Z_{KL}(\beta,+)$ is the partition function in the same volume but under purely + boundary conditions. For large β the main ingredient is the energy of the interface and thus $\tau(\beta,n) > 0$. Whenever $\mathcal{U}$ is a domain in R^2 of unit area with piecewise smooth boundary (we shall use $\mathcal{D}_1$ to denote the set of all of them) we define the surface tension functional:

$$\mathcal{U} \in \mathcal{D}_1 \longrightarrow \mathcal{W}_\tau(\mathcal{U}) = \oint_{\partial \mathcal{U}} \tau(\beta, n_\ell)\, d\ell . \quad (3)$$

It turns out that it reaches the minimal value for the domain $W^\beta \in \mathcal{D}_1$ determined by means of the following Wulff construction. Namely, one takes the half plane

$$L_{n,\lambda} = \{x \in \mathbb{R}^2 : (x,n) \geq \lambda \tau(\beta,n)\} \quad (4)$$

for each unit vector n and defines the set

$$W^\beta_\lambda = \bigcap_n L_{n,\lambda} = \{x \in \mathbb{R}^2 : (x,n) \leq \lambda \tau(\beta,n) \ \forall n \in \mathcal{S}\} .$$

Then $W^\beta = W^\beta_{\lambda_0}$ with λ_0 chosen so that the area of $W^\beta_{\lambda_0}$ equals one $(W^\beta_{\lambda_0} \in \mathcal{D}_1)$. If $\mathcal{U}$ is not of the ideal Wulff shape, we will evaluate its deviation from W^β by means of the Hausdorff distance

$$d(\partial \mathcal{U}, \partial W) = \max \{\sup_{x \in \partial \mathcal{U}} \mathrm{dist}(x, \partial W), \sup_{x \in \partial W} (x, \partial \mathcal{U})\} . \quad (5)$$

Now we are ready to formulate our main result.

<u>Theorem</u> Let us consider the small canonical ensemble of the Ising model in an L x L square V under the periodic boundary conditions that is characterized by fixing the magnetization $M_V(\sigma)=m_L\cdot|V|$ and suppose that $m_L \to m$ as $L\to\infty$ with $m=(1-2\alpha)m^*(\beta)$ where $\alpha\in(0,1/4)$ and $m^*(\beta) = \langle\sigma_0\rangle^{+,\beta}$. Consider further the set $A(L,\varepsilon,\delta)$ of all those configurations that contain a single big contour Γ such that

$$d(\Gamma_t, \partial W^{\beta}_{\sqrt{\alpha}L\lambda_0}) < \varepsilon L^{1/2+\delta}$$

with a suitable shift Γ_t of Γ. ($W^{\beta}_{\sqrt{\alpha}L\lambda_0}$ is $W^{\beta}_{\lambda_0}$ rescaled so that its area is αL^2.)

Then, if β is sufficiently large, the probability of $A(L,\varepsilon,\delta)$ tends for all ε,δ to 1 as $L\to\infty$.

<u>Remarks:</u>

1. The value α is chosen from the interval $\langle 0,1\rangle$ to fall into the coexistence region $(-m^*(\beta), m^*(\beta))$ (the vertical plateau of the corresponding thermodynamical potential as function of magnetization m). The restriction $\alpha < 1/4$ is related to our use of periodic boundary conditions. Namely, our theorem is not valid for $m\in(-\frac{1}{2}m^*(\beta),\frac{1}{2}m^*(\beta))$ since then the energetically most favourable configuration consists of two parallel contours winding around the torus and enclosing a strip.

2. In spite of the unpleasant limitation $\alpha < 1/4$ we are using the periodic boundary conditions since for fixed + or - boundary conditions the above theorem does not seem to hold for any α. The point is that the boundary of the square V influences the shape Γ if it passes near ∂V. There is only a small fraction of such configurations; this fraction, however does not vanish as $L\to\infty$

3. The shape $W^{\beta}_{\lambda_0}$ for $\beta = \infty$ is exactly the square of unit area. It may be shown that $\tau(\beta, n)$ is continuous in β on $[\beta_0, \infty)$. Hence the droplet $W^{\beta}_{\lambda_0}$ is for large β close to the square.

4. For the particular case of planar Ising model the surface tension $\tau(n, \beta)$ may be exactly calculated by duality to an angle-dependent correlation length[7]. It turns out that at $\beta \neq \infty$ there are no cusps in the polar plot of τ implying that the corresponding Wulff shape is everywhere rounded with no facets presents. This is presumably a general feature shared by two-dimensional models that is related to the nonstability of interfaces of any given direction[8].

5. Though a generalization of our statement to other two-dimensional models should be possible in principle, attempt at generalization to dimensions larger then two meets several problems.

a) The structure of the set of Gibbs states at low temperatures is up to now not known for the dimensions $\nu \geq 3$. For example it is not even known whether the ground state (i.e. Gibbs state at zero temperature) $\langle \ \rangle^{\mathrm{diag}, \beta=\infty}$ corresponding to the boundary conditions

$$\sigma_t^{\pm \mathrm{diag}} = \begin{cases} +1 & t^1 + t^2 + t^3 \geq 0 \\ -1 & t^1 + t^2 + t^3 < 0 \end{cases}$$

($\nu = 3$) is extremal state or it may be decomposed into other ground states! E.g., is the equality

$$\langle \ \rangle^{\mathrm{diag}, \beta=\infty} = \frac{1}{2}\left[\langle \ \rangle^{+, \beta=\infty} + \langle \ \rangle^{-, \beta=\infty} \right]$$

valid? (It seems that results of [8,9] using equivalance with the Six-vertex model suggest that it is so.)

b) We have not a sufficient understanding of the surface tension $\tau(n,\beta)$(existence, bounds, ...). Only for interfaces of type (1,0,0) with boundary conditions

$$\sigma_t^{\pm,i} = \begin{cases} +1 & t^i \geq 0 \\ -1 & t^i < 0 \end{cases}$$

i = 1,2,3, we get sufficient knowledge from the results of Dobrushin[10].

c) Unlike in the two-dimensional case, the surface tension functional $\mathcal{W}$ does not satisfy in three dimensions the condition of stability at the point $W_{\lambda_0}^{\beta}$ (at least not for Hausdorff distance): for every d, $\varepsilon > 0$ there exist a domain $\mathcal{U}(\varepsilon, d) \in \mathcal{D}_1$ such that

$$\mathcal{W}(\mathcal{U}(\varepsilon,d)) - \mathcal{W}(W_{\lambda_0}^{\beta}) < \varepsilon$$

and in the same time

$$\min_{t\in\mathbb{R}^3} d(\partial\mathcal{U}(\varepsilon,d)+t, \partial W_{\lambda_0}^{\beta}) > d$$

(we can always attach long "hairs" of negligible surface). A cure here might be to measure the distance of $\mathcal{U}$ and W by, e.g., evaluating the volume of their symmetric difference (a concept related to the "flat norm" from[11]).

6. Our Theorem is a statement on the level of the law of large numbers. The methods used in the proof allow, however, to obtain more precise statements the level of the local central limit theorem, i.e. to study the character of fluctuations of the contour Γ around the droplet $W_{\sqrt{a}L\lambda_0}^{\beta}$.

REFERENCES

1. Rottman, C. and Wortis, M., Physics Reports 103, 59 (1984).
2. Abraham, D.B., Surface structures and phase transitions - Exact results, in "Phase Transitions and Critical Phenomena", Vol. 10, eds. Domb, C. and Lebowitz J. L., Academic Press, London 1987.
3. van Beijeren, H. and Nolden, I., The Roughening Transition, in "Current Topics in Physics", eds. Schommers and van Blankenhagen, Springer Verlag 1987.
4. Dobrushin, R. L., Kotecký R. and Shlosman S.B.; in preparation
5. Minlos, R. A. and Sinai, Ya. G., Tr. Mosk. Mat. Obshch. 19, 113 (1968). (English trans.: Trans. Moscow Math. Soc. 19, 121 (1968).)
6. Wulff, G., Z. Kristallogr. 34, 449 (1901).
7. Abraham, D. B. and Reed, P., J. Phys. A10, L 121 (1977).
8. Dobrushin, R. L. and Shlosman, S. B., Sov. Sci. Rev. C5, 54 (1985).
9. Blöte, H. W. and Hilhorst, H. J., J. Phys. A15, L 631 (1982); Nienhuis, B., Hilhorst, H. J. and Blöte, H.W.J., J. Phys. A17, 3559 (1984).
10. Dobrushin, R. L., Theory Prob. Appl. 17, 582 (1972).
11. Federer, H., "Geometric Measure Theory", Springer Verlag 1969.

FIRST ORDER PHASE TRANSITIONS: THE RENORMALIZATION GROUP POINT OF VIEW

R.Kotecký

Dept. Math. Physics
Charles University
V Holešovičkách 2, Praha 8
Czechoslovakia

ABSTRACT

The aim of the lecture is to present a rather non-technical account of the coarse graining approach to the description of the first-order phase transitions. After a brief discussion of the main idea of the Pirogov-Sinai theory, it is explained how, (for models at low temperatures) the renormalization group can be naturally introduced in the contour formulation and what is the resulting flow diagram. Recent application to the random field Ising model by Bricmont and Kupiainen is mentioned.

1. FROM THE PEIERLS ARGUMENT TO THE PIROGOV-SINAI THEORY

The coarse graining approach introduced in ref. 1 contains two ingedients. Namely, the renormalization group ideology and the use of contours for expressions of relevant variables. It shares the latter aspect with the Pirogov-Sinai theory[2)] to which it presents an alternative.

The idea of using contours, i.e. geometrical objects with a direct relevance to properties of the system, for a study of first-order phase transitions comes back to

Peierls with his celebrated argument[3]. We recall it briefly to set the stage for further considerations.

Consider thus the Ising model on, say, a square lattice:

$$i \in \mathbb{Z}^2, \ \sigma_i \in \{+1,-1\}, \ H = -J\sum_{\langle i,j\rangle} \sigma_i\sigma_j - h\sum_i \sigma_i, \ J>0. \quad (1)$$

There are two ground states, $\sigma^+ = \{\sigma_i = +1, i \in \mathbb{Z}^2\}$ for $h \geqslant 0$ and $\sigma^- = \{\sigma_i = -1, i \in \mathbb{Z}^2\}$ for $h \leqslant 0$. For vanishing external field, $h = 0$, these two ground states are in competition. First order phase transitions may be viewed as an unstability with respect to the boundary conditions. Thus to show that a transition exists, one has to prove that fixing the configuration to equal σ^+ outside a finite volume $V \subset \mathbb{Z}^2$, the probability that the spin at a fixed site, say = 0, equals -1, σ_0 = -1, is small (uniformly in V). We replace now configurations by corresponding collections of disjoint contours that form boundaries between regions with spin + and those with spin - . The condition σ_0 = -1 means in the language of corresponding contour ensemble that a contour γ must exist that encircles the site 0. While the probability of presence of such a contour of the "length" $|\gamma|$ = k is bounded by $e^{-2\beta J k}$ (we pay for a contour by the energy proportional to its length), the number of contours of length k encircling the site 0 is bounded by $k \cdot 3^k$. The latter factor ("the entropy of contours") is, for β large enough, overwhelmed by the energy term yielding, after the summation over all contours that encircle the lattice site 0, the bound

$$P_{V,+}(\sigma_0 = -1) < \varepsilon \quad (2)$$

uniformly in V. The + boundary conditions thus lead to the

phase that may be described as the sea of pluses with small islands of minuses.

There are two points to be noticed about the Peierls argument. Both of them are implied by the symmetry of the Ising model. First, we know from the very beginning that a phase transition should be expected for $h = 0$. The second point concerns the proof of the bound

$$P_{V,+}(\gamma) \leq e^{-2\beta J|\gamma|} \tag{3}$$

on the probability of presence of a fixed contour γ in our contour ensemble. We have

$$P_{V,+}(\gamma) = \frac{\sum_{\partial;\gamma\in\partial} \exp\left[-2\beta J\left(\sum_{\bar\gamma\in\partial}|\bar\gamma|\right)\right]}{\sum_{\partial} \exp\left[-2\beta J\left(\sum_{\bar\gamma\in\partial}|\bar\gamma|\right)\right]}$$

with the sum in the nominator taken only over those collections ∂ of contours that contain the given contour γ. Now the desired bound follows from the observation, and this is the point to be noticed, that dropping any contour from an admissible family of contours leads again to an admissible family. Indeed, all collections of mutually disjoint contours are admissible - to all of them there corresponds a spin configuration and the partition function with + boundary conditions is

$$Z_{V,+} = \sum_{\partial\subset V} \prod_{\bar\gamma\in\partial} e^{-2\beta J|\bar\gamma|} \tag{4}$$

with the sum over all collections ∂ of disjoint contours in V.

Both points may turn out not to be true for more

general models preventing us to use the simple minded Peierls argument. To see it we consider a simple example of a nonsymmetric perturbation of the Ising Hamiltonian, say,

$$H = -J\sum_{\langle i,j\rangle}\sigma_i\sigma_j - K\sum_{\Delta}\prod_{i\in\Delta}\sigma_i - \sum_i h\sigma_i \qquad (5)$$

Here the sum is over all triplets of lattice sites $\Delta = ((i_1, i_2),(i_1 + 1, i_2),(i_1 + 1, i_2 + 1))$. The weight of a contour now depends not only on its geometrical shape but also on the sign of the configuration outside. For example, the relative energy of a single - in the sea of + is 8J + 6K + 2h, while it is 8J - 6K - 2h if we interchange the signs. Correspondingly, one has to distinguish + and - contours and replace (4) by

$$Z_{V,+} = \sum_{\partial}\prod_{\gamma\in\partial_+}\phi_+(\gamma)\prod_{\gamma\in\partial_-}\phi_-(\gamma)\exp\left[-h_+V_+(\partial)-h_-V_-(\partial)\right] \qquad (6)$$

Here the sum is over all collections of disjoint contours with matching signs: inside a + contour we first meet - - contours, inside them again + contours, etc.; $\partial_{+(-)}$ is the collection of all $+(-)$ contours in ∂, $\phi_{+(-)}$ the respective weights, $V_{+(-)}$ is the area occupied by pluses (minuses) in the corresponding configuration, and $h_+ = h + K$, $h_- = -h-K$. The phases + and - coexist at the vanishing temperature, $\beta = \infty$, whenever $h_+ = h_-$ (i.e. h = -K). But, since the lowest possible excitations at a nonzero temperature have different weights for the two phases, we do not know a priori for what particular value $h = h_t$ the transition takes place. Also, the straightforward way of proving (3) will not go through any more: extracting a contour from a compatible family ∂ of signed contours we get a family that is no more compatible. (It cannot be cured by changing the signs of all contours inside that in question - this would change their weights.)

The solution to all these problems presents the Pirogov-Sinai theory[2]. Not entering into the technical details, we only mention that the main idea is to look for two distinct contour functionals (collection of contour weights) Ψ_+ and Ψ_- that, considering e.g. Ψ_+, not only yields the partition function (6) by means of the sum of disjoint + contours of the form (4) with $e^{-2\beta J|\gamma|}$ replaced by $\Psi_+(\gamma)$ (and without any condition on matching signs), but that also recovers correctly the probabilities of collections of external contours implying the existence of distinct + and - phases. Indeed, an equation (in a suitable Banach space) can be written down, whose solution yields such contour functionals Ψ_+, Ψ_-, and the value h_t of the transition external field in the same time.

2. THE RENORMALIZATION GROUP APPROACH

We will see now how the same nonsymmetric situation is tackled with help of the renormalization group. For the standard Ising model (1), the expected flow diagram is in the low temperature region governed by the fixed point $T = 0$, $h = 0$:[4]

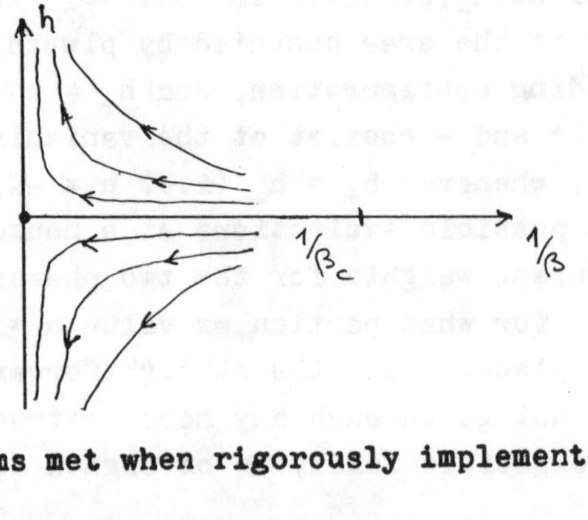

The problems met when rigorously implementing the renormalization group are of two types. One is the problem of parametrization. Starting with a Hamiltonian of the form (5), new further reaching many-body interactions are introduced as a result of the renormalization group transformation. It is necessary to have a control over such additional terms so that they do not blow up and do not spoil the picture after many interactions. Second problem is presented by so called Griffith-Pearce pecularities[5]. Roughly speaking, it is not clear a priori that the flow does not cross the transition line. If this happened, the discontinuity at the transition line would make it difficult even to define unambiguously the renormalization group transformation.

We solve both problems by using a parametrization and introducing a renormalization group transformation that are appropriate and natural just for the low temperature region. Namely, we take as our starting point the formula (6) with one parameter $h_o = h_+ - h_-$ (this will eventually turn out to be the only relevant parameter) and with an infinite number of parameters $\Phi_+(\gamma)$, $\Phi_-(\gamma)$ instead of J, K, &c. (these parameters will be irrelevant and well controled even after many iterations). The renormalization group transformation is defined so that it leads to a model of the same form (6) with new weights Φ_+', Φ_-' and external fields h_+', h_-'. The definition sketched on the next page was split up into three steps - coarse graining, blocking, and rescaling. Choosing a fixed length L, the "small contours" are those of linear size smaller that L and block contours consists of L^2 - - blocks.

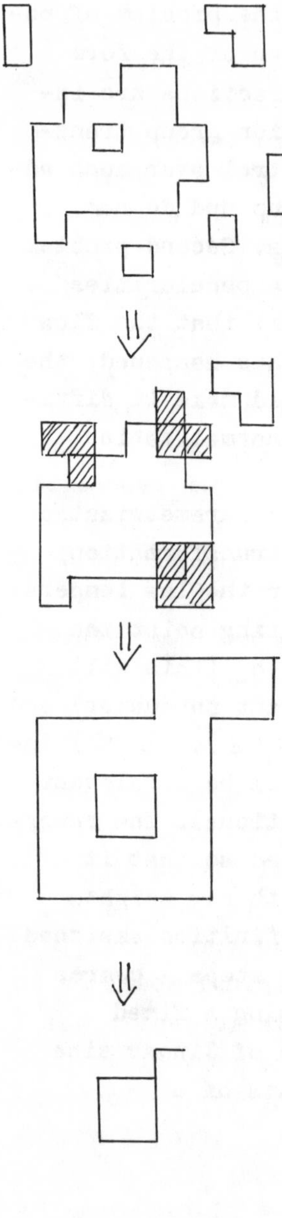

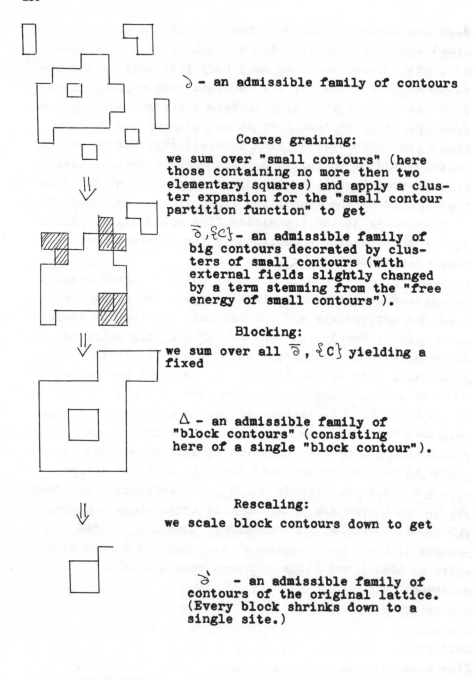

∂ - an admissible family of contours

Coarse graining:

we sum over "small contours" (here those containing no more then two elementary squares) and apply a cluster expansion for the "small contour partition function" to get

$\bar{\partial}, \{C\}$ - an admissible family of big contours decorated by clusters of small contours (with external fields slightly changed by a term stemming from the "free energy of small contours").

Blocking:

we sum over all $\bar{\partial}$, $\{C\}$ yielding a fixed

Δ - an admissible family of "block contours" (consisting here of a single "block contour").

Rescaling:

we scale block contours down to get

∂' - an admissible family of contours of the original lattice. (Every block shrinks down to a single site.)

Even without filling up all the technical details it is clear that the external field dramatically expands: $h'_+ - h'_- \approx L^2(h_+ - h_-)$. Indeed, every site outside contours from ∂' represents an L^2 - block contained fully in $V_+(\partial)$ or $V_-(\partial)$ for every ∂ contributing to ∂'. Realizing also that the length $|\gamma|$ of a typical γ contributing to γ' is $L|\gamma'|$, we get $\phi'_{\pm}(\gamma') \sim e^{-\beta'|\gamma'|}$ if we choose $\beta' = L\beta$. For the standard Ising model this verifies the flow diagram drawn above (with the $\frac{1}{\beta}$ axis actually representing infinitely many directions $\phi_{\pm}(\ )$).

In the nonsymmetrical situation the flow diagram may be actually used to place the transition line h_t. Notice first that if $h_+ - h_-$ is large enough, one can settle everything in a single step. Indeed, new contours may be introduced by including all regions with - spins into them. Even such "thick" contours are dumped (due to the condition on $h_+ - h_-$) at low temperatures and the existence of the single + phase may be proved by means of cluster expansions.

Similarly the existence of the single-phase for $h_- - h_+$ large enough can be proved. For intermediate values of $h_+ - h_-$ we apply our renormalization group transformation leading to a model with external fields that yield much larger difference, $h'_+ - h'_- \approx L^2(h_+ - h_-)$. If $h'_+ - h'_-$(or $h'_- - h'_+$) falls into the + phase (or - phase) region, we can conclude that also the original model is in the corresponding phase. Otherwise we have to repeat the renormalization group transformation. As a result we split up the axis of the original field h into two open halflines the values from which eventually, after a finite number of steps, lead to the manifestly pure phases. The transition value h_t is the common boundary of these two half lines. Of course, starting at the line h_t, we will flow along it under renormalization group transformations.

We conclude with a few remarks. First we notice that, unlike the original Pirogov-Sinai theory, our approach works also for complex external fields (and complex Hamiltonians in general). This might prove useful for a study of metastable phases,and to extend the Pirogov-Sinai theory to the complex case was actually our original motivation.

Let us also observe that our approach might be used to put the theory of finite size scaling at first order phase transitions on a rigorous footing. Indeed, if we start with a finite volume V, we get after k steps a system in the volume of the area $L^{-2k}|V|$. Choosing k such that L^{2k} is of the order $|V|$, we might compute some relevant variables of the resulting model "by hand" and extrapolate for the original model in the finite volume V.

Finally, let us mention that an important generalization was reached by including random fields[6]. In that way J.Bricmont and A.Kupiainen succeeded recently to settle a long-termed dispute by proving that there is a long range order in a random field Ising model at low temperatures and $\nu \geqslant 3$.

We will not try to present here a summary of their technically involved paper. Let us only reproduce the main new ingredient. Namely, if we start with a fixed, site dependent, external field h_i, we should avoid in the coarse graining step the region D of large fields $D = \{ i \,|\, h_i \geqslant \delta\beta \}$ where we have not a good control of the contour weights. The idea is to keep track of this region during iterations and to show that its area eventually diminishes. Actually, the probability of a very high peak, say, $h_i = L^k \beta$ is about $\exp\left[-\frac{1}{\varepsilon^2} L^{2k}\right]$ (ε is the variance of the singhle site random field). Thus it typically appears at distance $\exp\left[\frac{1}{\varepsilon^2} L^{2k}\right]$ from other such peaks and the site i may be after k iterations extracted from the region D (the iterated field $\sum_{j \in L^k} h_j$ with the sum over an L^k - block is not large any more if compared with the iterated temperature). As a result, when including the variance ε^2 into the flow picture, the behaviour is governed by the zero temperature zero field fixed point. Finally we remark that while Bricmont and Kupiainen are considering only the symmetric Ising case, a generalization (by different methods) to unsymmetric cases is under investigation[7].

REFERENCES

1. Gawędzki, K., Kotecký, R. and Kupiainen, A., Journ. Stat. Phys. 47, 701(1987).
2. Pirogov, S. A. and Sinai, Ya. G., Teor. Mat. Fiz. 25, 358 (1975) and 26, 61 (1976); [Theor. Math. Phys. 25, 1185 (1975) and 26, 39 (1976)];
   Sinai, Ya. G., "Theory of Phase Transitions: Rigorous Results", Akademiai Kiado, Budapest, 1982.
3. Peierls, R. E., Proc. Camb. Phil. Soc. 32, 477 (1936); Griffiths, R., Phys. Rev. A136, 437 (1964); Dobrushin, R., Theor. Prob. Appl. 10, 193 (1965).
4. Klein W., Wallace,D.J.,and Zia,R.K.P., Phys. Rev. Lett. 37, 639 (1976).
5. Griffiths, R. and Pearce, P.A., Phys. Rev. Lett. 41, 917 (1978) and Journ. Stat. Phys. 20, 499 (1979).
6. Bricmont, J. and Kupiainen, A., Phys. Rev. Lett. 59, 1829 (1987) and to be published.
7. Zahradník, M., in preparation.

ONE-DIMENSIONAL SCATTERING PROBLEM FOR RANDOM POTENTIAL AND SOME KINETIC PROPERTIES OF DISORDERED SYSTEMS

L.A. Pastur
Physics Technical Institute of Low Temperature
Ukrainian Academy of Science, Kharkov
Lenin's Prospect 47, USSR

1. ONE-DIMENSIONAL RANDOM SCATTERING

Let us consider one-dimensional quantum mechanical scattering problem for the Schrödinger equation

$$- \psi'' + v(x)\psi = k^2\psi \tag{1}$$

with the potential which is equal to zero outside of the interval (0,L). It means that we are looking for solution of (1) which satisfies the condition

$$\psi(x) = \begin{cases} e^{ikx} + \tau_L e^{-ikx}, & k \leqq 0 \\ t_L e^{ikx}, & x \geqq L \end{cases} \tag{2}$$

The transmission and reflection coefficients T_L and R_L are

$$T_L = |t_L|^2 \ , \quad R_L = |\tau_L|^2 \ , \quad R_L = 1 - T_L \tag{3}$$

In paper we shall suppose that the potential v(x) is the

restriction on the interval (0,L) of a random ergodic process. In such a case transmission coefficient T_L is a random variable and we are interested in its statistical properties for large L $(L \to \infty)$.

We mention two typical examples of the random potential.

$$\text{(i)} \quad v(x) = \sum_j u(x - x_j) \tag{4}$$

where U(x) is a function with compact support and $\{x_j\}$ is an ergodic point process in $\mathbb{R}$. In particular a number of results in the theory of disordered systems was obtained for the so-called Frisch-Lloyd model

$$u(x) = k_0 \sum \delta(x - x_j) + v_0, \tag{5}$$

where v_0 is nonrandom nonnegative constant, $k_0 > 0$ and $\{x_j\}$ is the Poisson point process. This potential models the field of short-range chaotically distributed impurities (see e.g. [1]).

(ii) v(x) is the Markov or Markov-type process with sufficiently good ergodic properties (e.g. with exponential mixing). In particular

$$v(x) = V(X(x)) \tag{6}$$

where X(x) is the random value on the smooth Riemannian manifold K and V(X) is a smooth function on K which has only finite order critical points (see e.g. [3,8] for using this potential in the spectral theory of the random Schrödinger operator).

In [1] (see also [2]) it is shown that for any ergodic v(x) with probability 1 there exists the limit

$$-\lim_{L\to\infty} L^{-1} \ln T_L \equiv \bar{\gamma} \leq 0 \tag{7}$$

$$\bar{\gamma} = 2\gamma \tag{8}$$

where γ is the so-called Lyapunov exponent of (1).

Using the terminology of the disordered systems theory [1] we can say that $L^{-1}\ln T_L$ is a self-averaging quantity, just like density of states, conductivity etc.

Consider now the mean value $\langle T_L \rangle$ of T_L. In [2] it is shown that for the same potential as in (7) there exists nonnegative limit

$$-\lim_{L\to\infty} L^{-1} \ln \langle T_L \rangle \equiv \gamma_T. \tag{9}$$

According to Jensen inequality

$$\bar{\gamma} \geq \gamma_T \tag{10}$$

and, as we shall see later, this inequality is strict at least for the high energy of the incident wave (see formula (17)).

2. QUASI ONE-DIMENSIONAL CASE

One-dimensional scattering problem describes the process of wave or particle propagation in the sufficiently thin filament or wire containing the disordered segment (0,L). Consider now the bundle of M such isolated wires. Then the transparency $\Sigma_{L,M}$ of this bundle is

$$\Sigma_{L,M} = M^{-1} \sum_{i=1}^{M} T_L^{(i)} \tag{11}$$

where $T_L^{(i)}$ is the transmission coefficient of i-th wire. These random variables are independent and identically distributed. Writing $T_L^{(i)}$ in the form

$$T_L^{(i)} = e^{-L\bar{\gamma} + L\xi_L^{(i)}} \tag{12}$$

we conclude from (7) that the distribution function $F_L(dx)$ of $\xi_L^{(i)}$ becomes degenerate in the limit $L \to \infty$ if $\gamma > 0$. More exactly if $\Delta = (a,b)$ is an interval then [5]

$$-\lim_{L\to\infty} L^{-1} \ln F_L(\Delta) = \phi(c) \tag{13}$$

where $c = \min\{|a|,|b|\}$ if $\Delta \not\ni 0$, $c=0$ if $\Delta \ni 0$, $\phi(x)$ is nonnegative convex function and $\phi(0)=0$ (the so-called large deviation type behaviour of F_L).

In [4] it is argued*) that if $M=e^{qL}$, $q>0$, then

$$p\text{-}\lim_{L\to\infty} L^{-1} \ln \Sigma_L = -\bar{\gamma} - \max_{\phi(x)\leq q} \{\phi(x) - x\} \equiv -\gamma_q \tag{14}$$

It is easy to see that when q changes from 0 to ∞ γ_q changes continuously from $\bar{\gamma}$ to γ_T. Besides γ_q has the second order derivative discontinuity at the point $q_0=\phi(x_0)$, where x_0 is the point of minimum of $\phi(x) - x$ $(\phi'(x_0) = 1)$, and the remainder term $-L^{-1}\ln\Sigma_L + \gamma_q$ has the order $O(L^{-1}\ln L)$ for $q < q_0$ and $O(e^{-L(q-q_0)})$ for $q>q_0$.

We notice also that essentially the same formula appears in the disordered statistical mechanics giving the expression for the free energy of some simple model of the spin glasses [6].

---

*) For rigorous proof see [5] .

3. CALCULATION OF THE γ_T.

We have already seen that the transmission coefficient has sufficiently complicated statistical properties. For example even for white noise which is the simplest random potential in the theory of disordered systems these properties are investigated only in high energy (semiclassical) region (see [1]and referencees therein). This region is singled out by the inequality

$$Dk^{-1} << 1 \tag{15}$$

where constant D determines the correlation function of white noise potential:

$$<v(x)v(x')>= D\delta(x-x')$$

In particular in this region for $L\to\infty$

$$<T_L> = \frac{\pi^{5/2}}{2}(2\gamma L)^{-3/2}\exp\{-L\gamma/2\}(1+o(1)), \tag{16}$$

From (8) and (16) we see that

$$\gamma_T = \frac{\bar{\gamma}}{4} \tag{17}$$

It turns out that this asymptotic relation is valid in all known cases of small γ_T [9].

If the inicident wave energy $E=k^2$ is smaller than the lower edge E_g of the spectrum of the Schrödinger operator in $L^2(\mathbb{R})$ with the potential (5), then by using the version of cluster expansion it can be shown [2], that γ_T is an analytic function of the parameter ck^{-1}*) in some circle cente-

---

*) Here c is the concentration of the Poisson point $\{x_j\}$ in(5)

red at the origin. The radius of this circle which is some analog of the Ruelle circle in statistical physics tends to zero as $E\uparrow E_g$.

Consider now the ergodic Markov type potential v(x) (i. e. the example (ii) above). Let y(x) be the Cauchy solution of (1) defined by conditions $y(0)=\sin\alpha$, $y'(0)=\cos\alpha$, $\alpha\in[0,\pi)$, and

$$\tau(x) = \left[y^2(x) + y'^2(x)\right]^{\frac{1}{2}}$$

be the envelope of this solution. Then for any real s there exists the limit

$$\lim_{L\to\infty} L^{-1} \ln \langle\tau^s(L)\rangle \equiv \mu(s) \tag{18}$$

which does not depend on α and has the properties [3,8]:

(a) $\mu(0) = \mu(-2) = 0,$ (19)

(b) $\mu(s)$ is convex,

(c) $\mu(s) = \mu(-2-s),$

(d) $\mu'(0) = \gamma.$

Using these properties and the formula [1]

$$T_L = \frac{4}{2+\tau_s^2(L)+\tau_c^2(L)} \tag{20}$$

where $\tau_s = \tau|_{\alpha=0}$, $\tau_c = \tau|_{\alpha=\pi/2}$, it can be shown that [9]

$$\gamma_T = -\mu(-1) \tag{21}$$

If γ_T is small it is reasonable to approximate $\mu(s)$ by parabola $\frac{\gamma}{2}(s+2)s$. Then (8) and (21) gives a heuristic proof of the asymptotic relation (17).

To obtain more information about γ_T let us introduce phase $\phi(x)$ of the solution $y(x)$ by relation $y'/ky = \cot\phi(x)$. Then from (1) we obtain that

$$\phi' = k - \frac{v(x)}{k}\sin^2\phi \tag{22}$$

and if $v(x)$ is the Markov process then the pair $(v(x),\phi(x))$ has the same property with that infinitesimal operator $-\frac{\partial}{\partial\phi}(k - \frac{v}{k}\sin^2\phi) + A_v$, where A_v is the infinitesimal operator of $v(x)$ (in the case of (6) A_v should be replaced by the Laplace-Beltrami operator on K).

Since [1]

$$\tau(x) = \exp\left\{\frac{1}{2k}\int_0^x v(t)\sin 2\phi(t)\,dt\right\}, \tag{23}$$

$\tau^{-1}(x)$ is the multiplicative functional of the pair $(v(x),\phi(x))$. Thus basing on (21)-(23) and on Feynman-Kac formula we can obtain γ_T as the lowest eigenvalue of the operator

$$A_v + \frac{\partial}{\partial\phi} - \frac{v}{k^2}\sin\phi\frac{\partial}{\partial\phi}\sin\phi .$$

For example using perturbation theory for this operator we can find asymptotic expansion of γ_T for small potential or high energy. The first terms of these expansions are the following

$$\gamma_T = \frac{1}{8k^2}\int_0^\infty B(x)\cos 2kx\,dx, \qquad \text{small potential} \tag{24}$$

$$\gamma_T = -\frac{1}{32k^4}B'(0), \qquad \text{high energy.} \tag{25}$$

Here $B(x) = v(0)v(x)$. Comparing these formulae with the respective formulae for Lyapunov exponent [1] and using (8) we can justify the asymptotic relation (17) in these cases.

The similar results (including (17)) can be obtained also for the potential (5) [9].

4. CONNECTION WITH KINETICS.

We start with mentioning the formulae well known in the theory of disordered systems for the electric conductance G_L and heat conductance K_L of the disordered segment (0,L) embedded in the ideal one-dimensional medium:

$$G_L = \frac{e^2}{h} \frac{\langle T_L \rangle_F}{\langle R_L \rangle_F} , \tag{26}$$

$$K_L = \frac{\pi^2 k_F T}{3h} \frac{\langle E^2 T_L \rangle_{F,B}}{\langle E^2 R_L \rangle_{F,B}} . \tag{27}$$

Here $\langle \ldots \rangle_{F,B}$ denotes the integration over E with the weigth functions $-\frac{\partial n_{F,B}}{\partial E}$, $n_{F,B}(E)$ are the Fermi and Bose distribution functions respectively. Formula (2) for zero temperature (T=0) was obtained by Landauer (see e.g. [1] for derivation and discussion of these formulae).

Now we discuss shortly two problems to which the results and methods described above can be applied.

The first of them is the problem of calculation of stochastic sound signal absorbption rate in superconductors after some quite realistic assumptions. This problem can be reduced (see e.g.[10] and references therein) to the scattering problem for Dirac equation in the frame which moves with the velocity s

$$-i(\sigma_z v_z - s)\Psi' + (\beta v(x) + \sigma_x \Delta)\Psi = E\Psi \tag{28}$$

where $\psi=(\psi_1,\psi_2)$ is two component wave function, σ_x and σ_z are the Pauli matrices, v_z is the component of the electron velocity along the sound propagation direction z, s is the sound velocity (which plays the role of the moving frame velocity here), $\beta=s/v_z$, v(x) is the form of sound signal (it is supposed that v(x) is not zero only if $0 \leqq x \leqq L$), Δ is the superconductor gap.

The rate of sound absorption Q is

$$Q = \frac{1}{4} n(E_F) \frac{s^2}{v_F^2} \int_0^\infty \frac{d\beta}{\beta^3} \int_{-\infty}^{\infty} dE (\varepsilon_+ - \varepsilon_-) \left[n_F(\varepsilon_+) - n_F(\varepsilon_-) \right] \langle R_L \rangle \qquad (29)$$

where $n(E_F)$ is the density of states at Fermi energy E_F, $\varepsilon_\pm(E,\beta)$ are branches of unperturbed ($v(x) \equiv 0$) dispersion law of (28) and $R_L=1-T_L$ is the reflection coefficient for this equation. The last formula is evidently similar to (26) and (27).

Note, that the inner integral in (29) diverges if we put in it limiting for $L=\infty$ value of R_L which is equal to 1 according to (7). For this reason it is important to take into account the regularising role of $\langle R_L \rangle$, which became sufficiently fast decreasing for high (semi-classical) energies. The results of corresponding calculations [11] depends on relative magnitudes of the parameters entering in the (28) and (29). The simplest case corresponds to $\beta<1$ and $T \ll \omega \ll \Delta$, where ω is the sound frequency and T is the temperature. In this case Q is proportional to the $(\log V_0^2 L/s)^{\frac{1}{2}}$ where V_0 is the amplitude of a signal ($|v(x)| \leqq V_0$). Note, that the parameter $V_0^2 L/s\Delta$ is large.

At zero temperature only region $\beta>1$ contributes to absorption which in the case $\omega \sim \beta$, when $V_0^2 L/s\Delta$ is again nonlinearity parameter, is proportional to $\ln^2(V_0^2 L/s\Delta)$.

Both last expressions for Q depend rather weakly on amplitude and length of signal. For comparison we note that linear theory (perturbation with respect to v) gives $Q \sim V_0^2 L$ and for periodic signal $Q \sim V_0$ [10].

It is also interesting to note that the spectrum of the operator defined by (29) on the whole axis is different for the $\beta<1$ and $\beta>1$. In the former case for sufficiently random potential (white noise, (5) or (6)) the spectrum is a pure point, dense and all eigenfunctions are exponentially decreasing with the rate which is equal to the Lyapunov exponent (just as for the Schrödinger operator with the same potential [3,8]). In the latter case ($\beta>1$) for any ergodic potential with $\langle |v(x)| \rangle < \infty$ the spectrum is pure absolutely continuous. Thus we have here the first example of one-dimensional differential operator whose spectrum is a pure point in one region of parameters ($\beta<1$) and is pure absolutely continuous in another one ($\beta>1$).

The second problem which we would like to mention here is the calculation of interband transition rate in semiconductors induced by stochastic electric signal v(x). In the simplest two-band approximation (one valence (v) and one conductance (c) band) and for the simplest form of the interband matrix element $\langle E,c|v(x)|E,v\rangle = 2\lambda u(x).\cos\Delta t$ where Δ is the gap between v and c bands we must solve the Dirac equation

$$i\Psi = \hat{\varepsilon}\sigma_z \Psi + u(x) \begin{pmatrix} 0 & \lambda \\ \lambda^* & 0 \end{pmatrix} \Psi$$

for the two component wave function $\Psi = (\psi_v, \psi_c)$ with the initial condition which corresponds to a completely filled valence band and then calculate transition rate Q:

$$Q \sim \int w(E)\,dE \quad , \qquad w(E) = |\psi_c(E,\tau)|^2$$

where τ is the duration of the signal.

In the linear theory which is based on the perturbation theory with respect to λ, $Q\sim\Delta\tau|\lambda|^2$ under the condition that $\tau|\lambda|<<1$. In nonlinear theory (where the case $\tau|\lambda|>>1$ is the most interesting) by using the methods described above we can show [12] that for periodic signal $Q_{per}\sim\Delta|\lambda|$ and for stochastic signal $Q_{stoch}\sim\Delta\tau_{cor}^{-1} = (|\lambda|\tau_{cor})^{-1}Q_{per}$. Here τ_{cor} is the correlation time of signal and we suppose that $|\lambda|\tau_{cor}<<1$. Thus $Q_{stoch}>>Q_{per}$ and $|\lambda|^2\tau_{cor}<<|\lambda|$ plays the role of transition frequency.

REFERENCES

1. Lifshitz, I.M,. Gredeskul,S.A., and Pastur, L.A. Intro duction to the Theory of Disordered systems, Moskow, Nauka, 1982, and New-York, Wiley&Son, 1988.

2. Marchenko, A.V., and Pastur, L.A., Teor.Mat.Fiz.1986, v.68,p.433.

3. Goldsheidt, I.Ja., Molchanov, S.A., and Pastur,L.A., Funk.Anal.i ego Pril.,1977,v.77,p.1-11.

4. Lifshitz, I.M., Gredeskul,S.A., and Pastur, L.A., Zurn. Eksp.and Teor.Fiz.,1982,v.83,p.2362.

5. Pastur, L.A., in preparation.

6. Derrida, B., Phys.Rev.1981,v.B.

7. Lifshitz, I.M., and Kirpichenkov, B.Ja. Zurn.Eks.and Teor.Fiz.,1979,v.77,p.989.

8. Molchanov, S.A., Izv.Ak.Nauk SSSR, ser.mat.1978,v.42,p. 7a.

9. Marchenko, A.V., Molchanov, S.A., and Pastur, L.A., Teor.Mat.Fiz. (in press).

10. Bratus, E.N., and Shumeiko, V.S., J.Low Temp.Phys. 1985 v.60.p.109.

11. Bratus, E.N., Gredeskul, S.A., Pastur,L.A., and Shumeiko, V.S., Fiz.Nizk.Temp.1986,v.12,p.322 and Teor.Mat. Fiz. (in press).

12. Bratus, E.N., Gredeskul, S.A., Pastur, L.A., and Shumeiko, V.S., Phys.Lett.A (in press).

ONE-DIMENSIONAL MARKOVIAN-FIELD ISING MODEL AND DISCRETE STOCHASTIC MAPPINGS

Ulrich Behn and Valentin A. Zagrebnov
Sektion Physik, Karl-Marx-Universität Leipzig, DDR-7010 Leipzig, German Democratic Republic, and Laboratory of Theoretical Physics, JINR, Dubna 141980, USSR

Dedicated to the memory of Raphael Hoegh-Krohn

ABSTRACT

The Ising chain in a Markovian field taking two values with nonzero expectation is investigated by analyzing a related discrete stochastic map. For zero temperature a finite Markov chain analysis yields a discontinuous behaviour of physical quantities. For nonzero temperature the map is characterized by the multifractal properties of its measure and the negative Lyapunov exponent.

1. INTRODUCTION

The calculation of the partition function of the 1D Ising model in a quenched random field h_n can be related to the problem of one spin in an auxiliary field ξ_n [1,2]

$$Z = \sum_{\{s_n\}} \exp\Big[\beta \sum_{n=1}^{N} \big(J s_n s_{n+1} + h_n s_n\big)\Big] = \sum_{s_N} \exp\Big[\beta\Big(\xi_N s_N + \sum_{n=1}^{N-1} B(\xi_n)\Big)\Big] \quad (1)$$

which is governed by the discrete stochastic mapping

$$\xi_n = h_n + A(\xi_{n-1}) = f(h_n, \xi_{n-1}), \quad \xi_0 = 0, \quad n = 1, \ldots, N \quad (2)$$

where $A(x) = (2\beta)^{-1} \ln[\mathrm{ch}\beta(x+J)/\mathrm{ch}\beta(x-J)]$ and $B(x) = (2\beta)^{-1} \ln[4\mathrm{ch}\beta(x+J)\mathrm{ch}\beta(x-J)]$. The Chapman-Kolmogorov equation for the joint probability density for (ξ_n, h_n) reads[3]

$$p_n(x,\eta) = \int d\eta' \int dx' \; T(\eta|\eta') p_{n-1}(x',\eta') \, \delta(x - f(\eta, x')) \quad (3)$$

where $T(\eta|\eta') = \alpha\delta(\eta+\eta'-2h_o) + (1-\alpha)\delta(\eta-\eta')$ is the transition probability for the field h_n taking the values $h_0 + \delta h$, $\delta = \pm$. α is the probability that δ changes from lattice site to lattice site.

The fixed point of (3) gives the invariant measure of the auxiliary field $p^*(x) = \int d\eta\, p^*(x,\eta)$ which can be used to calculate physical quantities[2,5].

The properties of (2) depend on the nature of the driving process h_n and on the shape of the function A.

For zero temperature $A(x)$ is piecewise linear with parts where $\partial_x A(x) = 0$. As a consequence (2) generates for given parameters only a finite number of possible states. Thus p^* can be determined with the theory of finite Markov chains[3]. There exist an infinite but countable number of critical parameter values where the number of possible states changes and a discontinuous behaviour of physical quantities is observed[5].

For nonzero temperatures $_xA(x)$ is nowhere zero. As a consequence (2) generates an uncountable number of possible states which can be labelled using the language of symbolic dynamics. The iteration of (3) generates in the nth step 2^n bands which are, possibly, separated by gaps[2-9]. In the limit $n \to \infty$ we have, due to the nonlinearity of A, a multifractal structure[5]. The fractal dimension undergoes as a function of the physical parameters continuous as well as discontinuous transitions similar to order parameters in phase transitions[3,10]. For high temperatures the fractal can be approximated by a strictly self-similar Cantor set which allowes analytic results[3].

The dynamical properties of (2) are characterized by the always negative Lyapunov exponent, so that we deal with a simple system showing frustration with a strange but nonchaotic attractor.

2. ZERO TEMPERATURE PROPERTIES

2.1. Finite Markov Chain Formalism

For $T=0$ the map is piecewise linear, $A(x) = \pm J$ for $x \gtrless \pm J$, and x otherwise. Therefore (2) generates only a finite number of states x_i which map exclusively into themselves[3,5]. For given (h_0, h, J) and $0<\alpha<1$ the x_i take the values

$$x(n_\sigma, \pm J) = \sum_\sigma n_\sigma h_\sigma \pm J \tag{4}$$

where the n_σ are chosen such that $x \in \bigcup_{\sigma=\pm} [h_\sigma - J,\ h_\sigma + J]$.

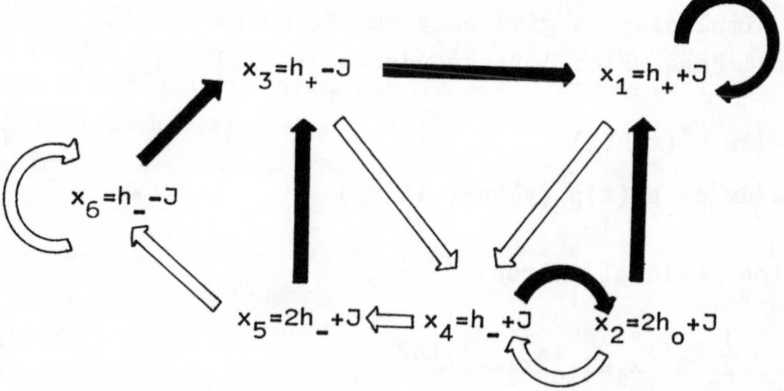

Fig. 1. Flow diagram for the six essential states in the range $-h_-/2 < J < h_+/2$. Full (open) arrows indicate the action of (2) with $h_n = h_+$ ($h_n = h_-$).

For fixed h these states build in the (x, h_0, J) space a complicated structure similar to a honeycomb.

The probability density is the sum of weighted δ functions located at the points $\{x_i, h_i\}$ forming the states of a finite Markov chain[11]. Introducing the vector of the weights $\{w_i^{(n)}\} = \vec{w}^{(n)}$ the Chapman-Kolmogorov equation (3) converts into the matrix equation

$$\vec{w}^{(n)} = D\, \vec{w}^{(n-1)} \tag{5}$$

where the elements D_{ij} are α if $x_i = f(h_i = h_\sigma, x_j = f(h_{-\sigma}, \cdot))$, and $1-\alpha=\gamma$ if $x_i = f(h_i = h_\sigma, x_j = f(h_\sigma, \cdot))$, and zero otherwise. The invariant measure corresponds to the unique ergodic fixed point given by $(1 - D)\vec{w}^* = 0$.

Obviously, for a given parameter range the construction of the space of states and the calculation of the invariant measure could be done on a computer using formula manipulation procedures.

2.2. Physical Properties

The invariant measure allowes us to calculate physical quantities such as free energy and magnetization[2] (for simplicity we give only the formulas for the uncorrelated case which corresponds to $\alpha = 1/2$)

$$f = -\int dx\ p^*(x) B(x) \tag{6}$$

$$\langle m \rangle = \int dx \int dy\ p^*(x) p^*(y) \operatorname{th}\beta(x + A(y)) \tag{7}$$

and the residual entropy[5]

$$s_{res} = \frac{1}{2} k_B (w^*_{x_i=J} + w^*_{x_i=-J}) \ln 2 \quad . \tag{8}$$

In Fig. 2 we present results for one cross section of the above mentioned honeycomb-like structure, namely for fixed exchange J [5]. Increasing the expectation of the magnetic field h_0 the number of states increases in a discontinuous way at the critical values $h_c^{(k)} = h - 2J/k$, $k = 1, 2, \ldots$ and the magnetization jumps by

$$\Delta m^{(k)} = k(1/2)^{k+1} \quad \text{with} \quad \sum_{k=1}^{\infty} \Delta m^{(k)} = 1 \quad . \tag{9}$$

The saturation value $\langle m \rangle = 1$ is reached for $h_0 \geqslant h$ after an infinite but countable number of jumps (phase transitions). At the critical values the magnetization is just in the middle between the steps and the residual energy is

$$s_{res} = (1/2)^{k+2} k_B \ln 2 \quad . \tag{10}$$

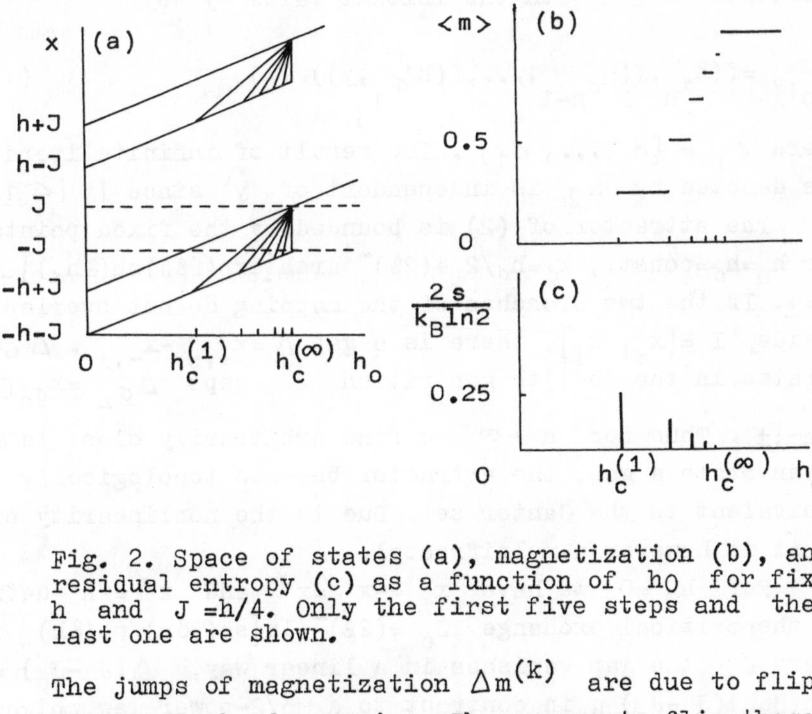

Fig. 2. Space of states (a), magnetization (b), and residual entropy (c) as a function of h_0 for fixed h and $J = h/4$. Only the first five steps and the last one are shown.

The jumps of magnetization $\Delta m^{(k)}$ are due to flips of clusters of k down spins. The energy to flip these clusters vanishes at $h_c^{(k)}$, and $\Delta m^{(k)}$ is simply $(s_\uparrow - s_\downarrow) \times$ length of the cluster $\times$ probability of the cluster. At the thresholds the clusters are frustrated and the groundstate is macroscopically degenerated.

For $0 < \alpha < 1$ the space of states, and therefore the location of discontinuities will be not changed. The measure and therefore the altitude of the steps depends however on α. A different cross section ($h = J/4$, J fixed) is considered in Ref. 5.

3. NONZERO TEMPERATURE PROPERTIES

3.1. The Fractal Measure

For $T>0$ the map is infinitely many times differentiable and generates an uncountable number of states which can be labelled by symbolic dynamics[12]. We denote the nth iteration of (2) with the initial value y by[3]

$$x_{\underset{\sim}{\sigma}_n;y} = f(h_{\sigma_n}, f(h_{\sigma_{n-1}}, \ldots, f(h_{\sigma_1}, y))\ldots) \tag{11}$$

where $\underset{\sim}{\sigma}_n = \{\sigma_n, \ldots, \sigma_1\}$. The result of infinite iterations denoted by $x_{\underset{\sim}{\sigma}}$ is independent of y since $|f'|<1$.

The attractor of (2) is bounded by the fixed points for $h_n = h_\sigma = \text{const.}$, $x_{\underset{\sim}{\sigma}} = h_\sigma/2 + (2\beta)^{-1}\text{arsh}[\exp(2\beta J)\text{sh}(\beta h_\sigma)]$, $\sigma = \pm$. If the two branches of the mapping do not overlap inside $I = [x_{\underset{\sim}{-}}, x_{\underset{\sim}{+}}]$, there is a gap $\Delta = x_{+,\underset{\sim}{-}} - x_{-,\underset{\sim}{+}}$. Δ generates in the $(n+1)$th generation 2^n gaps $\Delta_{\underset{\sim}{\sigma}_n} = x_{\underset{\sim}{\sigma}_n+;\underset{\sim}{-}} - x_{\underset{\sim}{\sigma}_n-;\underset{\sim}{+}}$. Thus for $n \to \infty$ we find arbitrarily close to a given state a gap, the attractor becomes topologically equivalent to the Cantor set. Due to the nonlinearity of A it is however a multifractal.

For $h_0 = 0$ we have $x_{\underset{\sim}{+}} = -x_{\underset{\sim}{-}} = x^*$ and $x^* = 2h$ defines the critical exchange $J_c = (2\beta)^{-1}\ln[\text{sh}(3\beta h)/\text{sh}(\beta h)]$. Near J_c the gap vanishes in a linear way, $\Delta(J-J_c) = 2\text{th}(3\beta h)(J-J_c)$, in contrast to a 5/2-power law which has been claimed to be observed numerically[13].

Only for high temperatures $A(x)$ can be replaced by $x(x^*-h)/x^*$ which gives a strictly self-similar Cantor set (the first gap of which has the exact value) with[3]

$$d_f^{\text{Cantor}} = \begin{cases} 1 & \text{if } \Delta \leqslant 0 \\ \ln 2/\ln[x^*/(x^*-h)] & \phantom{\text{if }} \Delta > 0 \end{cases} \tag{12}$$

In the general case d_f can be evaluated[10] considering an auxiliary repeller problem[14]. This approach was recently employed[15] to calculate the generalized fractal dimensions[16] D_q for $q = \pm\infty$.

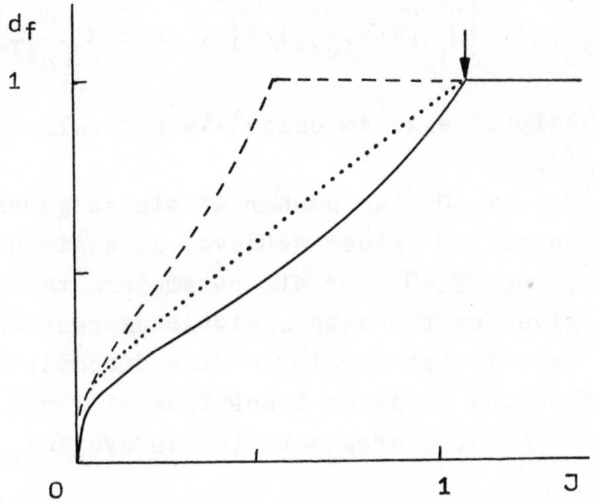

Fig. 3. Fractal dimension versus βJ for $\beta h = 1$ calculated (-----) in zeroth-order perturbation theory[2], (———) by the repeller method[10], and (·········) in the Cantor approximation[3]. The arrow indicates the value for which the gap vanishes.

Complementary to the hierarchy of gaps we find iterating the Chapman-Kolmogorov equation (3) a hierarchy of bands. A normalized initial measure $p_0(x)$ living on I generates in the first iteration two bands $p_\sigma(x) = p_0(y_{\sigma;x})W(y_{\sigma;x})/2$ living on the mappings of the initial support $x_{\sigma;I}$. Here

$$y_{\sigma;x} = f^{-1}(h_\sigma, x) = (2\beta)^{-1}\ln\left[\mathrm{sh}\beta(J+x-h_\sigma)/\mathrm{sh}\beta(J-x+h_\sigma)\right] \qquad (13)$$

denotes the monotonic inversion of (2) and

$$W(x') = \left|2\mathrm{ch}\beta(x'+J)\mathrm{ch}\beta(x'-J)/\mathrm{sh}(2\beta J)\right| \qquad (14)$$

is the prefactor which appears evaluating the δ function in (3). Denoting the result of n inversions by $y_{\underset{\sim}{\sigma}_n;x}$ we obtain for the 2^n bands in the nth generation the closed expression[5)]

$$p_{\underset{\sim}{\sigma}_n}(x) = p_0(y_{\underset{\sim}{\sigma}_n;x}) \prod_{\nu=1}^{n} \left[W(y_{\underset{\sim}{\sigma}_\nu;x})/2\right] , \quad x \in x_{\underset{\sim}{\sigma}_n;I} \qquad (15)$$

which may be helpful e.g. to calculate generalized scaling exponents[16,17)].

Whereas for $T=0$ the number of states jumps crossing critical parameter values we have, at a given step of the iteration, for $T>0$ for all parameters the same number of bands given by the same analytic expression (15). We therefore expect that for $T>0$ the discontinuities will be smoothed and no phase transition will occur in accordance with general arguments for 1D systems.

3.2. The Lyapunov Exponent

For $T>0$ we have $f'<1$ so that the Lyapunov exponent $\lambda = \int dx\, p^*(x) \ln f'(x)$ is negative, the mapping is nonchaotic. For $T=0$ we have $f'=1$ for $|x|<J$ and zero otherwise. The support however is $[-h-J, h+J]$ so that $\lambda = -\infty$.

The Lyapunov exponent is a measure of the convergence of two trajectories corresponding to the same realization of the driving process starting with different initial values. $\lambda < 0$ corresponds to $\lim_{n\to\infty} |x_{\underset{\sim}{\sigma}_n;y} - x_{\underset{\sim}{\sigma}_n;y'}| = 0$. Roughly speaking, the convergence is the faster the 'fewer' states are at our disposal. For $T \to 0$ the space of states collapses to a finite number and $\lambda \to -\infty$.

For high temperatures we have $\lambda^{\text{Cantor}} = \ln[(x^*-h)/x^*]$ and for $d_f \leq 1$ we find an explicit expression between dynamic and geometric characteristic numbers $\lambda^{\text{Cantor}} = -\ln 2/d_f^{\text{Cantor}}$.

4. CONCLUDING REMARK

It is obvious that related models such as the random exchange Ising chain can be treated by adapting this formalism. We reproduced in this way easily the results for magnetization obtained by a different method in Ref. 18.

A further possible application could be the Ising chain in a quasiperiodic two-valued field[19].

REFERENCES

1. Rujan,P., Physica A91,549(1978).
2. Györgyi,G. and Rujan,P., J.Phys. C17,4207(1984).
3. Behn,U. and Zagrebnov,V.A., J.Stat.Phys. 47,939(1987).
4. Behn,U. and Zagrebnov,V.A., JINR Dubna E17-87-138 (1987).
5. Behn,U. and Zagrebnov,V.A., J.Phys. A21(1988).
6. Bruinsma,R. and Aeppli,G., Phys.Rev.Lett. 50,1494(1983).
7. Aeppli,G. and Bruinsma,R., Phys.Lett. 97A,117(1983).
8. Normand,J.M.,Mehta,M.L. and Orland,H., J.Phys. A18, 621(1985).
9. Andelman,D., Phys.Rev. B34,6214(1986).
10. Szepfalusy,P. and Behn,U., Z.Phys. B65,337(1987).
11. Kemeny,J.G. and Snell,J.L., Finite Markov Chains, Springer, Berlin 1983.
12. Guckenheimer,J. and Holmes,P., Nonlinear Oscillations, Dynamical Systems and Bifurcations of Vector Fields, Springer, Berlin 1986.
13. Satija,I.I., Phys.Rev. B35,6877(1987).
14. Szepfalusy,P. and Tel,T., Phys.Rev. A34,2520(1986).
15. Evangelou,S.N., J.Phys. C20,L511(1987).
16. Halsey,T.C.,Jensen,M.H.,Kadanoff,L.P.,Procaccia,I. and Shraiman,B.I., Phys.Rev.A 33,1141(1986).
17. Eykholt,R. and Umberger,D.K., Phys.Rev.Lett. 57,2333 (1986).
18. Derrida,D., Vannimenus,J. and Pomeau,Y., J.Phys. C11, 4749(1978).
19. Luck,J.M., J.Phys. A20,1259(1987).

THE PARISI SPIN-GLASS ORDER PARAMETER FOR A FRUSTRATED ISING MODEL

B. Nachtergaele* and L. Slegers

Instituut voor Theoretische Fysica
Universiteit Leuven
Celestijnenlaan 200 D
B-3030 Leuven (Belgium)

ABSTRACT

In this contribution we calculate rigorously the overlap distribution for a low-temperature approximation of a frustrated Ising model (the so-called PUD model). It is argued that the properties of the overlap distribution as an order parameter to detect spin-glass transitions, should be investigated in more models which can be treated mathematically.

1. INTRODUCTION

The Edwards-Anderson model[1)] for a spin glass is defined as follows. It is an Ising-spin model on Z^d ($x_i \in \{-1,1\}$, $i \in Z^d$) with a random Hamiltonian : for any finite volume $\Lambda \subset Z^d$

$$H_\Lambda(x) = \sum_{\substack{|i-j|=1 \\ i,j \in \Lambda}} J_{ij}\, x_i\, x_j \tag{1}$$

where the J_{ij} are i.i.d. random variables with $\mathbb{E}[J_{ij}] = 0$ and $0 < \mathbb{E}[J_{ij}^2] < +\infty$. The disorder (randomness) has to be considered as quenched, i.e. one looks for properties which hold a.e. The interesting physical quantities are expected to be self-averaging.

The most studied spin glass however, is a mean-field model (Sherrington-Kirkpatrick[2)]). For each N one has the Hamiltonian

*Onderzoeker I.I.K.W. Belgium

$$H_N(x) = \frac{1}{N} \sum_{\substack{i,j=1 \\ i \neq j}}^{N} J_{ij}\, x_i\, x_j \tag{2}$$

where the J_{ij} are again i.i.d., $\mathbb{E}[J_{ij}] = 0$ but now $\mathbb{E}[J_{ij}^2] = NJ^2$. From the heuristic treatment of Parisi the following picture arises[3]: there is a critical temperature $0 < T_c$ such that for $T < T_c$ there exist infinitely many "pure phases", the number of which increases with the number of spins as $\exp(\gamma N)$, $\gamma > 0$. The equilibrium state can be decomposed formally into these pure phases :

$$< \cdot > = \sum_{\alpha \in I} P_\alpha < \cdot >_\alpha \tag{3}$$

The similarity between two pure phases indexed by α and β may be measured by the overlap $q_{\alpha\beta}$

$$q^N_{\alpha\beta} = \frac{1}{N} \sum_{i=1}^{N} \langle x_i \rangle_\alpha \ \langle x_i \rangle_\beta \tag{4}$$

The order parameter for the transition proposed by Parisi is then the overlap distribution defined by [4]

$$P(q) = \lim_{N\to\infty} P^N(q) = \lim_{N\to\infty} \sum_{\alpha,\beta\in I} P_\alpha\, P_\beta\ \delta(q - q^N_{\alpha\beta}) \tag{5}$$

The main hypothesis of Parisi can be reduced to the assumption that there is a natural ultrametric structure on the set of all pure phases (see e.g. [5]). But the mathematical treatment of the SK-model is far from complete, although some progress was made by several authors[6-7]. Almost all important questions about this and other spin-glass models are unanswered. Among them : the proof of the existence of infinitely many pure phases (disjoint extremal states) and how they can be obtained; the question whether or not P(q) is self-averaging (takes a definite value a.e.). More recently doubts were raised in the literature about the qualities of P(q) as an order parameter for spin-glass models other than the SK-model[8-10].

In view of all this it seems to be very important to search for rigorous results in simplified models with[11-13] or without randomness[14-15]. In the next section we present the results of a calculation of P(q) for a non-random frustrated Ising model. There is certainly no spin-glass transition for this model as it is one-dimensional and finite-ranged. But one could expect that some special kind of ordering occurs at T = 0, just as the one-dimensional Ising ferromagnet has non-zero magnetisation in its groundstate.

2. CALCULATION OF P(q)

We consider the one-dimensional lattice Z with an Ising variable $x_i \in \{-1,1\}$ attached to each site $i \in Z$. For any $n \geqslant 2$ define the local Hamiltonians

$$H_N^{(n)}(x) = - \sum_{i=-N}^{N-n} (x_i\, x_{i+1} - \frac{1}{n} x_i\, x_{i+n}) \tag{6}$$

All these models have non-zero residual entropy, due to the frustration which causes an exponentially large degeneracy of the groundstate[16]. For convenience we will treat the case n = 2 here, but everything can be done for the general case. This case n = 2 is a low-temperature approximation of a two-dimensional frustrated Ising model (the so-called PUD model, see ref. 15)). Here we only state the main results concerning the calculation of P(q) for this model, for all temperatures, including T = 0. For proofs and more details we refer the reader to ref. 17).

The equilibrium state ω_β at inverse temperature β for the local Hamiltonians $H_N^{(2)}$ (6), has a quasi-product structure, i.e. its N-point density functions are completely determined by its 3-point density $\rho_\beta(\cdot,\cdot,\cdot)$. Denote

$$\rho'_\beta(x_0,x_1) = \sum_{x_2} \rho_\beta(x_0,x_1,x_2) \tag{7}$$

Then

$$\omega_\beta^N(x_0,..,x_N) = \frac{\rho_\beta(x_0,x_1,x_2)\rho_\beta(x_1,x_2,x_3)\ldots\rho_\beta(x_{N-2},x_{N-1},x_N)}{\rho'_\beta(x_1,x_2)\rho'_\beta(x_2,x_3)\ldots\rho'_\beta(x_{N-2},x_{N-1})} \tag{8}$$

The ρ_β itself can be determined by a variational principle[18]. The result is :

$$\rho_\beta(x_0,x_1,x_2) = \lambda_\beta^{-1}\phi_\beta(x_0,x_1)\psi_\beta(x_1,x_2)k_\beta(x_0,x_1,x_2) \tag{9}$$

where for $\beta < \infty$

$$k_\beta(x_0,x_1,x_2) = \exp -\beta h(x_0,x_1,x_2) \tag{10}$$

with

$$h(x_0,x_1,x_2) = - x_0\, x_1 + \frac{1}{2}\, x_0\, x_2 \tag{11}$$

and ϕ_β and ψ_β are positive eigenvectors of the transfer matrix L_β and its adjoint $L_\beta^\star$ respectively, corresponding to their largest eigenvalue λ_β and suitably normalized. The transfermatrix L_β itself is defined by :

$$(L_\beta\phi)(x_1,x_2) = \sum_{x_0} k_\beta(x_0,x_1,x_2)\phi(x_0,x_1) \tag{12}$$

For $\beta = +\infty$ the same structure is present but with a slightly modified definition of k_β (see ref. 16-17)).

We generalize the definition of the overlap (4) a little bit. The overlap of two configurations can be measured by any two observables which depend on n + 1 sites.

$$q_{x,y}^N = \frac{1}{2N}\sum_{i=-N}^{N-n} A(x_i,\ldots,x_{i+n})B(y_i,\ldots,y_{i+n}) \tag{13}$$

So we have to calculate :

$$P_\beta(q) = \lim_{N\to\infty} P_\beta^N(q)$$

$$= \lim_{N\to\infty} \sum_{\substack{x_{-N},\dots,x_N \\ y_{-N},\dots,y_N}} \delta(q - q_{x,y}^N)\omega_\beta^{2N}(x_{-N},..,x_N)\omega^{2N}(y_{-N},..,y_N) \tag{14}$$

First one has to show that the characteristic function $\tilde{P}_\beta^N$ of P_β^N can be written as

$$\tilde{P}_\beta^N(k) = \sum_{\substack{x_0,..,x_n \\ y_0,..,y_n}} f(x_0,..,x_n)f(y_0,..,y_n)$$

$$\times \exp\{(ik/2N)A(x_0,\dots,x_n)B(y_0,\dots,y_n)\}$$

$$\times \left(T_\beta(\tfrac{k}{2N})^{2N-n} g \right)(x_0,..,x_n,y_0,..,y_n) \tag{15}$$

where f and g are some specific functions and $T_\beta(k)$ is an operator defined as :

$$(T_\beta(k)a)(x_0,..,x_N,y_0,..y_N)$$

$$= \sum_{x_{N+1},y_{N+1}} \exp\{ik\ A(x_0,..,x_N)B(x_0,..,x_N)\}$$

$$k_\beta(x_0,x_1,x_2)\ \dots\ k_\beta(y_{N-1},y_N,y_{N+1})$$

$$a(x_1,..,x_{N+1},y_1,..y_{N+1}). \tag{16}$$

This is then the starting point for proving the following theorem :

Theorem

If the largest eigenvalue of L_β is non-degenerate then the overlap distribution $P_\beta(q)$ converges properly to the Dirac distribution as $N \to \infty$, i.e. $P_\beta(q) = \delta(q - Q(\beta))$, concentrated in $Q(\beta) = \omega_\beta(A)\omega_\beta(B)$. This also holds for $\beta = +\infty$ with $\omega_\infty = \lim_{\beta\to\infty} \omega_\beta$.

■

This theorem is actually more general; it holds for all one-dimensional finite-range models. In our case

$$Q(\beta) = \frac{1 - 2e^{-2\beta} + e^{-4\beta}}{5 - 2e^{-2\beta} + e^{-4\beta}} \tag{17}$$

One can check that $Q(\beta)$ is continuous as a function of β also at $\beta = \infty$.

3. CONCLUSION

We established the fact that $P(q)$ is trivial for a class of frustrated Ising models, in the sense that it is a Dirac distribution. But from a single calculation as this it is of course not possible to decide which of the following explanations holds :

- There is no spin-glass ordering in these models, also not at $T = 0$.
- The overlap-distribution is in general not a good order-parameter to determine such an ordering.

To answer this question $P(q)$ should be calculated for more models which can be treated mathematically.

REFERENCES

1. Edwards S.F., Anderson P.W., J. Phys. F5, 1965 (1975).
2. Sherrington D., Kirkpatrick S., Phys. Rev. Lett. 35, 1792 (1975).
3. Parisi G., Phys. Rev. Lett. 43, 1754 (1979);
Parisi G., J. Phys. A13, L115, 1101, 1887 (1980);
Mézard M., Parisi G., Virasoro M.A., Europhys. Lett. 1, 77 (1986).
4. Parisi G., Phys. Rev. Lett. 50, 1946 (1983).
5. Ramal R., Toulouse G., Virasoro M., Rev. Mod. Phys. 58, 765 (1986).

6. Aizenman M., Lebowitz J.L., Ruelle D., Comm. Math. Phys. 112, 3 (1987).

7. Fröhlich J., Zegarlinski B., Comm. Math. Phys. 112, 553 (1987).

8. Fisher D.S., Huse D.A., J. Phys. A 20, L1005 (1987).

9. Huse D.A., Fisher D.S., J. Phys. A 20 L997 (1987).

10. Bovier A., Fröhlich J., J. Stat. Phys. 44, 347 (1987).

11. Derrida B., Phys. Rev. B 24, 2613 (1981).

12. Ruelle D., Comm. Math. Phys. 108, 225 (1987).

13. Chayes J.T. et al, Comm. Math. Phys. 106, 41 (1986).

14. Villain J., J. Phys. C 10, 1717 (1977).

15. André G. et al, J. Physique 40, 479 (1979).

16. Nachtergaele B., Slegers L., "The groundstate and its entropy for a class of one-dimensional classical lattice systems", preprint KUL-TF-87/19, to appear in Physica A.

17. Nachtergaele B., Slegers L., J. Phys. A 21, 529 (1988).

18. Nachtergaele B., Slegers L., "Construction of equilibrium states for one-dimensional classical lattice systems", preprint KUL-TF-87/33, to appear in Il Nuovo Cimento B.

KINETIC EQUATION IN THE PRIGOGINE THEORY OF IRREVERSIBILITY

M. Rybaczuk, K. Weron
Institute of Physics Technical University of Wrocław
50-370 Wrocław, POLAND
and
Z. Suchanecki, A. Weron
Institute of Mathematics, Technical University of Wrocław
50-370 Wrocław, POLAND

ABSTRACT

By employing stochastic integrals with respect to the operator valued martingales, a mathematically rigorous and simple construction of the Λ operator in the Prigogine theory of irreversible dynamical system is given. As a consequence of the martingale technique a linearized kinetic equation is derived.

1. INTRODUCTION

Prigogine et al.[1-4] have related reversible evolution of dynamical systems to Markov processes without loss of information. The basic idea of this approach is that for unstable dynamical systems it is possible to find an irreversible operator Λ that converts the unitary group associated with classical dynamics into a strict contraction semigroup of a Markov process describing a random motion of points in the phase space. The constructions of the Λ operator presented in these papers illustrate the basic idea of the Prigogine theory of irreversibility. Namely, the reversible dynamical evolution and the irreversible evolution obeying the second law of thermodynamics are connected by a nonunitary operator acting on a distribution function ρ so that the original deterministic Liouville equation should be transformed by it to a dissipative equation describing the irreversible evolution of the dynamical system.

In the recent paper[5], to the construction of the operator Λ, we have employed the stochastic integrals with respect to the operator valued martingales. Our approach has an explicit random character, and also provides a mathematically rigorous and simple construction. Moreover, the martingale technique [5,6] allows us to obtain here the kinetic equation in the explicit form.

2. STOCHASTIC INTEGRAL APPROACH

A dynamical flow is defined as a measure space $(\Gamma, \mathcal{A}, \mu)$ on which acts a group $\{S_t\}$, $t \in \mathbb{R}$ of one-to-one transformations of Γ. For Hamiltonian systems Γ may be identified with a constant energy surface in the phase space, $\{S_t\}$ with a group of canonical transformations of Γ, and the existence of the invariant measure μ is guaranted by Liouville's theorem. We assume that all S_t are μ invariant, i.e., $\mu(S_t^{-1}A) = \mu(A)$ for each $A \in \mathcal{A}$. In this paper we assume that a dynamical flow is also a K-flow. By K-flow we mean such a dynamical flow in which there exists an increasing family $\{\mathcal{A}_t\}$ of sub - σ - fields of $\mathcal{A}$ that satisfies:

$$\sigma(S_t \mathcal{A}_s) = \mathcal{A}_{s+t} \, , \tag{1}$$

$$\bigcup_{t=-\infty}^{\infty} \mathcal{A}_t \text{ generates the } \sigma\text{-field } \mathcal{A}, \tag{2}$$

$$\bigcap_{t=-\infty}^{\infty} A_t = \{\phi, \Gamma\} \equiv A_{-\infty} . \qquad (3)$$

For the practical purposes, it is convenient to consider the following (equivalent) realization of K-flow. Let $L^2(\Gamma,\mu)$ be the space of all square integrable functions ρ, and $\{U_t\}$ a family of unitary operators on this Hilbert space defined by

$$U_t\rho(\omega) = \rho(S_{-t}\omega) , \quad \omega \in \Gamma . \qquad (4)$$

The family $\{A_t\}$ corresponds to the family $\{E_t\}$ of linear operator on $L^2(\Gamma,\mu)$, where $E_t \equiv E^{A_t}$ denotes the operator of conditional expectation with respect to the σ-field A_t.

Let's recall that $\{M_t\}$ is an operator-valued martingale on $L^2(\Gamma,\mu)$ if $\{M_t\rho\}$ is a martingale w.r.t. A_t for each $\rho \in L^2(\Gamma,\mu)$, i.e., $E^{A_s}M_t\rho = M_s\rho$ for $s < t$. In addition, we assume that

$$(M_{s_2} - M_{s_1})(M_{t_2} - M_{t_1}) = 0 \qquad (5)$$

and

$$\int_\Gamma (M_{s_2} - M_{s_1})\rho(M_{t_2} - M_{t_1})\rho d\mu = 0$$
$$\text{for each} \quad \rho \in L^2(\Gamma,\mu), \qquad (6)$$

where $s_1 < s_2 \leq t_1 < t_2$, and M_∞ is a positive, one-to-one operator such that

$$M_\infty 1 = 1, \quad M_\infty U_t = U_t M_\infty \quad \text{for each} \quad t \in \mathbb{R} . \qquad (7)$$

It is easy to check that the family $\{E_t\}$ is an example

of such operator valued martingale.

In order to construct the operator Λ, we need the concept of a stochastic integral with respect to the operator valued martingale $\{M_t\}$ (introduced above). Let remind the definition, as well as, same properties of the stochastic integral[6]. If f is a simple function of the form $f(t) = a_i$ for $t \in (s_i, s_{i-1})$, where $s_1 < s_2 < \ldots < s_{n+}$ then put

$$\int f dM = \sum_{i=1}^{n} a_i (M_{s_{i+1}} - M_{s_1}) . \tag{8}$$

This notion of the integral extends on a wider class of Borel measurable functions on $\mathbb{R}$ which contains, in particular, bounded and measurable functions. For such functions f the integral $\int f dM$ is a linear operator on $L^2(\Gamma,\mu)$, and if M_t are bounded so is $\int f dM$. Moreover, it can be shown that for any Borel measurable function f the integral is a densely defined operator on $L^2(\Gamma,\mu)$, and has the following properties:

$$\int (af + bg) dM = a\int f dM + b\int g dM , \tag{9}$$

$$\int f(s) dM_s \int g(t) dM_t = \int f(s) g(s) dM_s , \tag{10}$$

$$U_t\left(\int f(s) dM_s\right) = \left(\int f(s-t) dM_s\right) U_t . \tag{11}$$

Now, we can define the operator Λ by the formula

$$\Lambda\rho = \int f dM + M_{-\infty}\rho , \qquad \rho \in L^2(\Gamma,\mu) , \tag{12}$$

for an appropriate function f. Namely, we assume that: f is a positive nonicreasing function on $\mathbb{R}$, $f(-\infty) = 1$ $f(+\infty) = 0$, end such that $\log f(s)$ is concave on $\mathbb{R}$. Under these assumptions Λ is a positive and one-to-one operator on $L^2(\Gamma,\mu)$, and although Λ^{-1} has not any of the properties of Λ, the operators $\Lambda U_t \Lambda^{-1}$ are "good operators" again. Precisely, let $W_t^* = \Lambda U_t \Lambda^{-1}$ then under the above assumptions on f, we have

(i) W_t^* is a linear non-negative contraction on $L^2(\Gamma,\mu)$,

(ii) W_t^* is densely defined and $W_t^* 1 = 1$,

(iii) $\{W_t^*\}$, $t \geq 0$, forms a semigroup and $\|W_t^*(\rho - 1)\|$ decreases strictly monotonically to 0, as $t \to +\infty$ for all densities $\rho \neq 1$.

It is not difficult to show[5)] that W_t^* has the form

$$W_t^* = \left(\int \frac{f(s)}{f(s-t)} \, dE_s \right) U_t + E_{-\infty} , \tag{13}$$

which does not depend on the initial martingale $\{M_t\}$.

3. LINEARIZED KINETIC EQUATION

The explicit form of the Λ operator, and consequently the contraction semigroup $W_t^* = \Lambda U_t \Lambda^{-1}$, allows us to discuss an analogue of the Liouville equation, i.e., the equation of motion of the transformed probability density $\tilde{\rho}_t = W_t^* \rho_o$. Let $\tilde{\rho}_{t+\delta t} = W_{t+\delta t}^* \rho_o$, then

$$\frac{d\tilde{\rho}_t}{dt} = \frac{1}{\delta t}(W_{t+\delta t}^* - W_t^*)\rho_o \Big|_{\delta t \to 0} , \tag{14}$$

where the contraction semigroup W_t^* has the form (13). Taking into account that

$$U_{t+\delta t} = U_t(1 - iL\delta t) ,$$

where L denotes the Liouville operator, and using some

approximation in series expansion of the function f, the operator $W^*_{t+\delta t}$ can be transformed to the more convenient form

$$W^*_{t+\delta t} \cong \int \frac{f(s)}{f(s-t)[1-\delta t\frac{f'(s-t)}{f(s-t)}]} dE_s U_t (1 - iL\delta t) + E_{-\infty}$$

$$\cong \int \frac{f(s)}{f(s-t)}[1 + \delta t\frac{f'(s-t)}{f(s-t)}] dE_s U_t (1 - iL\delta t) + E_{-\infty}$$

Hence, after simple calculations, we obtain

$$\frac{1}{\delta t}(W^*_{t+\delta t} - W^*_t)\Big|_{\delta t\to 0} = \int \frac{f(s)}{f(s-t)} \frac{f'(s-t)}{f(s-t)} dE_s U_t - i\int \frac{f(s)}{f(s-t)} dE_s U_t L$$

$$= \int \frac{f(s)}{f(s-t)} dE_s U_t \int \frac{f'(s)}{f(s)} dE_s - i\int \frac{f(s)}{f(s-t)} dE_s U_t L$$

where we have used the property (11).
Denoting

$$\phi = \int \frac{f'(s)}{f(s)} dE_s ,$$

we have

$$\frac{1}{\delta t}(W^*_{t+\delta t} - W^*_t)\Big|_{\delta t\to 0} = (W^*_t - E_{-\infty})\phi - i(W^*_t - E_{-\infty})L$$

$$= [(W^*_t - E_{-\infty})\phi W^{*-1}_t - i(W^*_t - E_{-\infty})LW^{*-1}_t]W^*_t$$

$$= (\tilde{\phi} - i\tilde{L})W^*_t ,$$

where $\tilde{\phi} = (W^*_t - E_{-\infty})\phi W^{*-1}_t$ and $\tilde{L} = (W^*_t - E_{-\infty})LW^{*-1}_t$.
Finally, the equation of motion has the form

$$\frac{d\tilde{\rho}_t}{dt} = -i\tilde{L}\tilde{\rho}_t + \tilde{\phi}\tilde{\rho}_t \,, \tag{15}$$

which is just analogous to the classical kinetic equation with $\tilde{\phi}\tilde{\rho}_t$ as a scattering integral [7]. Note that the scattering integral in (15) is responsible for the irreversible dynamics.

In order to see the physical sense of $\tilde{\phi}$ and $\tilde{L}$ operators, let's transform them using the properties (9) - - (11).

$$\tilde{\phi} = \int \frac{f(s)}{f(s-t)}\, dE_s U_t \int \frac{f'(s)}{f(s)}\, dE_s [U_{-t} \int \frac{f(s-t)}{f(s)}\, dE_s + E_{-\infty}]$$

$$= \int \frac{f(s)}{f(s-t)} \frac{f'(s-t)}{f(s-t)}\, dE_s [\int \frac{f(s-t)}{f(s)}\, dE_s + U_t E_{-\infty}]$$

$$= \int \frac{f'(s-t)}{f(s-t)}\, dE_s = -\int \frac{d}{dt}\,[\ln f(s-t)] dE_s \,,$$

where the second term vanishes as a consequence of the fact that the value of the stochastic integral with respect to the constant process is zero. In a similar manner, we obtain

$$\tilde{L} = \int \frac{f(s)}{f(s-t)}\, dE_s U_t L[U_{-t} \int \frac{f(s-t)}{f(s)}\, dE_s + E_{-\infty}]$$

$$= \int \frac{f(s)}{f(s-t)}\, dE_s L \int \frac{f(s-t)}{f(s)}\, dE_s + \int \frac{f(s)}{f(s-t)}\, dE_s U_t L E_{-\infty}$$

$$= (1 - E_{-\infty}) L \,.$$

REFERENCES

1. Prigogine, I. and Stengers, I., "Order Out of Chaos", Bantam Books, Toronto, 1984.
2. Misra, B., Prigogine, I. and Courbage, M., Physica 104A, 359 (1980).

3. Goldstein, S., Misra, B. and Courbage, M., J. Stat. Phys. 25, 111 (1981).
4. Courbage, M. and Prigogine, I., Proc. Natl. Acad. Sci. USA 80, 2412 (1983).
5. Suchanecki, Z., Weron, A., Rybaczuk, M. and Weron, K., Submitted to Physica.
6. Suchanecki, Z. and Weron, A., to appear, Exp. Math.
7. Cergignani, C., "Theory and Application of the Boltzmann Equation", Scottish Academic Press, Edinburgh, 1975.

MATHEMATICS

Department of [illegible]
[illegible] of [illegible]
[illegible]

10. Intro[illegible]

This [illegible] while the [illegible] random [illegible]

We shall [illegible] section [illegible] be given [illegible] have been [illegible] As for [illegible] now [illegible] can [illegible] approach [illegible] shall [illegible] where the [illegible] not go [illegible] our calc[illegible]

The [illegible] fields that [illegible] some preparatory discussions, we sh[illegible] random field [illegible] of plane [illegible] problems [illegible] random fie[illegible]

Our [illegible] course, [illegible] call them [illegible] some particular class of [illegible] in line with the [illegible]

WHITE NOISE ANALYSIS AND GAUSSIAN RANDOM FIELDS

Takeyuki HIDA

Department of Mathematics
Faculty of Science, Nagoya University
Chikusa-ku, Nagoya 464
JAPAN

§0. Introduction

This report consists of two parts; one is somewhat expository, while the other is presented to propose an attempt to study Gaussian random fields.

We shall start with some background of white noise analysis in Section 1, then two classes of generalized Brownian functionals will be given in Section 2. It should be emphasized that those classes have been introduced in a quite natural way under a lot of motivations. As for the recipes for how to define those generalized functionals we can imitate the methods in the finite dimensional case. Namely, the approach by L. Schwartz and that by I.M. Gel'fand and others. We shall briefly discuss, in Section 3, the so-called causal calculus, where the time development is always taken into account. We will not go into details on the infinite dimensional rotation group, however our calculus has been developed in connection with this group.

The second part of this paper is concerned with Gaussian random fields that may be represented in terms of a white noise. After having some preparatory discussions, we shall come to classes of Gaussian random fields depending on manifolds, in particular on a certain class of plane curves in Section 4. Having been motivated by the actual problems we are led, in Section 5, to discuss the variation of those random fields when the curve changes.

Our approach to such a stochastic variational calculus is, of course, far from a general theory, however we can appeal to the classical theory of functional analysis to carry on this study by choosing some particular class of fields. We also note that this work is still in line with the calculus using the infinite dimensional rotation group.

§1. Background

We start with a Gel'fand triple

$$E \subset L^2(R^d) \subset E^*,$$

where E is a nuclear space and E^* is the dual space of E. A characteristic functional $C(\xi) = \exp[-\frac{1}{2}\|\xi\|^2]$, $\xi \in E$, uniquely determines a probability measure μ on E^* such that

$$C(\xi) = \int_{E^*} \exp[i\langle x,\xi\rangle]\, d\mu(x).$$

This measure μ is called a white noise measure and is thought of as the probability distribution of a white noise $W(t)$, $t \in R^d$.

We shall work, in earlier sections, mostly with a white noise with one dimensional parameter, where $W(t)$ may be viewed as $\dot{B}(t)$ the time derivative of an ordinary Brownian motion $B(t)$. Therefore, μ-almost every x is viewed as a sample function of $\dot{B}(t)$ and hence a member $\Phi(x)$ of the complex Hilbert space $(L^2) = L^2(E^*,\mu)$ may be thought of as a functional of white noise or of Brownian motion. In view of this, $\Phi(x)$ is called a Brownian functional.

It is well known (e.g. [4]) that (L^2) admits a direct sum decomposition

$$(L^2) = \bigoplus_{n=0}^{\infty} H_n \tag{1}$$

and that the isomorphism

$$H_n \simeq \widehat{L^2}(R^{nd}, n!d^n t), \quad t \in R^d, \tag{2}$$

is established, where $\wedge$ denotes the symmetrization in n d-dimensional variables. The space H_0 is tacitly defined to be the one dimensional space $\mathbb{C}$. With the representation (2) any $\Phi(x)$ has a representation in terms of a sequence of functions on finite dimensional Euclidean spaces, and is ready to be analyzed.

Let us now introduce a useful transformation called S-transform on (L^2) (see [11]):

$$(S\Phi)(\xi) = \int_{E^*} \Phi(x + \xi)\, d\mu(x), \qquad \xi \in E. \tag{3}$$

The S-transform carries Φ_n, the H_n-component of Φ, to a homogeneous polynomial in ξ of degree n of the form

$$(S\Phi_n)(\xi) = \int_{R^{nd}} f_n(u)\, \xi^{\otimes n}(u)\, d^n u, \qquad u \in R^d,$$

where f_n is a function determined by the isomorphism (2).

Consider the particular case $d = 1$. Then we see that a version of a Brownian motion $B(t)$ can be realized in H_1 in such a way that $B(t) = B(t,x) = \langle x, I_{[0,t]}\rangle$, I being the indicator function. A formal relation $\dot{B}(t,x) = x(t)$, $x \in E^*$, is in agreement with the fact that x is viewed as a sample function of $\dot{B}(t)$.

§2. Generalized Brownian functionals

Here are presented two typical methods to define generalized Brownian functionals. The first one given in [I] below uses the Sobolev space technique, while the second one given in [II] employs the technique similar to the case of ordinary Schwartz space S.

Throughout this section we set $d = 1$.

[I] The space $(L^2)^-$.

As we have seen before, white noise $\{\dot{B}(t); t \in R\}$ or $\{x(t); t \in R\}$ is considered as the system of variables of Brownian functionals in question. With this choice of variables the most simple and basic functionals are no doubt polynomials in them, for instance $x(t)^n$, although it has no rigorous meaning. We can, however, make it a generalized functional applying the renormalization technique, with which its S-transform is given by $\xi(t)^n$.

We now define a class of generalized Brownian functionals of degree n. The isometry (2) can be generalized through the diagram

$$\begin{array}{ccccc} \mathcal{H}_n^{+} & \hookrightarrow & \mathcal{H}_n & \hookrightarrow & \mathcal{H}_n^{-} \\ \updownarrow & & \updownarrow & & \updownarrow \\ \widehat{H}^{(n+1)/2}(R^n, n!d^nu) & \hookrightarrow & \widehat{L}^2(R^n, n!d^nu) & \hookrightarrow & \widehat{H}^{-(n+1)/2}(R^n, n!d^nu), \end{array}$$

where the double headed vertical arrow means an isometry and $\hookrightarrow$ denotes a continuous injection. A member of the space $\mathcal{H}_n^-$ thus obtained is called a generalized Brownian functionals of degree n.

When we discussed the Feynman integrals (see [8],[13]) we came across the idea of using Donsker's delta function. To make this function a generalized Brownian functional we have formed a weighted sum of the $\mathcal{H}_n$'s. More precisely, take a positive decreasing sequence c_n and denote the $\mathcal{H}_n^+$-norm by $\| \ \|_n$. Set

$$\| \ \|_+ = \{ \sum_{n=0}^{\infty} c_n^{-2} \| \ \|_n^2 \}^{1/2} .$$

The completion of the algebraic sum $\sum_n \mathcal{H}_n^+$ with respect to the norm $\| \ \|_+$ is denoted by $(L^2)^+$. It is a Hilbert space and plays the role of the space of test functionals. The dual space of $(L^2)^+$ relative to the (L^2)-norm is denoted by $(L^2)^-$.

Definition. The Hilbert space $(L^2)^-$ is called the space of generalized Brownian functionals.

Remark. The space $(L^2)^-$ is often written as

$$(L^2)^-_{\{c_n\}} \quad \text{or} \quad \bigoplus_{n=0}^{\infty} c_n \mathcal{H}_n^- .$$

We have thus obtained another triple of large spaces:

$$(L^2)^+ \hookrightarrow (L^2) \hookrightarrow (L^2)^-$$

The bilinear form that links $(L^2)^+$ and $(L^2)^-$ is denoted by $\langle\ ,\ \rangle_\mu$.

It is easy to see that the S-transform is naturally extended to $(L^2)^-$ and it is denoted by the same symbol S.

Examples of an $(L^2)^-$-functional.

i) Polynomials in $x(t)$'s.

Let $\Phi(x)$ be such that

$$(S\Phi)(\xi) = \prod_j \xi(t_j)^{n_j} \qquad \text{(finite product).}$$

Then, $\Phi(x)$ is denoted by $:\prod x(t_j)^{n_j}:$ and is the renormalized functional of the product $\prod x(t_j)^{n_j}$. This $\Phi(x)$ is a member of H_n^- with $n = \Sigma\, n_j$.

ii) Donsker's delta function ([10],[12])

$$\Phi(x) = \delta_0(B(t) - p),$$

where $B(t)$ is a Brownian motion realized by $\langle x, I_{[0,t]}\rangle$. It is also a generalized Brownian functional. This generalized functional plays a role to pin a Brownian trajectory to a fixed point p when we discuss the Feynman integral.

[II] The space $(S)^*$.

To fix the idea we take a particular self-adjoint operator A:

$$A = -\left(\frac{d}{du}\right)^2 + u^2 + 1$$

acting on $L^2(R^1)$ and form the second quantized operator:

$$\Gamma(A) = \bigoplus_n A^{\otimes n}$$

that acts on (L^2). The domain of $\Gamma(A^p)$ is denoted by (S_p). The space of test functionals is given by

$$(S) = \bigcap_p (S_p)$$

and the dual space $(S)^*$ of (S) is the space of <u>generalized Brownian functionals</u>.

We shall not go into details here, but only refer to the literature [5], [8],$\cdots$.

§3. Causal calculus.

Once again we remind that $\dot{B}(t)$'s or $x(t)$'s are the variables of Brownian functionals. Based on them, differential and integral operators are introduced and we can carry out the causal calculus noting that t stands for a time. We still keep the assumption $d = 1$.

<u>Definition</u>. The operator

$$\partial_t = S^{-1} \frac{\delta}{\delta\xi(t)} S, \qquad t \in R^1,$$

is called a differential operator, where $\frac{\delta}{\delta\xi(t)}$ is the Fréchet derivative acting on functionals on E.

The domain of ∂_t is rich enough, including $(L^2)^+$. Following [11], the adjoint operator ∂_t^* for ∂_t is defined:

$$\langle\Phi,\partial_t F\rangle_\mu = \langle\partial_t^*\Phi,F\rangle_\mu, \qquad \Phi \in (L^2)^-, \quad F \in (L^2)^+.$$

The operator ∂_t acts as an annihilation operator, while ∂_t^* does as a creation operator. Multiplication by $x(t)$ is defined as

$$x(t)\cdot = \partial_t^* + \partial_t$$

and the operators ∂_t^*, $t \in R^1$, lead us to a generalization of Itô type stochastic integral (see [14]).

The infinite dimensional Laplace-Beltrami operator, the Lévy Laplacian and other operators can be introduced by using the operators

∂_t and ∂_t^*, $t \in R^1$. These operators have a close connection with the rotation group as is imagined from their name. The definition of the group in which we are interested is now given (after H. Yoshizawa): If g is a linear homeomorphism of E such that $\|g\xi\| = \|\xi\|$, then g is called a rotation of E. The collection of such g's forms a group under the usual product. We denote it by $O(E)$.

Definition. The group $O(E)$ is called the rotation group of E. In case E is unspecified, the group is called an infinite dimensional rotation group and is denoted by O_∞.

Many interesting subgroups of $O(E)$ have been discovered together with their own probabilistic or analytic meaning. For example, an inductive limit of finite dimensional rotation groups is a subgroup of O_∞ and characterizes the Laplace-Beltrami operator. We are, at present, particularly interested in one-parameter subgroups of $O(E)$ that come from diffeomorphisms of the time parameter space R^1. Such a one-parameter subgroup is often called a "whisker". The most important whisker is the shift $\{S_t\}$. To fix the idea, the basic space E is taken to be the Schwartz space $\mathcal{S}$. Define S_t by

$$S_t : \xi(u) \to (S_t\xi)(u) = \xi(u-t). \tag{4}$$

Then $\{S_t \,;\, t \in R^1\}$ forms a one-parameter subgroup of $O(E)$. The adjoint S_t^* is a transformation on E^* that keeps the measure μ invariant. To be more important, S_t^* shifts the time as much as t. Such an action can be represented by the unitary operator U_t:

$$U_t\Phi(x) = \Phi(T_t x), \quad T_t = S_t^*. \tag{5}$$

As is expected, we can prove

Proposition. The relations

$$\begin{cases} U_h\partial_t = \partial_{t+h}U_h, \\ U_h\partial_{t-h}^* = \partial_t^* U_h \end{cases} \tag{6}$$

hold.

Some other whiskers we refer to [4].

§4. Gaussian random fields

From now onward we shall discuss random fields depending on a manifold in a certain class. In most of the interesting cases, random fields in question are expressed as linear functionals of white noise. We therefore briefly discuss some basic properties of a white noise with R^d-parameter.

To avoid inessential complexity we set $d = 2$. Linear, but generalized functionals of white noise form a space H_1^- in the notation established in Section 2 [I], where d is taken to be two. As a generalization of the isometry established there we have

$$H_1^- \simeq H^{-3/2}(R^2). \tag{7}$$

If a plane curve C is smooth enough, we may restrict the white noise to C because of the property of a Sobolev space. Based on this fact, we can discuss a restriction of an R^2-parameter random field to C, provided that the field is in H_1^-.

We then come to the rotation group $O(E)$, where E should be taken to be $D_0 = \{\xi \in C^\infty ;\ (w\xi)(u) = \xi(\frac{u}{|u|^2}) \frac{1}{|u|^2} \in C^\infty\}$, w being the reflection, which is most convenient for us to find good whiskers. Here is a list of whiskers:

Two shifts $\{S_t^1\}$, $\{S_t^2\}$:

$$(S_t^1\xi)(u,v) = \xi(u-t,v),$$

$$(S_t^2\xi)(u,v) = \xi(u,v-t).$$

Isotropic dilation $\{\tau_t\}$:

$$(\tau_t\xi)(u,v) = \xi(ue^t,ve^t)e^t.$$

Rotations $\{\gamma_\theta\}$ on R^2.

Special conformal transformations $\{\kappa_t^1\}$, $\{\kappa_t^2\}$;

$$\kappa_t^i = wS_t^i w, \qquad i = 1,2.$$

It is known (see [6]) that the above whiskers form a six-dimensional Lie subgroup of $O(E)$ and they describe the projective invariance of white noise. The subgroup in often called the conformal group and is denoted by $C(2)$. It is isomorphic to the Lorentz group $SO(3,1)$.

This observation leads us to introduce the collection $\mathfrak{S}$ of all circles in R^2. $\mathfrak{S}$ is topologized so as to be homeomorphic to $R^2 \times (0,\infty)$. Consider now the group $\tilde{C}(2)$ acting on the plane R^2 and generating the group $C(d)$. For example $\tilde{S}_t^1(u,v) = (u-t,v)$, $\tilde{S}_t^1 \in \tilde{C}(2)$.

<u>Proposition</u>. The manifold $\mathfrak{S}$ is $\tilde{C}(2)$-invariant, and $\tilde{C}(2)$ acts transitively.

<u>Proof</u>. We only note the action of κ_t^i, and in reality it is enough to observe the result of the reflection $\tilde{w} : u \to \dfrac{u}{|u|^2}$ which is conformal.

The group $C(2)$ and the above assertion can give us plausible interpretation why we are interested in the random fields depending on a circle in $\mathfrak{S}$.

§5. <u>Variational calculus</u>

The variation of an ordinary (non random) functional depending on a plane curve C has been discussed by many authors; among others J. Hadamard and P. Lévy [1], [2]. It is natural to ask whether or not their methods could be applied to the case of random functionals.

Suppose we are given a family $\mathbb{C}$ of plane curves:

$$\mathbb{C} = \{C : \text{there is a diffeomorphism } S^1 \to C \text{ of } C^\infty\text{-class}\}$$

with a suitable topology. Let a functional

$$U(C) = \int_C f(s)\, ds, \qquad C \in \mathbb{C} \tag{8}$$

be given, where s is a parameter of arc length and where $f(s)$ is the restriction of $f(x,y)$ in the Sobolev space of order 2 to the curve C. Then the variation, when C deforms infinitesimally and within $\mathbb{C}$, is given by

$$\delta U = \int_C \{\frac{\partial f}{\partial n}(s)\delta n(s) - f(s)\delta(ds)\}, \tag{9}$$

where $\delta(ds)$ denotes the varaition of the line element along the curve.

Let us now find a counter part in a random field. If we assume it is in $\mathcal{H}_1^-$-space, then associated $\mathcal{S}$-transform, has an expression of the form

$$U(C) = U(C;\xi) = \int_C f(s)\xi(s)\, ds \tag{10}$$

and we are ready to apply the formula (9). Here one should note that $\delta U(C;\xi)$ involves a term

$$\int_C f(s)(\frac{\partial}{\partial n}\xi)(s)\ \delta n(s) ds.$$

The stochastic version of this formula must have non-manageable singularity, as we can guess from the result by Si Si [17],[18]. However, if C is a circle, as was done in [17], we can embed into a R^2-parameter generalized field.

Our present setup is as follows.

i) Let $X(t)$, $t \in R^2$, be an ordinary Gaussian random field, and let $X(C)$, $C \in \mathfrak{S}$, be given by

$$X(C) = \int_C F(s)X(s)\, ds. \tag{11}$$

ii) Deformation of C is permitted only by the action of $\tilde{C}(2)$.

iii) Derivatives of X are assumed to be well-defined.

iv) The variation $\delta(ds)$ can actually be computed as each member of $\tilde{C}(2)$ acts in the following manner.

1) Two kinds of shift and rotations of R^2 never change the line element ds.

2) Isotropic dilation. Infinitesimal deformation changes

$$ds \rightarrow ds' = (1 + dt)ds.$$

3) For $\tilde{\kappa}_t^1$,

$$ds \rightarrow ds' = ds + 4u(\frac{du}{ds})^2 \, dtds,$$

and for $\tilde{\kappa}_t^2$

$$ds \rightarrow ds' = ds + 4v(\frac{dv}{ds})^2 \, dtds.$$

We may choose a base consisting of six infinitesimal deformations denoted by α_j, $1 \leq j \leq 6$. Then, by the above computation, we may actually obtain their result, denoted by $\delta_j(ds)$. Summing up we have proved

<u>Theorem</u>. Suppose $X(C)$, $C \in \$$, be given by (11). Then, under the assumption iii) the variation δX of $X(C)$ due to $\tilde{C}(2)$ is expressed in the form

$$\delta X(C) = \sum_j dt_j \int_C \{\alpha_j(FX)(s)\, \delta_j(s)ds + (FX)(s)\, \delta_j(ds)\}.$$

More finer results we refer to [9].

<u>Concluding remarks</u>.

1) Remind that $\tilde{C}(2)$ acts effectively on $\$$. Hence we may fix the unit circle C_0, and for any given $C \in \$$ we can find $g \in \tilde{C}(2)$ such that $C = gC_0$. It may be fine if we can write

$$X(C) = X(gC_0) = \tilde{X}(g).$$

With the parameter space $\tilde{C}(d)$, we have a hope to discuss a spectral theory.

2) We can see many examples that can be expressed in the form (11).

3) Take the group $C(2)$. Then the list in Section 4 gives a unitary representation of $C(2)$. To be lucky, the Casimir operator is constant. This information is helpful when we discuss a generalization of the canonical representation of a Gaussian process.

REFERENCES

[1] P. Lévy, Sur la variation de la distribution de l'électricité sur un conducteur dont la surface se déforme, Bull Soc. mathématique de France, 46 (1918), 35-68.

[2] ______ , Problèmes concrets d'analyse fonctionnelle, Gauthier-Villars, Paris, 1951.

[3] T. Hida, Analysis of Brownian functionals, Carleton Math. Lecture Notes, no.13 (1975).

[4] ______ , Brownian motion, Japanese ed. Iwanami, 1975, English ed. Springer-Verlag, 1980.

[5] ______ , H.-H. Kuo, J. Potthoff and L. Streit, White noise: An infinite dimensional analysis, to appear.

[6] ______ , K.-S. Lee and S.-S. Lee, Conformal invariance of white noise, Nagoya Math. J., 98 (1985), 87-98.

[7] ______ , K.-S. Lee and Si Si, Multidimensional parameter white noise and Gaussian random fields, Balakrishnan volume, (1987) 177-183.

[8] ______ , J. Potthoff and L. Streit, White noise analysis and applications, Mathematics + Physics, Vol. 3, World Scientific, to appear.

[9] ______ , Si Si, A variational calculus of Gaussian random fields, to appear.

[10] I. Kubo, Itô's formula for generalized Brownian functionals,

Lecture Notes in Control and Information Sci., 49 (1983), 156-166.

[11] I. Kubo and S. Takenaka, Calculus on Gaussian white noise, Proc. Japan Academy, 56A (1980-1), 376-380, 411-416, 58A (1982), 186-189.

[12] H.-H. Kuo, Donsker's delta function as a generalized Brownian functional and its application, Lecture Notes in Control and Information Sci., 49 (1983), 167-178.

[13] ______, Brownian functionals and applications, Acta Applicandae Math., 1 (1983), 175-188.

[14] ______ and A. Russek, White noise approach to stochastic integration, J. Multivariate Analysis, 24 (1988), 218-236.

[15] P.A. Meyer et J.A. Yan, A propos des distributions sur l'espace de Wiener, Séminaire de Probabilités XXI, Lecture Notes in Math., 1247 (1987), 8-26.

[16] Si Si, A note on Lévy's Brownian motion, Nagoya Math. J., 108 (1987), 121-130.

[17] ____, A note on Lévy's Brownian motion II, Nagoya Math. J., to appear.

[18] ____, Topics on Gaussian random fields, RIMS Proc., (1988) ed. A. Noda.

FLOWS IN INFINITE DIMENSIONS AND ASSOCIATED TRANSFORMATIONS OF GAUSSIAN MEASURES

Ana Bela Cruzeiro
Centro de Matemática e Aplicações Fundamentais,
I.N.I.C., Av. Prof. Gama Pinto, 2, 1699 Lisboa Codex
PORTUGAL

ABSTRACT

The existence of a flow associated with a vector field on a (infinite-dimensional) Wiener space, under some L^p assumptions on the vector field as well as its divergence is proved. The divergence describes the action of the flow on the Wiener neasure and it's quasi-invariance is shown. An application in statistical hydrodynamics is discussed.

0. INTRODUCTION

In several mathematical and physical situations one is lead to solve ordinary differential equations $\frac{du}{dt} = A(u)$ on infinite-dimensional spaces, namely in abstract Wiener spaces in the sense of L. Gross (cf.[5]). Very often the vector fields we want to integrate are not smooth and resources of classical differential calculus are not sufficient. On the other hand, it is important to look for invariants of the motions described and therefore to study the action of the flows on the underlying Wiener measures.

An analysis of Wiener functionals was introduced by P. Malliavin (cf. [7], [8]) and is now know as the Malliavin calculus. In this framework, we proved in [4] the existence of global flows associated with vector fields defined almost everywhere and belonging to certain Sobolev classes. As we work in infinite dimensions, the hypothesis do not imply any continuity or lipchitz-regularity. These results are obtained assuming that the vector field "conserves" the gaussian measure, that is, using the hypothesis that it's divergence is exponentially integrable. Moreover, the image of the Wiener measure by the global flows obtained has a density with respect to the Wiener measure, for which we gave an explicit formula involving the divergence. In the case of a constant vector field it reduces to the Cameron-Martin formula ([2]).

The aim of this work is to show that, by weakening the assumptions on the vector field and, on the other hand, imposing the condition that the divergence is bounded, we still obtain a global flow as well as the conservation of the measure. This is proved using different techniques than the ones used in [4]. Finally, we refer to an exemple where these methods have had an application - the case of Euler equations for a two-dimensional fluid, where the divergence of the vector field is zero, which means that the measure is invariant ([1]).

Although we are using the classical Wiener space, we remark that everything can be done in the framework of a general abstract Wiener space.

The author wishes to express her gratitude to professor Paul Malliavin, whose encouragements and suggestions made this work possible. She is also gratefull to the organizers of the XXIV Winter School in Karpacz for their kind invitation.

1. NOTATIONS

Let X be the classical Wiener space, that is, the space of continuous real functions x defined on $[0,1]$ and verifying $x(0) = 0$. X is a Banach space for the norm $||x||_\infty = \max |x(\tau)|$. We denote by H the Cameron-Martin (Hilbert) space $H = \{x \in X : \int_0^1 |x'(\tau)|^2 \, d\tau < +\infty\}$. The brownian motion defines on X a gaussian measure μ, the Wiener measure, and (X,H,μ) is a Wiener space in the sense of Gross. We also consider an orthonormal basis of H, $e_1,\ldots,e_k,\ldots$, such that the e_k belong to the space $H_0 = \{h \in H : h'' \text{ is a measure}\}$. The scalar product $\langle x,e_i\rangle$ defined for $x \in H$ can be extended to the space X and the notation will be kept for the extension.

We shall consider the differential calculus on the Wiener space in the sense of Malliavin (cf. [8] and references therein). We recall some notions that will be necessary in the sequel. In this framework, the directions of derivation are the vectors in H. A classical result of Cameron and Martin (cf. [2]) characterizes the vectors in H as the directions that leave μ invariant under translation; moreover, if $\tau_h(x) = x + h$, $x \in X$, we have the explicit formula:

$$\frac{d(\tau_h)_* \mu}{d\mu} = \exp\left(\int_0^1 h'(\tau)\,dx(\tau) - \frac{1}{2}\int_0^1 |h'(\tau)|^2 d\tau\right) \quad (1.1.)$$

and this density belongs to $\bigcap_p L^p_\mu$. For a Wiener functional, that is, a function defined on X with values on a Banach space E, we define the gradient of ϕ as the linear operator $\nabla\phi(.) \in \mathcal{L}(H;E)$ given by:

$$\nabla\phi(x)(h)=D_h\phi(x)=\lim_{\varepsilon\to 0}\frac{1}{\varepsilon}[\phi(x+\varepsilon h)-\phi(x)],\ h\in H\ , \qquad (1.2.)$$

the limit being taken in the almost-everywhere sense. Supposing $\nabla\phi$ is an Hilbert-Schmidt operator, we can define $\nabla^2\phi$ and, in general, $\nabla^i\phi(.)\in\mathcal{L}^i(H;E)$, where $\mathcal{L}^i(H;E)$ denotes the class of i-linear operators on H with values in E. Let W_r^p be the corresponding Sobolev space:

$$W_r^p=\{\phi:X\to E:\ \phi\in L^p(\mu) \text{ and } \forall 1\le i\le r\ \|\nabla^i\phi\|\in L_\mu^p(X;\mathcal{L}_{(2)}^i(H;E))\},$$

where the index (2) means we are considering the Hilbert-Schmidt operators with the corresponding norms. We note:

$$\|\phi\|_{r,p}=\|\phi\|_{L^p}+\sum_{i=1}^r\|\nabla^i\phi\|_{L^p(X;\mathcal{L}_{(2)}^i(H,E))} \qquad (1.3.)$$

For a functional $A\in L^2(X;H)$ we recall that the divergence of A, noted δA, is defined as the element of $L^2(X;\mathbb{R})$, when it exists, that verifies:

$$\int f\ \delta A\ d\mu=\int(A|\nabla f)\ d\mu \qquad \forall f\in W_1^2(X;\mathbb{R}) \qquad (1.4.)$$

and we refer to [6] for conditions on its existence.

We shall use the test-function space defined by $\mathcal{D}=\{f:X\to\mathbb{R}:\exists d,\exists F\in C_0^\infty(\mathbb{R}^d)\ :\ f(x)=F(x_1,\dots,x_d)\}$, where $x_i=\langle x,e_i\rangle$. Finally, let X_α be the subspace of X consisting of the Hölderian functions of order α, with the following norm: $\|x\|_\alpha=\sup_{\tau\neq\tau'}|x(\tau)-x(\tau')|/|\tau-\tau'|^\alpha$. It is well known that, for $0<\alpha<1/2$, $\mu(X_\alpha)=1$.

2. A NON-LINEAR CAMERON-MARTIN FORMULA

We recall the following result that was proved in [4]

for vector fields defined on X.(a vector field on X is a Wiener functional with values in the space of derivations, that is, the Cameron-Martin subspace).

2.1. THEOREM - Let A be a vector field on X satisfying the following hypothesis:

$$\text{(i)}\quad A\in \bigcap_p W_4^p \text{ and } \forall\lambda>0 \quad \int_X \exp(\lambda \|A(x)\|_H) d\mu(x) < +\infty$$

$$\text{(ii)}\quad \forall\lambda>0 \quad \int_X \exp(\lambda \|\nabla A(x)\|) d\mu(x) < +\infty$$

$$\text{(iii)}\quad \forall\lambda>0 \quad \int_X \exp(\lambda |\delta A(x)|) d\mu(x) < +\infty$$

Then there exists a flow $U_t^A(x)$ defined μ-a.e-x for all $t\in \mathbb{R}$ such that

$$\text{(iv)}\quad U_t^A(x) = x + \int_0^t A(U_\xi^A(x))d\xi \qquad \mu\text{-ae.-}x \quad \forall t$$

Furthermore, for all $t\in\mathbb{R}$, the image of the measure μ under the action of the flow, $(U_t^A)_*\mu$, has a density $k_t(x)$ w.r.t. μ given by:

$$\text{(v)}\quad k_t(x) = \exp\left(\int_0^t \delta A(U_{-\xi}^A(x))d\xi\right), \quad \text{with } k_t \in \bigcap_p L_\mu^p$$

We also recall that the hypothesis of this theorem do not imply any sort of continuity for the vector field.

If we take $A(x) = h\in H$, then $U_t(x) = x + th$. In this case it is not difficult to verify hypothesis (i)-(iii) and, because the divergence is given by the stochastic integral in the Itö sense of h', namelly $\delta A(x) =$ $= \int_0^1 h'(\tau)dx(\tau)$, we obtain the Cameron-Martin formula (1.1.) for $t = 1$.

3. INFINITE-DIMENSIONAL FLOWS: THE MAIN RESULT

In this paragraph we show that, by imposing the boundness of the divergence, we can weaken the hypothesis on the vector field and its derivative and still have a globally defined (in some sense) associated flow. As is theorem 2.1., the study of the divergence of the vector field allow to control the explosion of the flow and gives it's action on the Wiener measure.

Let V_n be the finite-dimensional subspace of H_0 generated by $\{e_1,...,e_n\}$, where the e_k are the elements of the basis of H introduced in paragraph 1. The orthogonal projection of H on V_n can be extended to X by the formula $\Pi_n(x) = \sum_{k=1}^{n} <x,e_k>e_k$; if we write $V_n^{\perp}=\Pi_n^{-1}(0)$, we have a pseudo-direct sum decomposition of H, namely $X = V_n \oplus V_n^{\perp}$. The Wiener measure can be decomposed in $\mu=\mu \otimes \mu_n^{\perp}$, where μ_n is the standard gaussian n-dimensional measure and $\mu_n^{\perp}$ is supported by $V_n^{\perp}$.

For a functional $A \in L^2(X;H)$, and if E^{V_n} denotes the orthogonal projection from $L^2(X;H)$ to $L^2(V_n;H)$, we define:

$$A_n(x) = \Pi_n[E^{V_n}A(x)] \qquad (3.1.)$$

We proved in [4] that the finite-dimensional vector fields A_n verify the estimations $\|A_n\|_{r,p} \leq \|A\|_{r,p}$ $\forall r,p : 0 \leq r < +\infty,\ 1 < p < +\infty$ and that $\delta A_n = E^{V_n}\delta A$, where δA_n is taken in the $\mathbb{R}^n$ sense, with respect to the measure μ_n.

We shall use a finite-dimentional result proved in [3]:

3.1. LEMMA - Let B be a vector field on $\mathbb{R}^n$, $B \in C^4$, such that $B \in W_1^4(\mathbb{R}^n)$ and $\delta B \in L^\infty_{\mu_n}$. Then there exists a flow $U_t^B(x)$ associated with B, defined μ_n-a.e. in x for all $t \in \mathbb{R}$. Furthermore, if $k_t^B(x) = \exp \int_0^t \delta B(U_{-s}^B(x))\, ds$, we have $\dfrac{d(U_t^B)_*\mu_n}{d\mu_n}(x) = k_t^B(x)$.

We are now able to prove the following:

3.2. THEOREM - Let $A \in \bigcap_p W_5^p(X;H)$ and be such that $\delta A \in L^\infty_\mu$. Then there exists a probability space $(\Omega, \mathcal{F}, P)$ and a flow $U_t(\omega)$ defined P-a.e. in ω for all $t \in \mathbb{R}$ taking values in X, $U_t(\omega) \in C(\mathbb{R};X)$, that verifies:

(i) $U_t(\omega) = U_0(\omega) + \int_0^t A(U_s(\omega))ds$ P-a.e.-w, $\forall t \in \mathbb{R}$

Moreover, there exists a function $k_t \in L^2$, such that:

(ii) $\int f(U_t(\omega))dP(\omega) = \int f\, k_t\, d\mu$ $\forall f \in \mathcal{D}$ $\forall t \in \mathbb{R}$

Proof: By the estimations $\|A_n\|_{r,p} \leq \|A\|_{r,p}$ and Sobolev's immersion lemma, the vector fields belong to the class C^4. Furthermore $\delta A_n \in L^\infty$ and, by lemma 3.1., we can associate to A_n a flow U_t^n defined globally in time for μ_n-almost every initial condition. Let k_t^n denote the density of the image of the measure μ_n under the action of U_t^n given by lemma 3.1.

We will consider the "modified flows" $\tilde{U}_t^n$ defined on X by $\tilde{U}_t^n(x) = U_t^n(\Pi_n x) + y_n$, where $y_n = x - \Pi_n x$. Because of the decomposition of the measure μ, $\mu = \mu_n \otimes \mu_n^\perp$, it is easy to see that

$$\int f(\tilde{U}_t^n(x))\, d\mu\, (x) = \int f\, k_t^n\, d\mu \tag{3.2.}$$

where $k_t^{\,n}(x) = \exp \int_0^t \delta A_n(\tilde{U}_{-s}^{\,n}(x))\,ds$.

On the other hand, for $t \in \mathbb{R}^+$, $\tilde{U}_t^{\,n}$ can be looked upon as a random variable with values un the space $C(\mathbb{R}^+;X)$. With respect to the uniform norm topology, $C(\mathbb{R}^+;X)$ is equivalent to the space $Y = C(\mathbb{R}^+ \times [0;1];\mathbb{R})$. Let ν_n be the law of $\tilde{U}_t^{\,n}$ on Y, i.e., $\nu_n(\Gamma) = \mu(\{x : \tilde{U}^n_{\cdot}(x) \in \Gamma\})$, for $\Gamma \subset Y$. Then, if we prove that:

a) $$\lim_{R\to+\infty} \sup_n \nu_n\,(|y(0,0)| \geq R) = 0 \tag{3.3.}$$

b) $$\lim_{\delta\to 0} \sup_n \nu_n\,(\sup_{\substack{0\leq t\leq t'\leq t_0 \\ 0\leq\tau\leq\tau'\leq\tau_0 \\ (t'-t)+(\tau'-\tau)\leq\delta}} |y(t,\tau)-y(t',\tau')|\geq\rho)=0 \quad \forall\rho,t_0>0\,, \ \forall 0<\tau_0\leq 1$$

we know (cf. [9]) that $\{\nu_n\}$ is a precompact set on the space of measures on Y endowed with the weak topology.

a) follows immediatly from the definition of X; with respect to b), we have:

$$I_n = E_{\nu_n}(\sup_{t,t',\tau,\tau'} |y(t,\tau)-y(t',\tau')|) \leq \int \sup|\tilde{U}_t^n(x)(\tau)-\tilde{U}_{t'}^{\,n}(x)(\tau')|\,d\mu$$

$$\leq \int (\sup\|\tilde{U}_t^n(x)-\tilde{U}_{t'}^{\,n}(x)\|_X + |\tau-\tau'|^\alpha\, \|\tilde{U}_{t'}^{\,n}(x)\|_\alpha)\,d\mu$$

$$\leq \int [\delta^{1/2}(\int_0^{t_0} \|\tilde{U}_s^{\,n}(x)\|_X^2\,ds)^{1/2} + \delta^\alpha(\|x\|_\alpha + \int_0^{t_0} \|A^n(\tilde{U}_s^{\,n}(x))\|_H\,ds)]\,d\mu$$

Now, applying the change of variables formula (3.2.) and using de L^∞ uniform boundness of δA_n, the L^1 boundness of A, as well as the fact that $\int \| x \|_\alpha d\mu < +\infty$, we have an uniform bound of I_n and condition b) follows. Therefore we obtain a (weak) limit measure ν. By Skorohod's theorem, there exists a probability space $(\Omega, \mathcal{F}, P)$ and a family of functions $\tilde{\tilde{U}}^n_t(\omega)$, $U^A_t(\omega)$, $\omega \in \Omega$, having laws, respectively, ν^n and ν on the space $C(\mathbb{R}^+;X)$; furthermore, $\tilde{\tilde{U}}^n_t(\omega) \dashrightarrow U_t(\omega)$ P-a.e. in ω. By taking limits in equality (3.2.) we obtain (ii) for $t \in \mathbb{R}^+$, k_t being a L^2 weak limit of the function k^n_t.

We prove now that $U^A_t(\omega)$ is in fact a flow associated with A. For the finite-dimensional approximations, we have:

$$\tilde{U}^n_t(x) = x + \int_0^t A_n(\tilde{U}^n_s(x))ds \qquad \mu\text{-a.e.-}x \quad \forall t \qquad (3.4.)$$

The difficulty in taking a limit of this expression is caused by the fact that A is not necessarily continuous. Let us write $\eta_n(t,x) = A_n(\tilde{U}^n_t(x))$, $t \in \mathbb{R}^+$ and note σ^n the law of the random variavle η_t on the space $C(\mathbb{R}^+;X) \simeq C(\mathbb{R}^+ \times [0,1] ; \mathbb{R})$. Using the precompactness criterium (3.3.), we have:

$$E_{\sigma^n}\left(\sup_{t,t',\tau,\tau'} |y(t,\tau)-y(t',\tau')|\right) \leq \int \sup\left(\|A_n(\tilde{U}^n_t(x)) - A_n(\tilde{U}^n_{t'}(x))\|_H + |\tau-\tau'|^\alpha \, \|A_n(\tilde{U}^n_{t'}(x))\|_H\right)d\mu$$

$$\leq \int \left[\delta^{1/2}\left(\int_0^{t_0} \|\nabla A_n . A_n(\tilde{U}^n_s(x))\|_H ds\right)^{1/2} + \delta^\alpha \int_0^{t_0} \|A_n(\tilde{U}^n_s(x)\|_H ds\right] d\mu$$

Using (3.2.), the uniform bounds of the L^2 norms of A_n and ∇A_n, the uniform L^∞ bound of k^n_t, as well as

Hölder inequality, we obtain the condition b) of (3.3). Therefore there exists a limit measure σ and a function $\eta(t,\omega)$ defined on the probability space $(\Omega,\mathfrak{F},P)$, having σ for law. Now, for the $f \in \mathfrak{D}$, we have:

$$\int f(A^n(\tilde{U}^n_t(x)))\,d\mu(x) = E_{\sigma_n} f(y(t)) \longrightarrow E_\sigma f(y(t)) = \int f(\eta(t,\omega))\,dP(\omega)$$

and, on the other hand,

$$\int f(A^n(\tilde{U}^n_t(x)))\,d\mu(x) = \int f\circ A^n\, k^n_t\,d\mu \longrightarrow \int f\circ A\, k_t\,d\mu = \int f(A(U^A_t(\omega)))\,dP(\omega)$$

by (ii). Hence we have $\eta(t,\omega) = A(U^A_t(\omega))$ and, taking the limit in equality (3.4) we get (i) for $t \in \mathbb{R}^+$.

Finally, we can repeat all the proof for the negative values of t, by working with the random variables $\tilde{U}^n_{-t}(x)$ and $A_n(\tilde{U}^n_{-t}(x))$, $t \in \mathbb{R}^+$. □

4. AN APPLICATION

We remark that we can prove theorem 3.2. by working each "coordinate" $\langle U^A_t(x), e_k\rangle$ independently, a method which could be more easily generalized to any abstract Wiener space framework. We also remark that, if A is continuous, for example, the second part of the proof (namely, the verification of (ii)) is avoided and, in this case, we could allow the functional A to take values in a more general space, say X_α, by using the approximations $A_n(x) = E^{V_n}(\Pi_n A(x))$.

We shall now briefly describe an example coming from statistical hydrodynamics that was considered in [1].

Let us consider the Euler equation on a 2-dimensional torus Π^2, given by:

$$\frac{\partial u}{\partial t} = -(u.\nabla)u - \nabla p \qquad (4.1.)$$

$$\text{div } u = 0$$

where u is the velocity, p the pression, $u.\nabla = u_1\partial_1 + u_2\partial_2$. We consider the equations on the torus and with periodic boundary conditions. Because of the condition div $u = 0$, to look for a smooth solution of (4.1.) turns out to be equivalent to solve the following equation:

$$\frac{\partial \Delta\phi}{\partial t} = \nabla^{\perp}\phi \, . \, \nabla\Delta\phi \qquad (4.2.)$$

where $u = \nabla^{\perp}\phi = (-\partial_2\phi, \partial_1\phi)$.

Let $e_k(x) = \frac{1}{\sqrt{2\pi}} e^{ik.x}$, $k \in Z^2$, $k.x = k_1x_1 + k_2x_2$ be a complete orthonormal (with respect to L^2) set of functions which are eigenfunctions of the operator $-\Delta$. For $p > 0$, let us consider the Sobolev spaces on the forms:

$$H^p = \{\phi = \sum_{k>0} \phi_k e_k : \sum k^{2p}|\phi_k|^2 < +\infty\},$$

with inner product given by $\langle u,v\rangle_p = \sum k^{2p}\phi_k\bar{\phi}_k$ ($k > 0$ means $k_1 > 0$ or $k_1 = 0$ and $k_2 > 0$). We can prove that there exists a gaussian measure μ, supported by $H^{1-\alpha}$, $\alpha > 0$, and such that:

$$\int \exp(i\langle \ell,\phi\rangle_2)\, d\mu(\phi) = \exp(-\tfrac{1}{2}||\ell||_2^2) \quad \forall \ell \in (H^{1-\alpha})^* \subset H^2$$

The structure $(H^{1-\alpha}, H^2, \mu)$ is a (complex) Wiener space. If we take the decomposition of ϕ into the sum $\phi(x,t) = \sum_{k>0} \phi_k(t)e_k$ (ϕ is real and we can assume $\int \phi\, dx = 0$), equation (4.2.) takes the form $\frac{d\phi}{dt} = B(\phi)$, where

$B \in L^2 (H^{1-\alpha};H^{1-\alpha})$ $\alpha > 3/2$ (see [1] for the details). Furthermore, due to the fact that the Euler system is conservative, we can prove that $\delta_\mu B = 0$. We are in presence of an analogous situation to the one described in the last sections, and where the techniques used in the proof of theorem 3.2. can be applied.

REFERENCES

[1] S. Albeverio and A.B. Cruzeiro, Global flows with invariant (Gibbs) measures for Euler and Navier-Stokes two dimensional fluids, to appear,(Bochum preprint 1988).

[2] R. H. Cameron and W.T. Martin, Transformation of Wiener integrals under translations, Ann. of Math. 45 (1944), 386-396.

[3] A.B. Cruzeiro, Équations différentielles ordinaires: non explosion et mesures quasi-invariantes, J. Funct. Anal. 54 (1983), 193-205.

[4] A. B. Cruzeiro, Équations différentielles sur l'espace de Wiener et formules de Cameron-Martin non linéaires, J. Funct. Anal. 54 (1983), 206-227.

[5] H.H. Kuo, "Gaussian Measures in Banach Spaces", Lect. Notes Math. 463, Springer - Verlag, Berlin, 1979.

[6] M. Krée and P. Krée, Continuité de la divergence dans les espaces de Sobolev, C. R. Acad. Sci. Paris, Ser I (1983).

[7] P. Malliavin, Stochastic calculus of variation and hypoelliptic operators, Proc. Int. Conf. on S.D.E., Kyoto (1976), 195-264, Kinokuniya, Tokyo, 1978.

[8] P. Malliavin, Analyse différentielle sur l'espace de Wiener, Proc. Int. Congress of Mathematicians, Warszawa (1983).

[9] D. W. Stroock and S.R.S. Varadhan, "Multidimensional Diffusion Processes", Grundlehren Math., 233, Springer-Verlag, Berlin, 1979.

ON POLYGONAL MARKOV FIELDS

Taivo Arak
Department of Mathematics, Tartu University,
Vanemuise 46, Tartu, 202400, ESTONIAN SSR

Donatas Surgailis
Institute of Mathematics and Cybernetics,
Academy of Sciences of the Lithuanian SSR,
Akademias 2 ,Vilnius,232001,LITHUANIAN SSR

ABSTRACT

A construction of a class of Markov random fields with finite number of values and two-dimensional time parameter is given. The constancy domains of these fields have polygonal form. There exists a subclass of fields satisfying all the Nelson axioms except the reflection property and a subclass with simple expressions for 1- and 2-dimensional correlation functions.

1. INTRODUCTION

The theory of Gaussian Markov fields has been developed by Wong, Pitt, Rozanov, Molchan and others. Non Gaussian fields can be obtained from Gaussian ones by using multiplicative density functionals. But in this manner it is impossible to construct fields with finite number of values. We come to an interesting class of such a fields by some sort of combination of basic ideas of three different areas: integral geometry, Ising model and the theory of interacting Markov processes. The class of fields is called polygonal for their realizations have constancy domains of polygonal form.

The polygonal fields seem to be a reasonable model for some real spatial data. In section 6 a subclass of the so called consistent fields will be considered with simple explicit expressions for one- and two-dimensional correlation functions. Consistent fields can be described in terms of random Markov evolution of a one-dimensional particle system and that makes them very convenient for computer simulation.

The following notations will be used in this paper: $\mathcal{T}$ - the family of all convex bounded open sets $T \subset \mathbb{R}^2$, $\mathcal{L}$ - the space of straight lines in $\mathbb{R}^2$, $\mathcal{L}_T$ - the set of straight lines which intersect T. For $l \in \mathcal{L}$ we introduce polar coordinates: $l = l(p,\alpha)$ where p is the length of the perpendicular from the origin to l and α is the angle between the perpendicular and abscissa-axes.

To make it easier to understand the constructions given below we start with a somewhat unusual but very natural definition of the Poisson point process.

2. POISSON POINT PROCESS

Let μ be a finite atomless measure on a measurable space $(X,\mathcal{F})$, Denote by $(X^n,\mathcal{F}_n,\mu^n)$ the n-fold direct product on $(X,\mathcal{F},\mu)$ with itself and set

$$\tilde{X}_n = \{(x_1,\ldots,x_n) \in X^n \colon x_1 < x_2 < \ldots < x_n\}, \quad \Phi = \bigcup_{n=0}^{\infty} \tilde{X}_n,$$

where $<$ is an arbitrary measurable order relation on X. The set ϕ can be interpreted as the space of all finite subsets of X. We introduce the σ-algebra $\mathcal{G}$ of all subsets $A \subset \Phi$ such that $A \cap \tilde{X}_n \in \mathcal{F}_n$ for n=0,1,2,... . Now the Poisson point process on X with intensity μ can be defined as a probability measure π on $(\Phi,\mathcal{G})$ with

$$\pi(A) = c^{-1} \sum_{n=0}^{\infty} \mu^n (A \cap \tilde{X}_n) ,$$

where c is the normalizing constant

$$c = \sum_{n=0}^{\infty} \mu^n(\tilde{X}_n) = \sum_{n=0}^{\infty} (\mu(X))^n/n! = e^{\mu(X)} .$$

The Poisson line process on $T \subset \mathbb{R}^2$ is defined as the Poisson point process on L_T.

3. POLYGONAL FIELD

3.1. The Space of Realizations. Let $T \subset \mathcal{T}$ and let J be a finite ordered set. For any function ω: $T \to J$ we denote by $\partial\omega$ the set of its discontinuity points. For any $(l_1,\ldots,l_n) \in L_T^n$ we introduce the set $\Omega_T(l_1,\ldots,l_n)$ of functions ω such that there exist closed intervals $[l_i] \subset l_i$ $(i=1,\ldots,n)$ satisfying the following conditions:

(i) $[l_i] \subset l_i \cap \bar{T}$, where $\bar{T}$ is the topological closure of T;

(ii) $\partial\omega = \bigcup_{i=1}^{n} l_i \cap T$;

(iii) if $l_i = l_j$, $i \neq j$ then $[l_i] \cap [l_j] = \emptyset$;

(iv) $\omega(z) = \lim_{\varepsilon\to 0} \sup_{|x-z|} \omega(x)$ for $z \in \partial T$.

Let us define

$$\Omega_T^{(n)} = \bigcup_{(l_1,\ldots,l_n)\in L_T^n} \Omega_T(l_1,\ldots,l_n), \quad \Omega_T = \bigcup_{n=0}^{\infty} \Omega_T^{(n)} ,$$

where $\Omega_T^{(0)} = \{\omega : \omega(z) \equiv \text{const} \in J\}$. The set Ω_T is the space of realizations of polygonal fields on T. A "typical" realization is represented in Fig.1.

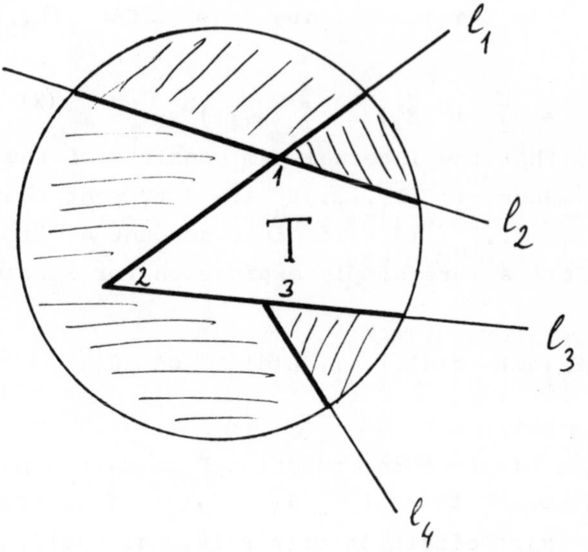

Fig.1. A "typical" realization $\omega\in\Omega_T(l_1,l_2,l_3,l_4)$. The "nodes" 1,2,3 are of type X,Λ and T correspondingly.

A topology on Ω_T can be introduced by claiming ω' to be "close" to $\omega\in\Omega_T(l_1,\ldots,l_n)$ if ω' can be obtained from ω by "small" movements of the lines $l_1,\ldots,l_n$. Let $\mathcal{B}_T$ be the Borel σ-algebra corresponding to this topology. It is easy to show that $\mathcal{B}_T$ coincides with the σ-algebra generated by the functions $\omega\rightarrow\omega(z)$, $z\in T$.

3.2. <u>The Basic Measure. Potentials</u>. Let μ be an atomless measure on L_T finite on compacts. For any measurable function $F:\Omega_T\rightarrow \mathbb{R}\cup\{+\infty\}$ we introduce a measure $Z_T=Z_{T,F,\mu}$ on

$(\Omega_T, \mathcal{B}_T)$ by the equality

$$Z_T(A) = \sum_{n=0}^{\infty} \frac{\mu(dl_1)\ldots\mu(dl_n)}{n!} \sum_{\omega \in A \cap \Omega_T(l_1,\ldots,l_n)} e^{-F(\omega)} \tag{1}$$

Notice that the internal sum consists of the finite number of summands. If $l_1,\ldots,l_n$ are different then any function $\omega \in \Omega_T(l_1,\ldots,l_n)$ is taken into account n! times in (1) and therefore a more simple expression for Z_T can be given:

$$Z_T(d\omega) = \mu(dl_1) \ldots \mu(dl_n) \quad \text{on} \quad \Omega_T^{(n)}. \tag{2}$$

If $Z_{T,F,\mu}(\Omega_T) < \infty$ the function F is called potential.

3.3. <u>Final Definition of a Polygonal Field</u>. The polygonal field on T corresponding to F and μ is the probability measure $P_T(\cdot) = Z_T(\cdot)/Z_T(\Omega)$ on $(\Omega_T, \mathcal{B}_T)$.

4. MARKOV PROPERTY

For any $S \subset T$ we denote by $\mathcal{B}_T(S)$ the σ-algebra generated by the random variables $\omega(z)$, $z \in S$.

A random field P_T on T is called Markovian if for any open sets $S, U \subset T$ such that $S \cup U = T$ and for any $A \in \mathcal{B}_T(S)$

$$P_T(A|\mathcal{B}_T(U)) = P_T(A|\mathcal{B}_T(S \cap U)).$$

A potential F is called additive if for any open sets $S, U \subset T$ with $S \cup U = T$ there exist a $\mathcal{B}_T(S)$-measurable function F_S and a $\mathcal{B}_T(U)$-measurable function F_U with $F = F_S + F_U$.

<u>Theorem 1</u>. If F is an additive potential then the

field $P_{T,F,\mu}$ is Markovian.

The validity of Theorem 1 is a consequence of the multiplicative form of the measure P_T (see (2)) and the following fact: all those intervals of $\partial\omega$ which have common parts both with S and U intersect the set $S \cap U$ and therefore the sets

$$B_\xi = \{\omega|_S : \omega \in \Omega_T, \ \omega|_U = \xi|_U\} \ (\xi \in \Omega_T)$$

depend only on $\xi_{|U \cap S}$

A detailed proof is given in [1].

5. THREE MAIN PROBLEMS

5.1. Existence Problem. A question arises for any given T,F,μ whether the field $P_{T,F,\mu}$ exists i.e. whether the quantity $Z_{T,F,\mu}(\Omega)$ is finite. The following result has been obtained by P. Mikkov (Tallinn).

Theorem 2. Let $\mu_0(dl) = dl$, $F_0(\omega) \equiv 0$ if ω does not contain nodes of type X and $F_0(\omega) = +\infty$ otherwise. Then

$$(M_1/\log n)^n \leq Z_{T,F_0,\mu_0}(\Omega_T^{(n)}) \leq (M_2 \ln\ln n / \ln n)^n,$$

where $M_i = C_i \, \mathrm{card}\, J \, \mathrm{diam}\, T$ $(i=1,2)$ and C_1, C_2 are absolute constants.

It follows from Theorem 2 that for any nonnegative functional F with excluded nodes of type X the field P_{T,F,μ_0} exists.

5.2. Extension Problem. As to the problem concerning extension of the field P_T on the whole plane the following theorem has been obtained by Surgailis .

Theorem 3. Assume that natural consistency conditions on the family $\{F_T\}_{T \subset \mathcal{T}}$ of additive potentials are satisfied and $F_T(\omega) > \gamma L(\partial\omega)$, where $L(\partial\omega)$ is the total length of $\partial\omega$ and

γ is a sufficiently large number (depending on μ). Then there exists a "thermodynamical limit"

$$P_{\mathbb{R}^2} = \lim_{T \uparrow \mathbb{R}^2} P_{T,F_T,\mu} .$$

In section 6 an example of a field will be presented for which both the existence and the extension problem are almost trivial.

5.3. Uniqueness Problem. The question about uniqueness of the limit $P_{\mathbb{R}^2}$ is more complicated than in lattice Gibbsian models. Namely, in utmost cases there exist limits with realizations containing (infinite) straight lines separating different values. But if we exclude such kind of limit distributions the uniqueness problem seems to have a sence.

6. CONSISTENT FIELDS

6.1. Definition. A family of fields $\{P_T\}_{T\in\mathcal{T}}$ is said to be consistent if

$$P_T\left\{\omega \in \Omega_T : \omega\Big|_S \in A\right\} = P_S(A)$$

for any $S,T\in\mathcal{T}$, $S\subset T, A \in \mathcal{B}_S$.

According to the Kolmogorov's theorem a consistent family of fields defines a random field P_{R^2} on the whole plane.

6.2. The Simpliest Example.

Theorem 4. Let card $J=2$, $\mu_0(dl)=dl$, $F(\omega)=2L(\partial\omega)$ with excluded X-modes. Then

(i) the field $P_T=P_{T,F,\mu_0}$ exists for any $T\in\mathcal{T}$,

(ii) $Z_T(\Omega_T) = e^{2L(\partial T) + \pi \mathrm{vol}(T)}$

(iii) the family $\{P_T\}_{T \in \mathcal{T}}$ is consistent;

(iv) the corresponding field $P_{\mathbb{R}^2}$ is invariant under Euclidean movements;

(v) for any fixed $l \in \mathcal{L}_T$ the lines l_i with segments $[l_i]$ intersecting l form a Poisson line process on $\mathcal{L}_1$ with intensity μ_0;

(vi) in particular, restriction of the field to any straight line l is a symmetric two-valued Markov process with transition intensity 2 ds, where ds is the element of length on l.

The proof of the theorem is based on the description of the field P_T in terms of random evolution of a one-dimensional particle system.

6.3. Generalizations. As to the case card $J \geq 3$ in the paper [1] sufficient conditions on a symmetric probability matrix $P=(p_{ij})$ $i,j \in J$ will be given for the existence of polygonal Markov field whose restriction on any straight line is a jump - type Markov process with transition probabilities p_{ij}. The case of nonsymmetric P will be considered in a subsequent paper.

REFERENCES

1. Arak, T., and Surgailis, D., "On Markov Fields with Polygonal Realizations", Probability and related fields, to appear.

NONDEMOLITION STOCHASTIC CALCULUS IN FOCK SPACE AND NONLINEAR FILTERING AND CONTROL IN QUANTUM SYSTEMS

V.P. Belavkin
Moscow Institute of Electronic Engineering (M.I.E.M.)
B. Vusovski 3/12, Moscow 109028 USSR

ABSTRACT

The class of continuous nondemolition measurements in quantum stochastic systems is characterized in terms of Hudson-Parthasarathy stochastic calculus. Two types of such measurements of a quantum stochastic process are described: a Poissonian counting measurement and a Brownian direct observation. The corresponding nonlinear filtering equations are derived in semi-martingal and density-matrix form, and a posterior Schrödinger equation is found.
A quantum continuous Bellman equation is derived for the solution of the quantum stochastic process optimal control problem with nondemolition measurements. The solution of this equation $u^0(t,u_t,\rho)$ together with the solution of the corresponding nonlinear filtering problem $\rho=\hat{\rho}_t(w_t,v(t))$, $w_t=(u_t,v_t)$ defines the optimal control strategy $d^0(t,w_t,v(t))=$ $=u^0(t,u_t,\hat{\rho}_t(w_t,v(t)))$.

INTRODUCTION. QUANTUM NONDEMOLITION PRINCIPLE.

The problem of description of continuous observation in quantum dynamical systems can be effectively solved in the framework of quantum stochastic calculus in the Fock space recently developed in sufficiently general form by Hudson and Parthasarathy [1], Maassen [2] and Meyer [3]. The stochastic calculus for general linear models of such quantum continual measurements in Boson systems and for some nonlinear models has been developed by Belavkin [4-6] and Barchielli and Lupieri [7], respectively. As shown in [5,6] such models can be naturally described as the models of indirect nondemolition measurements in open quantum dynamical systems even if they are obtained as a continual limit [8] of the sequences of inaccurate direct measurements partially demolishing an initially closed quantum system. Here we develop the general non-Markovian stochastic calculus of quantum processes and measurements in the Fock space, elaborating the idea of quantum nondemolition (QND) principle intensively discussed in connection with the problem of detection and estimation of gravitation waves [9-11]. We apply it for the solution of a quantum nonlinear filtering problem first formulated in [4], the partial cases of which are nondemolition detection and estimation problems of some (quantum) stochastic process.

A general mathematical formulation of the QND principle as given in [12] consists in the commutativity condition $[X_s, Y_t]=0$ $\forall s \geq t$, for a real-valued measurement process $Y=\{Y_t\}$ with respect to a quantum stochastic process $X=\{X_t\}$ described in a Hilbert space $\mathcal{H}$ by self-adjoint operators X_t, Y_t, $t \in \mathbb{R}_+$. (In the case of unbounded operators X_t, Y_t the QND condition should be understood as commutativity of J_t spectral projectors with the von Neumann algebra $\mathcal{A}_{[t}=\{X_s\}''_{s \geq t}$ for each t.) Due to this commutativity condition the sequential measurements of Y_t do not demolish, either at present or in future, the quantum dynamical subsystem, described in the Heisenberg picture at any instant, $t \in \mathbb{R}_+$, by a subalgebra $\mathcal{A}_t$ of operators X_t. It means the existence for any prior state vector $\xi \in \mathcal{H}$ of a posterior state, described by Umegaki conditional expectation [13] on the von Neumann algebra $\mathcal{A}_s \vee \mathcal{B}_t$, $s \geq t$, generated by X_s and Y_t with respect to the Abelian subalgebra $\mathcal{B}_t = Y_t''$. (The Umegaki conditional expectation on $\mathcal{A}_s \vee \mathcal{B}_t$ with respect to a non-central Abelian subalgebra $\mathcal{B}_t \not\subseteq \mathcal{A}_s'$ exists only for reduced states, described for irreducible $\mathcal{A}_s \vee \mathcal{B}_t$ only by eigenvectors ξ of Y_t.). Note, that usually the QND principle is used only in the context of self-nondemolishing measurements $Y_t=X_t$ meaning not more than their commutativity: $Y_t \in \{Y_s\}'_{s\geq t}$, which is a consequence of the general QND condition $Y_t \in \mathcal{A}_{[t}'$ with respect to a given nonicreasing family $\{\mathcal{A}_{[t}\}$ only in the case of direct measurements $Y_t \in \mathcal{A}_t$, $\bigvee_{s\geq t} \mathcal{A}_s = \mathcal{A}_{[t}$ in the corresponding dynamical subsystem $\mathcal{A}_t = X_t''$, described by the quantum stochastic process $X=\{X_t\}$. We shall distinguish the general QND processes described by the arbitrary nondecreasing families $\mathcal{B}_] = \{\mathcal{B}_{t]}\}$ of von Neumann algebras $\mathcal{B}_{t]} \subseteq \mathcal{A}_{[t}'$ on $\mathcal{H}$ with respect to non-increasing families $\mathcal{A}_[= \{\mathcal{A}_{[t}\}$ and QND observations with respect to $\mathcal{A}_[$, which are supposed to be also self-nondemolition, and hence are described by commutative families $\{\mathcal{B}_{t]}\}$ of Abelian algebras $\mathcal{B}_{t]} = \{Y_s\}''_{s\leq t}$.

Note that any quantum stochastic process is a QND process with respect to a trivial subsystem $\mathcal{A}_{[t} = c\hat{I}$, with respect to which a QND observation is a QND measurement process in the trivial traditional sense [9-11].

A linear stochastic calculus of QND processes with respect to a linear Markovian boson subsystem developed in [5, 6] helped us to derive a linear stochastic differential equation for a quantum Kalman-Busy filter which allows to find recurrently a posteriori mean $\varepsilon\{X_t | \mathcal{B}_{t]}\}$ value for the amplitude X_t of a continuously observed open quantum oscillator with Gaussian initial state. In the next section we shall explain the obtained results using the example of a simple stochastic quantum subsystem continuously observable by indirect QND measurements in the standard Fock space.

1. QND STOCHASTIC CALCULUS IN THE FOCK SPACE.

Let us consider an open quantum dynamical system over the algebra $\mathcal{A}=\mathcal{B}(\mathcal{E})$ of bounded operators X in a Hilbert space $\mathcal{E}$, which is described by a family $\tau=(\tau_t)$ of non-degenerated representations $\tau_t : \mathcal{A} \to \mathcal{B}(\mathcal{H})$, $t \in \mathbb{R}_+$, in the Hilbert space $\mathcal{H} = \mathcal{E} \otimes \mathcal{F}$, where $\mathcal{F} = \Lambda(L^2(\mathbb{R}_+))$ is the standard Fock space over $L^2(\mathbb{R}_+)$. We shall denote $\mathcal{A}_t = \tau_t(\mathcal{A})$, $\mathcal{F}_{[r,s[} = \Lambda(L^2[r,s[)$, $\mathcal{F}_t = \mathcal{F}_{[0,t[}$, $\mathcal{F}_{[t} = \mathcal{F}_{[t,\infty[}$, $\mathcal{N}_{[r,s[} = \mathcal{B}(\mathcal{F}_{[r,s[}) \otimes \hat{\mathbf{1}}_{[s}$, $\mathcal{N}_t = \mathcal{N}_{[0,t[}$, where $\mathbf{1}_{[t}$ is the identity operator on $\mathcal{F}_{[t}$, and we shall call the system $(\mathcal{A}, \tau)$ adapted if $\mathcal{A}_t \subseteq \mathcal{A} \otimes \mathcal{N}_t$ for all $t \in \mathbb{R}_+$.

We assume, that the family τ is generated by a quantum Ito equation [1-3]:*)

$$d\tau_t(X) = \alpha_t(X) dN_t + 2\mathrm{Re}\,\beta_t(X) dB_t^* + \gamma_t(X) dt, \qquad (1.1)$$

with respect to the standard gauge N_t, Bose annihilation B_t and Bose creation B_t^* processes in $\mathcal{E} \otimes \mathcal{F}$, the increaments of which satisfy the multiplication table

$$dN \cdot dN = dN, \quad dBdN = dB, \quad dNdB^* = dB^*, \quad dBdB^* = \hat{I} dt,$$

with zero other products, where $\hat{I} = I \otimes \mathbb{1}$ is the identity operator on $\mathcal{H}$.

THEOREM 1.1. Equation (1.1) with the initial condition $\tau_0(X) = X \otimes \hat{\mathbf{1}}$ for all $X = X^*$ generates an open adapted dynamical system $(\mathcal{A}, \tau)$, $\tau_t(X) \in \mathcal{A} \otimes \mathcal{N}_t$, iff the adapted weakly measurable and locally square-integrable [1] maps $t \to \alpha_t(X)$, $\beta_t(X)$, $\gamma_t(X)$ depend on $X \in \mathcal{A}$ in the following way:

$$\alpha_t(X) = \kappa_t(X) - \tau_t(X), \quad \beta_t(X) = \kappa_t(X)\hat{Z}_t - \hat{Z}_t \tau_t(X),$$
$$\gamma_t(X) = \hat{Z}_t^* \kappa_t(X) \hat{Z}_t - \tau_t(X)\left(\tfrac{1}{2}\hat{Z}_t^* \hat{Z}_t + i\hat{H}_t\right) - \left(\tfrac{1}{2}\hat{Z}_t^* \hat{Z}_t - i\hat{H}_t\right)\tau_t(X). \qquad (1.2)$$

) $\mathrm{Re}A = (A + A^*)/2$ denotes a self-adjoint part of an operator A, defined on the common for A and A domain in $\mathcal{H}$.

Here $\kappa_t(X)=\hat{V}_t^*\tau_t(X)\hat{V}_t$, $\hat{V}_t^*=\hat{V}_t^{-1}$, $\hat{Z}_t$ and $\hat{H}_t=\hat{H}_t^*$ are defined uniquely by some adapted operator-valued functions $V_t^*=V_t^{-1}$, Z_t and $H_t=H_t^*$ up to operators $\hat{v}_t$, $\hat{z}_t$ and $\hat{h}_t$ from the commutant A_t' of $\tau_t(A)$:

$$\hat{V}_t = \hat{v}_r^* V_t, \quad \hat{Z}_t = V_t^*(Z_t + \hat{z}_t), \quad \hat{H}_t = H_t + \mathrm{Im}\hat{z}_t^* Z_t + \hat{h}_t . \tag{1.3}$$

The proof of the theorem is based on the quantum Ito formula [1]. Note, that in the Markovian stationary case α_t, β_t and γ_t are the maps $A \to A_t$, and V_t, Z_t, H_t can be defined as images of some operators $V^*=V^{-1}$, Z and $H=H^*$ from A:

$$V_t = \tau_t(V), \quad Z_t = \tau_t(Z), \quad H_t = \tau_t(H) . \tag{1.4}$$

Let us consider a quantum process $Y=(Y_t)$ in H, which is adapted in the sense that $Y_t=Y_t^*$ is affiliated with $A \otimes N_t$, and nondemolishing (QND) with respect to (A,τ): $X_s Y_t = Y_t X_s$ for all $t \leq s$ and $X_s \in \tau_s(A)$. We assume, that the process Y is generated by the quantum Ito equation:

$$dY_t = D_t dN_t + 2\mathrm{Re}F_t dB_t^* + G_t dt, \tag{1.5}$$

with the corresponding (adapted, local square-integrable [1]) operator-valued functions $t \to D_t$, F_t, G_t. The following theorem is a consequence of the QND principle and Ito formula.

THEOREM 1.2. The adapted process Y_t, satisfying equation (1.4) is QND with respect to the system (1.1), iff

1) it is QND with respect to the processes $\hat{Y}_t$, $\hat{Z}_t$ and $\hat{H}_t$ up to $\hat{v}_t, \hat{z}_t$ and $\hat{h}_t \in A_t'$, i.e. if

$$[V_s, Y_t] = 0, \quad [Z_s, Y_t] = 0, \quad [H_s, Y_t] = 0, \quad s \geq t \qquad (1.6)$$

for the operators V_t, Z_t and H_t, defining (1.2) by (1.3), $Y_0 = y\hat{I}$ for $y \in \mathbb{R}$, and

2) $V_t D_t V_t^* = \hat{a}_t, \qquad V_t F_t = \hat{a}_t Z_t + \hat{b}_t,$

$$G_t = Z_t^* \hat{a} Z_t + 2\mathrm{Re}\hat{b}_t^* Z_t + \hat{c}_t, \qquad \forall t \in \mathbb{R}_+ \qquad (1.7)$$

where $\hat{a}_t^* = \hat{a}_t$, $\hat{b}_t$ and $\hat{c}_t = \hat{c}_t^*$ are adapted and commuting with $X_t \in A_t$ for all t:

$$[\hat{a}_t, X_t] = 0, \quad [\hat{b}_t, X_t] = 0, \quad [\hat{c}_t, X_t] = 0 \qquad (1.8)$$

Condition (1.6) for the Markovian case (1.4) can be omitted. The discovered structure (1.6) of QND processes (1.5) allows us to consider the following three basic types of QND observations:
a QND counting observation $Y_t = M_t$, where

$$dM_t = V_t^* v_t dN_t + 2\mathrm{Re} V_t^* Z_t dB_t^* + Z_t^* Z_t dt, \qquad M_0 = 0 \qquad (1.9)$$

($\hat{a} = \hat{I}$, $\hat{b} = 0$, $\hat{c} = 0$), a QND diffusion observation $Y_t = Q_t$, $Q_t = 2\mathrm{Re}A_t$ where

$$dA_t = V_t dB_t + Z_t dt, \quad dA_t^* = V_t^* dB_t^* + Z_t^* dt, \quad A_0 = 0 = A_0^* \qquad (1.10)$$

$(\hat{a}=0, \hat{b}=I, \hat{c}=0)$ and QND time observation $Y_t=t\hat{I}$ $(\hat{a}=0,\hat{b}=0,\hat{c}=I)$. As it follows from the next theorem, the basic QND processes M_t, A_t, A_t^* are commutative, but not mutually commutative. The multiplication table for their increaments reads

$$dM_t dM_t = dM_t, \quad dA_t dM_t = dA_t, \quad dM_t dA_t^* = dA_t^*, \quad dA_t dA_t^* = \hat{I}dt$$

and zero other products.

THEOREM 1.3. A family $Y=\{Y_1,\dots,Y_m\}$ of QND processes $Y_i(t)$ satisfying Ito equations (1.4) is mutually commutative $[Y_i(t),Y_k(s)]=0$, iff the operators $\hat{a}_i$, $\hat{b}_i$, $\hat{c}_i$ defining by (1.7) the coefficients D_i, F_i, G_i satisfy the QND conditions:

$$[\hat{a}_i(s),Y_k(t)] = 0, \; [\hat{b}_i(s),Y_k(t)] = 0 \;, \; [\hat{c}_i(s),Y_k(t)] = 0, \qquad s \geq t \qquad (1.11)$$

$$\hat{a}_i\hat{a}_k = \hat{a}_k\hat{a}_i, \quad \hat{a}_i\hat{b}_k = \hat{a}_k\hat{b}_i, \quad \hat{b}_i^*\hat{b}_k = \hat{b}_k^*\hat{b}_i, \quad \forall t \in \mathbb{R}_+ \qquad (1.12)$$

Moreover, the adapted nondemolated process Z_t in (1.7) defining (1.2) by (1.3), can be chosen in such a way, that $\hat{a}_i(t)\hat{b}_i(t)=0$ for all i and t. If $\hat{a}_i(t)$, $\hat{b}_i^*(t)$ and $\hat{c}_i(t)$ commute with Z_t for all t, as it can be done in Markovian case (1.4), then the processes $Y_i(t)$ satisfy the equivalent Ito equation:

$$dY_i(t) = \hat{a}_i(t)dM_t + 2\mathrm{Re}\hat{b}_i(t)dA_t^* + \hat{c}_i(t)dt, \qquad (1.13)$$

$Y_i(0)=y_i\hat{I}$ with respect to the canonical QND processes (1.9), (1.10).

The family $Y(t)=\{Y_1(t),\dots,Y_n(t)\}$ of mutually commutative QND observations (1.13) is called QND filter, if the operators $\hat{a}_i(t)$, $\hat{b}_i(t)$ and $\hat{c}_i(t)$ are affiliated with the Abelian algebra $\mathcal{B}_{t]}=\{Y(s)\}''_{s\leq t}$ for each t ($\hat{a}_i(t)$, $\hat{b}_i(t)$, $\hat{c}_i(t)\in\mathcal{B}_{t]}$ in the case of bounded operators. In the case of $\hat{a}_i(t)$, $\hat{b}_i(t)$ and $\hat{c}_i(t)$ defined by corresponding functions $a_i(v_{t]})$, $b_i(v_{t]})$ and $c_i(v_{t]})$ on the trajectory space $V_{t]}$ of the observed values v(s) of Y(s), $s\leq t$:

$$\hat{a}_i(t) = \int^{\oplus} a_i(v_{t]})\hat{I}_{v_{t]}}, \quad \hat{b}_i(t) = \int^{\oplus} b_i(v_{t]})\hat{I}_{v_{t]}},$$

$$\hat{c}_i(t) = \int^{\oplus} c_i(v_{t]})\hat{I}_{v_{t]}},$$

where $\hat{I}=\int^{\oplus}\hat{I}_{v_{t]}}$ is the orthogonal identity resolution for $Y_{t]}=\{Y(s)|s\leq t\}$. The condition $\hat{a}_i(t)\hat{b}_i(t)=0$ meaning either $a_i(v_{t]})=0$ or $b_i(v_{t]})=0$ decompose a QND filter on a jumping and continuous orthogonal parts:

$$dY_i(v_{t]}) = a_i(v_{t]})dM_t(v_{t]}) + c_i(v_{t]})dt, \quad v_t\in V'_{t]}$$

$$dY_i(v_{t]}) = 2\mathrm{Re}\,b_i^*(v_{t]})dA_t(v_{t]}) + c_i(v_{t]})dt, \; v_{t]}\in V^{\perp}_{t]}$$

where $V_{t]}=V^{|}_{t]}\cup V^{\perp}_{t]}$ is a measurable disjunction for each i and t and the increaments $dM_t(v_{t]})$, $dA_t(v_{t]})$, $dA_t^*(v_{t]})$ are defined in (1.9), (1.10) by the corresponding components of the decompositions

$$V_t = \int^{\oplus} V(v_{t]}), \quad Z_t = \int^{\oplus} Z(v_{t]}), \quad H_t = \int^{\oplus} H(v_{t]}) .$$

Note, that as it follows from (1.12), $b_i(v_{t]})=0$ for all i and $v_{t]}$, provided that $a_i(v_{t]})\neq 0$ for each $v_{t]}$ and some i, and $a_i(v_{t]})=0$ for all i and $v_{t]}$, provided that $b_i(v_{t]})\neq 0$ for each $v_{t]}$ and some i, so that the subsets $V^{\|}$ and $V^{\perp}$ do not depend on i.

2. QND FILTERING AND OPTIMAL CONTROL IN FOCK SPACE.

Let us consider an initial normal state ρ on $A=B(E)$ and vacuum state ω on $N=B(F)$, defined by normalized vectors $\psi\in E$ and $\upsilon\in F$, denote by $\omega_{[r,s[}$ the corresponding state on $N_{[r,s[}$, $\omega_t=\omega_{[0,t[}$ and $\omega_{[t}=\omega_{[t,\infty[}$. The open dynamical system (A,τ) defined by (1.1)-(1.3) is a quantum stochastic system with respect to the filtration $\{A\otimes N_t\}$, and the nondemolition process, defined by (1.4)-(1.8) is a regular semimartingal with respect to $\{A\otimes N_t\}$. We denote by $<\hat{X}>_t$ the conditional expectations on the von Neumann decomposable algebras $B'_{t]}\supseteq A_{[t}$ with respect to their central subalgebras $B_{t]}$, which are defined by projectors $E_{t]}$ on subspaces $E_{t]}\subseteq E\otimes F_t$, generated by the action of $B_{t]}$ on $\psi\otimes\upsilon$: $\varepsilon_{t]}(\hat{X})\xi=E_{t]}\hat{X}\xi$, for all $\xi\in E_{t]}$ and $\hat{\rho}_t=\varepsilon_{t]}\circ\tau_t$, the posterior state on A, identified with density matrix due to its normality

THEOREM 2.1. Let (A,τ,M) be a quantum stochastic process (1.1) with counting nondemolition measurement (1.9) and $<Z_t^*Z_t>_t$ be invertable for all t. Then the posterior mean value $\hat{x}_t=<\tau_t(X)>_t$ satisfies the following stochastic filtration equation:

$$d<\tau_t(X)>_t = <\delta_t(X)>_t d\tilde{M}_t + <\gamma_t(X)>_t dt, \qquad (2.1)$$

where $d\tilde{M}_t=dM_t - <Z_t^*Z_t>dt$,

$$\delta_t(X) = Z^*{}_t\tau_t(X)Z_t<Z_t^*Z_t>_t^{-1} - \tau_t(X) \ . \qquad (2.2)$$

In the Markovian case (1.4) the posterior density matrix $\hat{\rho}_t$

satisfies the recursive stochastic equation in Ito sense:

$$d\hat{\rho}_t + (K\hat{\rho}_t + \hat{\rho}_t K^* - Z\hat{\rho}_t Z^*)dt = (Z\hat{\rho}_t Z^*\hat{\rho}_t)d\tilde{M}_t \qquad (2.3)$$

where $K=iH-Z^*Z/2$, $\hat{Z_t^*Z_t}=\hat{\rho}_t(Z^*Z)$, which has a solution $\hat{\rho}_t = \hat{\psi}_t \otimes \hat{\psi}_t^*$ for $\hat{\rho}_0=\psi\otimes\psi^*$, if $\hat{\psi}_t$ satisfies the following nonlinear Ito equation

$$d\hat{\psi}_t + (i\tilde{H}_t + \tilde{Z}_t^*\tilde{Z}_t/2)\hat{\psi}_t dt = \tilde{Z}_t\hat{\psi}_t d\tilde{M}_t / \|Z\hat{\psi}_t\| \hat{\psi}_0 = \psi, \qquad (2.4)$$

where $\tilde{H}_t=H-(\hat{\psi}_t|Z^*Z|\hat{\psi}_t)^{\frac{1}{2}}\mathrm{Im}Z$, $\quad \tilde{Z}_t=Z-(\hat{\psi}_t|Z^*Z|\hat{\psi}_t)^{\frac{1}{2}}$

THEOREM 2.2. Let (A,τ,Q) be a quantum stochastic process (1.1) with diffusion nondemolition measurement $Q_t=2\mathrm{Re}A_t$, defined by (1.10). Then the posterior mean value $\hat{x}_t=\langle\tau_t(X)\rangle_t$ satisfies the following stochastic filtration equation:

$$d\langle\tau_t(X)\rangle_t = \langle\theta_t(X)\rangle_t d\tilde{Q}_t + \langle\gamma_t(X)\rangle_t dt, \qquad (2.5)$$

where $d\tilde{Q}_t=dQ_t - \langle 2\mathrm{Re}Z_t\rangle_t dt$,

$$\theta_t(X) = 2\mathrm{Re}\tau_t(X)(Z_t - \langle Z_t\rangle_t). \qquad (2.6)$$

In the Markovian case (1.4) the posterior density matrix satisfies the recursive stochastic equation in Ito sense:

$$d\hat{\rho}_t + (K\hat{\rho}_t + \hat{\rho}_t K^* - Z\hat{\rho}_t Z^*)dt = 2\mathrm{Re}(Z - \hat{z}_t)\hat{\rho}_t d\hat{Q}_t \qquad (2.7)$$

where $K=iH+Z^*Z/2$, $\hat{z}=\hat{\rho}(Z)$, which has a solution $\hat{\rho}_t=\hat{\psi}_t\otimes\hat{\psi}_t^*$ for

$\hat{\rho}_0=\psi\otimes\psi^*$, if $\hat{\psi}_t$ satisfies the nonlinear Ito equation

$$d\hat{\psi}_t + (i\tilde{H}_t + \tilde{Z}_t^*\tilde{Z}_t/2)\hat{\psi}_t dt = \tilde{Z}_t\hat{\psi}_t d\tilde{Q}_t , \qquad \hat{\psi}_0=\psi$$

where $\tilde{H}_t=H - (\hat{\psi}_t|\mathrm{Re}Z_t|\hat{\psi}_t)\mathrm{Im}Z$, $\tilde{Z}_t = Z-(\hat{\psi}_t|\mathrm{Re}Z_t|\hat{\psi}_t)$.

The linear continuous filtration for the Gaussian ρ_0 and canonical Z was considered in [4-6], and the general formulation of quantum nonlinear filtration for a quantum Markovian partially observable controlled objects in operational approach was given in [14].

Let us consider a quantum controlled process over the algebra $A=B(E)$ described by the family of normal representations $\tau_t:A\otimes N_t\otimes C_t$, or $\tau_t(u_t)$: $A\otimes N_t$, where C_t is an abelian C*-algebra $C(U_t)$ of functions $U_t\to C$, $N_t=B(F_t)$ for the Fock space F_t. Let $U_{[r,s[}\subseteq \times_{r\leq t<s} U(t)$ be a Hausdorf space of controlling processes $u_{[r,s[}=\{u(t)|r\leq t<s\}$ such that $U_t\times U_{[t,s[}= = U_s$ for for all $t,s>0$, where $u(t)\in U(t)$, $U_t=U_{[0,t[}\ni u_t$ and $U_{[t}=U_{[t,\infty[}\ni u_{[t}$, $U=U_{[0}$. We consider a quantum controlled process $X_t(u_t)=\tau_t(u_t,X)$ over the algebra $A\ni X$, $u_t\in U_t$ described by the Hudson-Parthasarathy danamical equation [1]

$$dX_t = \hat{A}_t dN_t + 2\mathrm{Re}\hat{B}_t dB_t^* + \hat{C}_t(u(t))dt, \quad X_0=X\otimes\hat{1}, \qquad (2.8)$$

where $\hat{A}_t(u_t)=\alpha_t(u_t,X)$, $\hat{B}_t(u_t)=\beta_t(u_t,X)$, $\hat{C}_t(u_t,u(t))=\gamma_t(u_{t]},X)$ are defined in a standard way (1.2),(1.3) by the operator-valued functions $V_t(u_t)=\tau_t(u_t,V)$, $Z_t(u_t)=\tau_t(u_t,Z)$, $H_t(u_t)= =\tau_t(u_t,H)$ with unitary $V\in A$, $Z\in A$ and self-adjoint $H(u(t))\in A$. We shall assume, that the control process $u(t)$ is defined by strategies $u_{[t}=\{d_{[t}(s,w_t,v_{[t,s[},v(s))|s\geq t\}\equiv d_{[t}(w_t,v_{[t})$, where $w_t=(u_t,v_t)$, $v_t=v_{[0,t[}$, $v_{[r,s[}=\{v(t)|r<t\leq s\}$ are the results of nondemolition measurements $v(t)$ on the interval

$[r,s[$, $v_{[t}=v_{[t,\infty[}$, described by a commutative process $Y(t)$, satisfying the equations

either $dY(t) = \hat{a}(t)dM_t + \hat{c}(t)dt$, or $dY(t) = 2\mathrm{Re}\hat{b}(t)dB_t^* + \hat{c}(t)dt$

with invertable $\hat{a}(t,u_t)$, $\hat{b}(t,u_t)\in\mathcal{B}_{t]}=\{Y(s)\,|\,s\leq t\}''$ defined by corresponding real-valued functions $a(w_t,v(t))$, $b(w_t,v(t))$ and $c(w_t,v(t))$.

Let us consider the optimal control problem with the operator-valued risk $u\in\mathcal{U}\to R_t(u)\in\mathcal{A}_{[t}(u) =\{\tau_s(u_s,\mathcal{A})\,|\,s\geq t\}''$, satisfying the equation

$$R_{t_0}(u) = \int_{t_0}^{t_1} S_t(u_t,u(t))dt + R_{t_1}(u) , \tag{2.9}$$

where $S_t(u_t,u(t))=\tau_t(u_t,S(u(t)))$ for a continuous $\mathcal{A}$-valued self-adjoint function $S(u(t))=S(u(t))^*$. The optimal control strategy $d^0_{[t}$ is defined as a solution of the extremal problem

$$\langle\rho\otimes\omega,\ R_t(u_t,d_{[t}(w_t,v_{[t}))\rangle = \inf, \tag{2.10}$$

where ρ is an initial normal state on $\mathcal{A}=\mathcal{B}(\mathcal{E})$, and ω is the vacuum state on $\mathcal{N}=\mathcal{B}(\mathcal{F})$. This solution can be found by the quantum dynamic programming method as a solution of the following Bellman continuous inverse-time equation.

THEOREM 2.3. Let $R(t,w_t,d_{[t})$ be the averaged $\mathcal{A}$-valued risk uniquely defined for a strategy $d_{[t}$ by

$$\langle\rho\otimes\omega,\ \tau_t(u_t,R(t,w_t,d_{[t}))\rangle = \langle\rho\otimes\omega,R_t(u_t,d_{[t}(w_t,v_{[t}))\rangle$$

due to Markovian condition for $X_t(u_t)$ with respect to ω =

$=\omega_t \otimes \omega_{[t}$, and

$$\hat{r}_t(w_t, d_t) = \varepsilon_{t]}[R_t(u_t, d_{[t}(w_t, v_{[t}))] =$$

$$= \langle \hat{\rho}_t(w_t), R(t, w_t, d_{[t}) \rangle$$

by the posterior risk, corresponding to the strategy $d_{[t}$, where $\varepsilon_{t]}$ is the conditional expectation on $A_{[t} \vee B_{t]}$ with respect to the commutative algebra $B_{t]}$, and $\hat{\rho}_t(w_t, v(t)) = \varepsilon_t \circ \tau_t(u_t)(v_{t]})$ be controlled a posterior state on A for a $w_t = (u_t, v_t) \in U_t \times V_t$.

Then

$$\inf_{d_t} \langle \hat{\rho}_t(w_t, v(t)), R(t, w_t, d_{[t}) \rangle = s(t, \hat{\rho}_t(w_t, v(t)))$$

where the functional $s(t,\rho)$ satisfies the following Bellman equation:

$$-\partial_t s(t,\rho) = \inf_{u(t)} \langle \rho, S(u(t)) + \Lambda(u(t), \delta) s(f,\rho) \rangle + \\ + \langle \rho, Z^+ Z \rangle \Delta s(t,\rho) \qquad (2.11)$$

in the case of counting observation and

$$-\partial_t s(t,\rho) = \inf_{u(t)} \langle \rho, S(u(t)) + \Gamma(u(t), \delta) s(t,\rho) \rangle + \\ + \frac{1}{2} \langle \rho, \Theta(\delta) \rangle^2 s(t,\rho) \qquad (2.12)$$

in the case of diffusion observation.

Here $\partial_t = \partial/\partial t$, $\delta = \delta/\delta\rho$, $\Delta s(\rho) = s(Z\rho Z^+ / \langle \rho, Z^+ Z \rangle) - s(\rho)$,

$$\Lambda(u(t),\delta)s(\rho) = \delta s(\rho) - 2\mathrm{Re}K(u(t))^*\delta s(\rho),$$

$$\Gamma(u(t),\delta)s(\rho) = Z^+\delta s(\rho)Z - 2\mathrm{Re}K(u(t))^*\delta s(\rho),$$

$$\Theta(\rho,\delta)s(\rho) = 2\mathrm{Re}(Z - \langle\rho,Z\rangle)^*\delta s(\rho),$$

and $K(u(t)), K(u(t))^*$ are defined as $\pm iH(u(t)) + Z^*Z/2$, and $\hat{\rho}_t(w_t)$ is a posterior state on A, for controlled and observed data $w_t=(u_t,v_t)$, satisfying the nonlinear filtering equations either (2.3) or (2.7) correspondingly:

$$d\hat{\rho}_t + \hat{\rho}_t\Lambda(u(t)) = \hat{\rho}_t\Delta dM_t, \quad d\hat{\rho}_t + \hat{\rho}_t\Gamma(u(t)) = \hat{\rho}_t\Theta d\tilde{Q}_t$$

where $\hat{\rho}_0=\rho$.

The linear programming for Gaussian ρ, canonical Z and squared $S(u(t))$ was considered in [6]. The general formulation of quantum dynamical programming for the partially observable controlled quantum objects in operational approach was given in [14].

ACKNOWLEDGMENTS

I would like to thank Professor S.Albeverio for inviting me to talk in 24-th Winter School of Theoretical Physics "Stochastic Methods in Mathematics and Physics" and Professors W.Karwowski, R.Gielerak and other members of Organizing Committee for the warm hospitality in Karpacz.

REFERENCES

1. Hudson, R.S., Parthsarathy, K.R., Quantum Ito's formula and stochastic evolution, Commun.Math.Phys. 93, 301 (1984).

2. Maassen, H., Quantum Markov processes on Fock space described by integral kernels in Quantum Probability and Applications II, ed. L.Accardi and W.von Waldenfelds Springer LNM 1136 (1985).

3. Meyer, P.A., Elements de Probabilités Quantiques. Exposes I à IV, Institut de Mathematique, Université Louis Pasteur Strasbourg (1985).

4. Belavkin, V.P., Optimal quantum filtration of Markovian signals. Problems of Control and Information Theory, v. 7(5), p.345-360(1978).

5. Belavkin, V.P., Quantum filtering of Markovian signals with quantum white noises. Radiotechnika i Elektronika, v.25(7),p.1445-1453(1980) (in Russian).

6. Belavkin, V.P., Nondemolition measurement and control in quantum dynamical systems. In: Information complexity and control in quantum physics, ed.A.Blaquiere, S.Diner, G.Lochak, Springer-Verlag, Wien-New York, p. 311-336(1987).

7. Barchielli A,. and Lupiere, G., Quantum stochastic calculus, operation valued stochastic processes and continual measurements in quantum mechanics, J.Math.Phys. (1985).

8. Barchielli, A., Lanzand, L., and Prosperi, G.M., Statistics of continuous trajectories in quantum mechanics: Operator-valued Stochastic Processes. Foundations of Physics, 13, (18),p.779-812(1983)

9. Braginsky, V., Vorontsov, Y., and Thorne, S., Quantum Nondemolition Measurements, Science 209(4456), p.547-557 (1980).

10. Barchielli, A., Continuous observation in quantum mechanics: An application to gravitational-wave detectors. Phys.Rev.D,v.32(2),p.347-367(1984).

11. Holevo, A.S., On quantum nondemolition principle. Theor. Math.Phys.65(3),p.415-422(1985).

12. Belavkin, V.P., Reconstruction theory for a quantum stochastic process. Theor.Math.Phys.62(3),p.275-289(1985).

13. Umegaki, H., Conditional expectation in operator algebras, I.Tohoku Math.J.,(6),p.177-181(1954); II,(8),p.86--100(1956).

14. Belavkin, V.P., Theory of the control of observable quantum systems. Automatica and Remove Control, 44 (2), p.178-188(1983).

Nongaussian linear filtering, identification of linear systems, and the symplectic group

Michiel Hazewinkel
Centre for Mathematics and Computer Science
P.O. Box 4079, 1009 AB Amsterdam, The Netherlands

Consider stochastic linear dynamical systems, $dx = Axdt + Bdw, dy = Cxdt + dv, y(0) = 0, x(0)$ a given initial random variable independent of the standard independent Wiener noise processes w, v. The matrices A, B, C are supposed to be constant. In this paper I consider two problems. For the first one A, B and C are supposed known and the question is how to calculate the conditional probability density of x at time t given the observations $y(s), 0 \leq s \leq t$ in the case that $x(0)$ is not necessarily gaussian. (In the gaussian case the answer is given by the Kalman-Bucy filter). The second problem concerns identification, i.e. the A, B, C are unknown (but assumed constant so that $dA = 0, dB = 0, dC = 0$), and one wants to calculate the joint conditional probability density at time t of (x, A, B, C), again given the observations $y(s), 0 \leq s \leq t$. The methods used rely on Wei-Norman theory, the Duncan-Mortensen-Zakai equation and a "real form" of the Segal-Shale-Weil representation of the symplectic group $Sp_n(\mathbf{R})$.

AMS classification: 93E11, 93B30, 17B99, 93C10, 93B35, 93E12
Key words and phrases: nongaussian distribution, identification, non-linear filtering, DMZ equation, Duncan-Mortensen-Zakai equation, propagation of nongaussian initials, Wei-Norman theory, Segal-Shale-Weil representation, reference probability approach, unnormalized density, Kalman-Bucy filter, Lie algebra approach to nonlinear filtering.

1. Introduction

Consider a general nonlinear filtering problem of the following type:

$$dx = f(x)dt + G(x)dw \quad , \quad x \in \mathbf{R}^n, \; w \in \mathbf{R}^m \tag{1.1}$$

$$dy = h(x)dt + dv \quad , \quad y \in \mathbf{R}^p, \; v \in \mathbf{R}^p \tag{1.2}$$

where f, G, h are vector and matrix valued functions of the appropriate dimensions, and the w, v are standard Wiener processes independent of each other and also independent of the initial random variable $x(0)$. One takes $y(0) = 0$.

The general non-linear filtering problem is this setting asks for (effective) ways to calculate and/or approximate the conditional density $\pi(x,t)$ of x given the observations $y(s), 0 \leq s \leq t$; i.e. $\pi(x,t)$ is the density of $\hat{x} = E[x(t)|y(s), 0 \leq s \leq t]$ the conditional expectation of the state $x(t)$.

One approach to this problem proceeds via the socalled DMZ equation which is an equation of a rather nice form for an unnormalized version $\rho(x,t)$ of $\pi(x,t)$. Here unnormalized means that $\rho(x,t) = r(t)\pi(x,t)$ for some function $r(t)$ of time alone. A capsule description of this approach is given in section 2 below. Using this approach was strongly advocated by Brockett and Mitter (cf. e.g. their contributions in [6]), and initially the approach had a number of nontrivial successes, both in terms of positive and negative results (cf. e.g. the surveys [9] and [4]). Subsequently, the approach

Various subselections of the material in this article have formed the subject of various talks at different conferences; e.g. the 2nd conference on the road-vehicle system in Torino in June 1987, the 24-th Winter school on theoretical physics in Karpacz in January 1988, the present one, the 3rd meeting of the Bellman continuum in Valbonne in June 1988, and the special program on signal processing of the IMA in Minneapolis in the summer of 1988. As a result this article may also appear in the proceedings of these meetings.

became less popular; perhaps because a number of rather formidable mathematical problems arose, and because the number of systems to which the theory can be directly applied appears to be quite small. Cf [4] for a discussion of some aspects of these two points.

It is the purpose of this paper to apply this approach to two problems concerning linear systems, which do not fall within the compass of the usual Kalman-Bucy linear filtering theory. More precisely, consider a linear stochastic dynamical system

$$dx = Axdt + Bdw, \quad x \in \mathbf{R}^n, \; w \in \mathbf{R}^m \tag{1.3}$$

$$dy = Cxdt + dv, \quad y, v \in \mathbf{R}^p \tag{1.4}$$

where the A,B,C are matrices of the appropriate sizes. The first problem I want to consider is the filtering of (1.3)-(1.4) in the case that the initial condition $x(0)$ is a non-gaussian random variable. The second problem concerns the identification of (1.3)-(1.4); i.e. one assumes that the matrices A,B,C are constant but unknown and it is desired to calculate the conditional density $\pi(x,A,B,C,t)$ of the (enlarged) state (x,A,B,C) at time t. Technically this means that one adds to (1.3)-(1.4) the equations

$$dA = 0, \; dB = 0, \; dC = 0 \tag{1.5}$$

and one considers the filtering problem for the nonlinear system (1.3)-(1.5). Strictly speaking this problem is not well posed. Simply because A,B,C can not be uniquely identified on the basis of the observations alone. In the DMZ equation approach this shows up only at the very end in the form that $\rho(x,A,B,C,t)$ will be degenerate in the sense that $\rho(Sx,SAS^{-1},SB,CS^{-1},t)=\rho(x,A,B,C,t)$ for all constant invertible real matrices S. As a result the normalization factor $\int\rho(x,A,B,C,t)dxdAdBdC$ does not exist, and in fact $\pi(x,A,B,C,t)$ is also degenerate. One gets rid of this by passing to the quotient space (finite moduli space) $\{(x,A,B,C)\}/GL_n(\mathbf{R})$ for the action just given and/or by considering (local) canonical forms. The normalization factor can be calculated by integrating over this quotient space.

Besides the DMZ-equation, already mentioned, the tools used to tackle the two problems described above are Wei-Norman theory and something which could be called a real form of the Segal-Shale-Weil representation of the symplectic Lie group $Sp_n(\mathbf{R})$. These two topics are discussed in sections 3 and 4 below.

2. The DMZ approach to nonlinear filtering

Consider again the general nonlinear system (1.1)-(1.2). These stochastic differential equations are to be considered as Ito equations. Let $\pi(x,t)$ be the probability density of $E[x(t)|y(s), 0 \leqslant s \leqslant t]$, the conditional expectation of $x(t)$. (Given sufficiently nice f,G and h if can be shown that $\pi(x,t)$ exists.) Then the Duncan-Mortensen-Zakai result [1, 10, 12] is that there exists an unnormalized version $\rho(x,t)$ of $\pi(x,t)$, i.e. $\rho(x,t)=r(t)\pi(x,t)$, which satisfies an evolution equation

$$d\rho = \pounds\rho dt + \Sigma h_k \rho dy_k(t), \quad \rho(x,0) = \psi(x) \tag{2.1}$$

where $\psi(x)$ is the distribution of the initial random variable $x(0)$ and where $\pounds$ is the second-order partial differential equation

$$\pounds\phi = \frac{1}{2}\sum_{i,j} \frac{\partial^2}{\partial x_i \partial x_j}(GG^T)_{ij}\phi - \sum_i \frac{\partial}{\partial x_i} f_i\phi - \frac{1}{2}\sum_k h_k^2\phi. \tag{2.2}$$

Here $h_k, y_k(t), f_i$ are components of $h, y(t)$ and f respectively and $(GG^T)_{ij}$ is the (i,j)-entry of the product GG^T of the matrix G and its transpose.

Equation (2.1) is a Fisk-Stratonovič stochastic differential equation. The corresponding Ito differential equation is obtained by removing the $-\frac{1}{2}\Sigma h_k^2\phi$ term from (2.2).

As it stands (2.1) is a stochastic partial differential equation. However the transformation

$$\tilde{\rho}(x,t) = \exp(\Sigma h_k(x) y_k(t))\rho(x,t) \tag{2.3}$$

turns it into the equation

$$d\tilde{\rho} = (\pounds\tilde{\rho} + \Sigma\pounds_i\tilde{\rho}y_i + \frac{1}{2}\Sigma\pounds_{i,j}\tilde{\rho}y_iy_j)dt \tag{2.4}$$

where $\pounds_i$ is the operator commutator $\pounds_i=[h_i, \pounds]=h_i\pounds-\pounds h_i$ and $\pounds_{ij}=[h_i, [h_j, \pounds]]$. Cf. [4] for more details. In (2.4) I have explicity indicated the dependence of the various quantities on x,t to stress that here $h(x)$ should simply be seen as a known function of x and not as the time function $h(x(t))$. Equation (2.4) does not involve the derivatives dy_i anymore; it makes sense for all possible paths $y(t)$, and can be regarded as a family of PDE parametrized by the possible observation paths $y(t)$. Thus there is a robust version of (2.1) and we can work with (2.1) as a parametrized family of PDE parametrized by the $y(t)$. Note that knowledge of $\tilde{\rho}(x,t)$ (and $y(t)$) immediately gives $\rho(x,t)$ and that the conditional expectation of any function $\phi(x(t))$ of the state at time t can be calculated by

$$E[\phi(x(t))|y(s),\ 0\leqslant s\leqslant t] = (\int\rho(x,t)dx)^{-1}\int\phi(x)\rho(x,t)\,dx \tag{2.5}$$

Possibly the simplest example of a filtering problem is provided by one-dimensional Wiener noise linearly observed:

$$dx = dw,\ x,\ w\in\mathbf{R} \tag{2.6}$$

$$dy = xdt + dv,\quad y,\ v\in\mathbf{R}. \tag{2.7}$$

In this case the corresponding DMZ equation is

$$d\rho = (\frac{1}{2}\frac{d^2}{dx^2} - \frac{1}{2}x^2\ \rho dt + x\rho\ dy \tag{2.8}$$

an Euclidean Schrödinger equation for a forced harmonic oscillator.

3. Wei-Norman theory

Wei-Norman theory is concerned with solving partial differential equations of the form

$$\frac{\partial\rho}{\partial t} = u_1A_1\rho + \cdots + u_mA_m\rho \tag{3.1}$$

where the A_i, $i=1,...,m$ are linear partial differential operators in the space variables $x_1,...,x_n$, and the u_i, $i=1,...,m$ are given functions of time, in terms of solutions of the simpler equations

$$\frac{\partial\rho}{\partial t} = A_i\rho\ ,\ i=1,...,m \tag{3.2}$$

which we write as

$$\rho(x,t) = e^{A_it}\psi(x)\ ,\ \psi(x) = \rho(x,0) \tag{3.3}$$

Originally, the theory was developed for the finite dimensional case, i.e. for systems of ordinary differential equations

$$\dot{z} = u_1A_1z + \cdots + u_mA_mz \tag{3.4}$$

where $z\in\mathbf{R}^k$, and the A_i are $k\times k$ matrices. Both in the finite dimensional case (3.4) and the infinite dimensional case (3.1) it is well known that besides in the given directions $A_1\rho,...,A_m\rho$, the to be determined function or vector can also move (infinitesimally) in the directions given by the commutators $[A_i, A_j]\rho=(A_iA_j-A_jA_i)\rho$, and in the directions given by repeated commutators $[[A_i, A_j], A_k]$, $[[A_i, A_j], [A_k, A_l]]$, etc. etc.

Let Lie$(A_1,..,A_m)$ be the Lie algebra of operators generated by the operators $A_1,...,A_m$. This is the smallest vector space L of operators containing $A_1,..,A_n$ and such that if $A,B\in L$ then also $[A,B]:=AB-BA\in L$. In the finite dimensional case (3.4) L is always finite dimensional, a subvector space of $gl_k(\mathbf{R})$, the vectorspace (Lie algebra) of all $k\times k$ matrices. In the infinite dimensional case the Lie algebra generated by the operators $A_1,...,A_m$ in (3.1) can easily be infinite dimensional and it often is; also in the cases coming from filtering problems via the DMZ equation. Cf. [5] for a number of examples.

This is the essential difference between (3.1) and (3.4). Accordingly, here I shall assume that the Lie algebra $L = \mathrm{Lie}(A_1,...,A_m)$ generated by the operators $A_1,...,A_m$ in (3.1) is finite dimensional. For a discussion of various infinite dimensional versions of Wei-Norman theory cf. [4]. Hence, granting this finite dimensionality property, by setting, if necessary, some of the $u_i(t)$ equal to zero, and by combining other $u_j(t)$ in the case of linear dependence among the operators on the RHS of (3.1), without loss of generality, we can assume that we are dealing with an equation

$$\frac{\partial\rho}{\partial t} = u_1A_1\rho + \cdots + u_nA_n\rho \tag{3.5}$$

with the additional property that

$$[A_i, A_j] = \sum_k \gamma_{ij}^k A_k \quad ;i,j = 1,...,n \tag{3.7}$$

for suitable real constants γ_{ij}^k; $i,j,k = 1,...,n$.

The central idea of Wei-Norman theory is now to try for a solution of the form

$$\rho(t) = e^{g_1(t)A_1}e^{g_2(t)A_2}\cdots e^{g_n(t)A_n}\psi \tag{3.8}$$

where the g_i are still to be determined functions of time. The next step is to insert the Ansatz (3.8) into (3.5), to obtain

$$\begin{aligned}\dot\rho &= \dot g_1A\, e^{g_1A_1}\cdots e^{g_nA_n}\psi + e^{g_1A_1}\dot g_2A_2e^{g_2A_2}\cdots e^{g_nA_n}\psi + \cdots \\ &\quad + e^{g_1A}\cdots e^{g_{n-1}A_{n-1}}\dot g_nA_ne^{g_nA_n}\psi\end{aligned} \tag{3.9}$$

Now, for $i = 2,...,n$ insert a term

$$e^{-g_{i-1}A_{i-1}}\cdots e^{-g_1A_1}e^{g_1A_1}\cdots e^{g_{i-1}A_{i-1}}$$

just behind $\dot g_iA_i$ in the i-th term of (3.9). Then use the adjoint representation formula

$$e^ABe^{-A} = B + [A,B] + \frac{1}{2!}[A,[A,B]] + \frac{1}{3!}[A,[A,[A,B]]]] + \cdots \tag{3.10}$$

and (3.7)) repeatedly, and use the linear independence of the $A_1,...,A_n$ to obtain a system of ordinary differential equations for the $g_1,...,g_n$ (with initial conditions $g_1(0)=0=g_2(0)=...=g_n(0)$).

These equations are always solvable for small time. However they may not be solvable for all time, meaning that finite escape time phenomena can occur.

Let's consider an example, viz. the example afforded by the DMZ equation (2.8). One calculates that

$$[\tfrac{1}{2}\frac{d^2}{dz^2} - \tfrac{1}{2}x^2, x] = \frac{d}{dx}, [\tfrac{1}{2}\frac{d^2}{dx^2} - \tfrac{1}{2}x^2, \frac{d}{dx}] = x$$

$$[\frac{d}{dz}, x] = 1,\ [A,1] = 0$$

where A is any linear combination of the four operators $\tfrac{1}{2}\frac{d^2}{dx^2} - \tfrac{1}{2}x^2, x, \frac{d}{dx}, 1$. Applying the recipe sketched above to the equation

$$\dot\rho = (\tfrac{1}{2}\frac{d^2}{dx^2} - \tfrac{1}{2}x^2)\rho + x\rho u(t) + \frac{d}{dx}\rho 0 + 1\rho 0 \tag{3.11}$$

one finds the equations

$$\dot g_1 = 0\ ,\ \cosh(g_1)\dot g_2 + \sinh(g_1)\dot g_3 = u(t), \tag{3.12}$$

$$\sinh(g_1)\dot g_2 + \cosh(g_1)\dot g_3 = 0,\ \dot g_4 = \dot g_3 g_2$$

which are solvable for all time.

This fact and the form of the resulting equations: straightforward quadratures and one set of linear equations $B(t)g=b(t)$, with $B(t),b(t)$ known and $B(t)$ invertible, is typical for the case that the Lie algebra $L=\oplus\mathbf{R}A_i$ spanned by the $A_1,...,A_n$ is solvable. This means the following. Let $[L,L]$ be the subvectorspace of L spanned by all the operators of the form $[A,B],A,B\in L$. It is easily seen that this is again a Lie algebra. Inductively let $L^{(n)}=[L,L^{(n-1)}]$ be the subvectorspace of L spanned by all operators of the form $[A,B]$, $A\in L,B\in L^{(n-1)}$, $L^{(0)}=L$. These are all sub Lie algebras of L.

The Lie algebra of L is called nilpotent if $L^{(n)}=0$ for n large enough. It is called solvable if $[L,L]$ is nilpotent. The phenomenon alluded to above, i.e. solvability of the Wei-Norman equations for all time, always happens in case L is solvable [11]. (And it is no accident that these algebras have been called solvable. Though this is not the result which gave them that name.)

Note that the DMZ equation (2.1) corresponding to a nonlinear filtering problem (1.1)-(1.2) is of the type (3.1) (with $u_h(t)=dy_k(t)$). Thus the Lie Algebra generated by the operators £, $h_1(x),...,h_p(x)$ occuring in (2.1) clearly has much to say about how difficult the filtering problem is. This Lie algebra is called the *estimation Lie algebra* of the system (1.1)-(1.2) and it can be used to prove a variety of positive and negative results about the filtering problem [4, 5, 9].

4. The Segal-Shale-Weil representation and a 'real form'

Let J be the standard symplectic matrix $J=\begin{bmatrix}0 & I_n\\ -I_n & 0\end{bmatrix}$, where I_n the $n\times n$ unit matrix. Consider the vector space of $2n\times 2n$ real matrices defined by

$$sp_n(\mathbf{R}) = \{M\colon JM + M^TJ = 0\}. \tag{4.1}$$

Writing M as a 2×2 block matrix, $M=\begin{bmatrix}A & B\\ C & D\end{bmatrix}$, the conditions on the $n\times n$ blocks A,B,C,D become

$$B^T = B,\ C^T = C,\ D = -A^T. \tag{4.2}$$

As we shall see shortly below this set of matrices occurs naturally for filtering problems coming from linear systems (1.1)-(1.2).

The corresponding Lie group to $Sp_n(\mathbf{R})$ is the group of invertible $2n\times 2n$ matrices defined by

$$Sp_n(\mathbf{R}) = \{S\in\mathbf{R}^{2n\times 2n}\colon S^TJS = J\} \tag{4.3}$$

(This is a *group* of matrices in that if $S_1, S_2\in Sp_n(\mathbf{R})$ then also $S_1S_2\in Sp_n(\mathbf{R})$ and $S_1^{-1}\in Sp_n(\mathbf{R})$ as is easily verified.)

There is a famous representation of $Sp_n(\mathbf{R})$ (or more precisely of its two-field covering group $\tilde{S}p_n(\mathbf{R})$) in the Hilbert space $L^2(\mathbf{R}^n)$ called the Segal-Shale-Weil representation or the oscillator representation; cf. [8]. Here the word 'representation' means that to each $S\in Sp_n(\mathbf{R})$ there is associated a unitary operator U_S such that $U_{S_1S_2} = U_{S_1}U_{S_2}$ for all $S_1, S_2\in Sp_n(\mathbf{R})$.

For the purposes of this paper a modification of it is of importance. It can be described as follows by explicit operators associated to certain specific kinds of elements of $Sp_n(\mathbf{R})$:

(i) Let P be a symmetric $n\times n$ matrix; then to the element

$$\begin{bmatrix}I & P\\ 0 & I\end{bmatrix}\in Sp_n(\mathbf{R})$$

there is associated the operator $f(x)\mapsto \exp(x^TPx)f(x)$

(ii) Let $A\in GL_n(\mathbf{R})$ be an invertible $n\times n$ matrix. Then to the element

$$\begin{bmatrix} A & 0 \\ 0 & (A^{-1})^T \end{bmatrix} \in Sp_n(\mathbf{R})$$

there is associated the operator

$$f(x) \mapsto |\det(A)|^{1/2} f(A^T x)$$

(iii) let Q be a symmetric $n \times n$ matrix. Then to the element

$$\begin{bmatrix} I & 0 \\ Q & I \end{bmatrix} \in Sp_n(\mathbf{R})$$

there is associated the operator

$$f(x) \mapsto \mathfrak{F}^{-1}(\exp(x^T Q x) \mathfrak{F} f(x))$$

where $\mathfrak{F}$ denotes the Fourier transform.

(The operator corresponding to the element

$$\begin{bmatrix} 0 & I \\ -I & 0 \end{bmatrix} \in Sp_n(\mathbf{R})$$

is in fact the Fourier transform itself).

Except for one snag to be discussed below, this suffices to describe the operator which should be associated to any element $S \in Sp_n(\mathbf{R})$. Indeed let

$$S = \begin{bmatrix} S_1 & S_2 \\ S_3 & S_4 \end{bmatrix} \in Sp_n(\mathbf{R}) \tag{4.4}$$

then there is an $s > 0, s \in \mathbf{R}$ such that $S_1 + sS_2$ is invertible and we have a factorisation

$$\begin{bmatrix} S_1 & S_2 \\ S_3 & S_4 \end{bmatrix} = \begin{bmatrix} I & 0 \\ (S_3+sS_4)(S_1+sS_2)^{-1} & I \end{bmatrix} \begin{bmatrix} I & S_2(S_1+sS_2)^T \\ 0 & I \end{bmatrix} \begin{bmatrix} S_1+sS_2 & 0 \\ 0 & (S_1^T+sS_2^T)^{-1} \end{bmatrix} \begin{bmatrix} I & 0 \\ -s & I \end{bmatrix} \tag{4.5}$$

(It is easily verified that all four factors on the right are in fact in $Sp_n(\mathbf{R})$.

Now assign to the operator S the product of the four operators corresponding to the factors on the RHS of (4.5) according to the recipe (i)-(iii) given above. There is a conceivable second snag here in that it seems a priori possible that different factorisations could give different operators. This in fact does not happen precisely because the 'representation' described by (i)-(iii) is a 'real form' of the oscillator representation $Sp_n(\mathbf{R}) \to \mathrm{Aut}(L^2(\mathbf{R}^n))$. The relation between the oscillator representation and (i)-(iii) above is given by the substitution $x_k \mapsto \sqrt{i}\, x_k$ where $i = \sqrt{-1}$. (The possible sign ambiguity which could come from the fact that the oscillator representation is really a representation of the covering $\tilde{S}p_n(\mathbf{R})$ rather than $Sp_n(\mathbf{R})$ itself also seems not to happen; if would in any case be irrelevant for the applications dicussed below.)

It remains to discuss the first snag mentioned just above (5.4) and why the words 'representation' and 'real form' above have been placed in quotation marks. The trouble lies in part (iii) of the recipe. Taking a Fourier transform and than multiplying with a quadratic exponential may well take one out of the class of functions which are inverse Fourier transformable. Another way to see this is to observe that the operator described in (iii) assigns to a function ψ the value in $t = 1$ of the solution of the evolution equation

$$\frac{\partial \rho}{\partial t} = ((\frac{\partial}{\partial x})^T Q \frac{\partial}{\partial x})\rho\ , \ \rho(x,0) = \psi(x) \tag{4.6}$$

and if Q is not nonnegative definite this involves anti-diffusion components for which the solution at $t = 1$ may not exist. Additionally, - but this is really the same snag - applying recipe (i) to a function may well result in a function that is not Fourier transformable.

What we have in fact is not a representation of all of $Sp_n(\mathbf{R})$ but only a representation of a certain sub-semi-group cone in $Sp_n(\mathbf{R})$.

For the applications to be described below this means that we must be careful to take factorizations such that applying the various operators successively continues to make sense. The factorization (5.5) does not seem optimal in that respect and we shall for the special elements of $Sp_n(\mathbf{R})$ which come from filtering problems use a different one.

Incidentally, one says that two structures over $\mathbf{R}$ are real forms of one another if after tensoring with $\mathbf{C}$ (= extending scalars to $\mathbf{C}$) they become isomorphic (over $\mathbf{C}$). It is in this sense that the 'representation' described by the recipe (i)-(iii) is a 'real form' of the oscillator representation.

5. Propagation of non-Gaussian initials

Now, finally, after all this preparation, consider a known linear dynamical system

$$dx = Axdt + Bdw, \; Cxdt + dv_i, \; x \in \mathbf{R}^n, w \in \mathbf{R}^m, \; y,v \in \mathbf{R}^p. \tag{5.1}$$

with a known, not necessarily Gaussian, initial random variable $x(0)$ with probability distribution $\psi(x)$.

The *DMZ* equations in this case is as follows

$$d\rho = \pounds\rho dt + \sum_{j=1}^{p}(Cx)_j dy_j(t) \tag{5.2}$$

where $(Cx)_j$ is the j-th component of the p-vector Cx. The operator £ in this case has the form

$$\pounds = \tfrac{1}{2}\sum_{i,j}(BB^T)_{i,j}\frac{\partial^2}{\partial x_i \partial x_j} - \sum_{i,j}A_{ji}x_j\frac{\partial}{\partial x_i} - \mathrm{Tr}(A) - \tfrac{1}{2}\sum_{j}(Cx)_j^2 \tag{5.3}$$

Taking brackets of the multiplication operators $(Cx)_j$ with £ yields a linear combination of the operators

$$x_1,\dots,x_n; \; \frac{\partial}{\partial x_1},\dots,\frac{\partial}{\partial x_n}; \; 1. \tag{5.4}$$

This is a straightforward calculation to check. Moreover, the bracket (= commutator product) of £ with any of the operators in (5.4) again yields a linear combination of the operators listed in (5.4). It follows that for linear stochastic dynamical systems (5.1) the associated estimation Lie algebra (= the Lie algebra generated by £, $(Cx)_1,\dots,(Cx)_p$) is always solvable of dimension $\leqslant 2n+2$.

As a mather of fact it is quite simple to prove that the system (5.1) is completely reachable and completely observable if and only if the dimension of the estimation Lie algebra is precisely $2n+2$ so that a basis of the algebra is formed by the $(2n+1)$ operators of (5.4) and £ itself.

In all cases Wei-Norman theory is applicable (working perhaps with a slightly larger Lie algebra than strictly necessarily makes no real difference).

Thus we can calculate effectively the solutions of the unnormalized density equation (5.2) provided we have good ways of calculating the expressions.

$$e^{t\pounds}\psi, \; e^{tx_i}\psi, \; e^{t\frac{\partial}{\partial x_i}}\psi, \; e^t\psi \tag{5.5}$$

for arbitrary initial data ψ. The last three expressions of (5.5) cause absolutely zero difficulties $(\exp(t\frac{\partial}{\partial x_i})\psi = \psi(x_1,\dots,x_{i-1},x_i+t,x_{i+1},\dots,x_n))$. Thus it remains to calculate the $e^{t\pounds}\psi$ where £ is an operator of the form (5.3). It is at this point that the business of the Segal-Shale-Weil representation of the previous section comes in. As a matter of fact the Segal-Shale-Weil representation itself, not the 'real form' described in section 4 above, is a representation of the Lie algebra spanned by the operators

$$i\, x_k x_j,\ x_k \frac{\partial}{\partial x_j} + \tfrac{1}{2}\delta_{k,j},\ i\frac{\partial^2}{\partial x_k \partial x_j},\ i = \sqrt{-1} \tag{5.6}$$

and apart form multiples of the identity (which hardly matter) and the occurence of $\sqrt{-1}$ these are the constituents of the operators £ in (5.3). It is to remove the factors $\sqrt{-1}$ that we have to go to a real form. Cf. [3] for more details on the Segal-Shale-Weil representation itself, and what it, and its real form, have to do with Kalman-Bucy filters.

It is convenient not to have to worry about multiples of the identity. To this end note that if $£' = £ + aI$ then $\exp(t£')\psi = \exp(ta)\exp(t£)\psi$, so that neglecting multiples of the identity indeed matters hardly.

The first observation is now that, modulo multiples of the identity operator, if £ and £′ are two operators of the form (5.3) then their commutator difference $[£,£'] = ££' - £'£$ is again of the same form. (To make this exact replace £ in (5.3) by $£ + \frac{1}{2}\mathrm{Tr}(A)$ and similarly for £′.) Thus these operators actually form a finite dimensional Lie algebra and this is, of course, the symplectic Lie algebra $sp_n(\mathbf{R})$. The correspondence is given by assigning to $£(=£(A,B,C))$ the $2n \times 2n$ matrix

$$£(A,B,C) \to \begin{pmatrix} -A^T & -C^TC \\ -BB^T & A \end{pmatrix} \tag{5.7}$$

(If you want to be finicky it is the operator $£(A,B,C) + \frac{1}{2}\mathrm{Tr}(A)$ which corresponds to the matrix on the right of (5.7).)

In terms of a basis on the left and right side the correspondence (i.e. the isomorphism of Lie algebras) is given as follows. Let E_{ij} be the $n \times n$ matrix with a 1 in spot (i,j) and zero everywhere else. Then

$$\frac{\partial^2}{\partial x_i \partial x_j} \mapsto \begin{pmatrix} 0 & 0 \\ -E_{ij} - E_{ji} & 0 \end{pmatrix} \tag{5.8}$$

$$x_i \frac{\partial}{\partial x_j} + \tfrac{1}{2}\delta_{i,j} \mapsto \begin{pmatrix} E_{ij} & 0 \\ 0 & -E_{ji} \end{pmatrix} \tag{5.9}$$

$$x_i x_j \mapsto \begin{pmatrix} 0 & E_{ij} + E_{ji} \\ 0 & 0 \end{pmatrix} \tag{5.10}$$

It is now straightforward to check that this does indeed define an isomorphism of Lie algebras from the Lie algebra of all operators $£(A,B,C) + \frac{1}{2}\mathrm{Tr}(A)$ where £ is as in (5.3) and the algebra $sp_n(\mathbf{R})$ described and discussed in section 4 above. For example one has

$$[\frac{\partial^2}{\partial x_1 \partial x_2}, x_2 x_3] = x_3 \frac{\partial}{\partial x_1} \tag{5.11}$$

which fits perfectly with

$$[\begin{pmatrix} 0 & 0 \\ -E_{12} - E_{21} & 0 \end{pmatrix}, \begin{pmatrix} 0 & E_{23} + E_{32} \\ 0 & 0 \end{pmatrix}] = \begin{pmatrix} E_{31} & 0 \\ 0 & -E_{13} \end{pmatrix} \tag{5.12}$$

It is precisely the correspondence (5.8) - (5.10) or, modulo multiples of the identity, (5.7), plus the fact that 'real form' described in section 4 of the SSW representation is precisely the way to remove the $\sqrt{-1}$ factors, plus, again, the fact that the SSW is really a representation, which makes it possible to use finite dimensional calculations to obtain expressions for

$$\exp(t(£(A,B,C)+\tfrac{1}{2}\mathrm{Tr}(A))\psi \tag{5.13}$$

for arbitrary initial conditions.

Basically the recipe is as follows. Take $£(A,B,C)+\frac{1}{2}\mathrm{Tr}(A)$. Let $M\epsilon\ sp_n(\mathbb{R})$ be its associated matrix as defined by the RHS of (5.7). Calculate $\exp(tM)=S(t)$. Write $S(t)$ as a product of matrices as in (i), (ii), (iii) in section 4. Apply successively the operators associated to the factors. The result, if defined, will be an expression for (5.13). One factorisation which can be used is that of (4.5) above. It does not, however, seem to be very optimal and it is difficult to show that everything is well defined.

It is better and more efficient to use a preliminary reduction. Consider the algebraic Riccati equation

$$A^TP + PA - PBB^TP + C^TC = 0 \tag{5.14}$$

determined by the triple of matrices (A,B,C). It is easy to check that for any solution P one has

$$\begin{pmatrix} I & -P \\ 0 & I \end{pmatrix} M \begin{pmatrix} I & P \\ 0 & I \end{pmatrix} = \begin{pmatrix} -\tilde{A}^T & 0 \\ -BB^T & \tilde{A} \end{pmatrix} \tag{5.15}$$

where $\tilde{A}=A-BB^TP$. Given this it becomes useful to know when (5.14) has a solution and to know some properties of the solutions. These will also be important for the next section. In fact the function $\mathrm{rc}(A,B,C)$ that assigns to the triple (A,B,C) under suitable conditions the unique positive definite solution of (5.14) is important enough to be considered a standard named function which should be available in accurate tabulated form much as say the Airy function or Bessel functions. I know of no such tables. The symbol 'rc' of course stands for Riccati.

Let A^* be the adjoint of the complex $n\times n$ matrix A, i.e. the conjugated transpose of A, so, if A is real, $A^*=A^T$. Consider the equation (algebraic Riccati equation)

$$A^*P + PA = PBB^*P - C^*C \tag{5.16}$$

(Here A is an $n\times n$ matrix, B an $n\times m$ matrix, C an $p\times n$ matrix.) Some facts about (5.16) are then as follows:

(5.17) If (A,B) is stabilizable, i.e. if there exists an F such that $A-BF$ has all eigenvalues with negative real part, then there is a solution of (5.16) which is positive semidefinite ($P\geqslant 0$) (and for this solution $\tilde{A} = A - BB^*P$ is stable).

(So in particular if (A,B) is completely reachable there is a solution of (5.14).)

(5.18) Suppose (5.16) has a solution $P\geqslant 0$ and suppose that (A,C) is completely observable. Then P is the only nonnegative definite solution of (5.16) and $P>0$.

(5.19) If (A,B,C) is *co* and *cr* then there is a unique $P>0$ which solves (5.16).

This last property is the essential one for this section. For the next one we need something better. Let $L^{co,cr}_{m,n,p}(\mathbb{R})$ be the space of all triples of real matrices (A,B,C) such that (A,B) is completely reachable and (A,C) is completely observable. Let $\mathrm{rc}(A,B,C) := P$ be the unique solution P of (5.16) such that $P>0$ (the matrix P is positive definite and selfadjoint). Then

(5.20) The function $\mathrm{rc}(A,B,C)$ from $L^{co,cr}_{m,n,p}(\mathbb{R})$ to the space of selfadjoint matrices is real analytic (and so in particular C^∞ (= smooth)

Moreover

(5.21) $\mathrm{rc}(TAT^{-1},\ TB,\ CT^{-1}) = (T^*)^{-1}\mathrm{rc}(A,B,C)T^{-1}$

(5.22) $\mathrm{rc}(-A^*),\ \pm C^*, \pm B^*) = \mathrm{rc}(A,B,C)^{-1}$

Property (5.21) is important in section 6; more precisely it will be important when these results are really implemented for multi-input multi-output systems. The point is that the matrices (A,B,C) are not determinable from the observations alone, simply because the systems (A,B,C) and (TAT^{-1}, TB, CT^{-1}) for $T\in GL_n(\mathbb{R})$ produce exactly the same input-output behaviour. For completely reachable and completely observable systems this is also the only indeterminacy. Property (5.21) guarantees that the whole analysis of these two section 5 and 6 'descends' to the moduli space (quotient manifold) $L^{co,cr}_{m,n,p}(\mathbb{R})/GL_n(\mathbb{R})$.

Having all this available it is tempting (and natural) to play the trick embodied by (5.15) again, this time using conjugation by a 2×2 block matrix with identities on the diagonal, a zero in the upper right hand corner and a Riccati equation solution Q in the lower left hand corner. This, however, is no particular good because this will introduce both the two factors

$$\begin{bmatrix} I & 0 \\ -Q & I \end{bmatrix}, \quad \begin{bmatrix} I & 0 \\ Q & I \end{bmatrix}$$

in the factorisation of $S(t)=\exp(tM)$, and at least one will cause difficulties with inverse and direct Fourier transforms; cf. part (iii) of the recipe of section 4.

Instead, writing

$$\exp(t\begin{bmatrix} -\tilde{A}^T & 0 \\ -BB^T & \tilde{A} \end{bmatrix}) = \begin{bmatrix} \exp(-t\tilde{A}^T) & 0 \\ -R & \exp(t\tilde{A}) \end{bmatrix} \tag{5.23}$$

one uses the factorisation

$$\exp(t\begin{bmatrix} -\tilde{A}^T & 0 \\ -BB^T & \tilde{A} \end{bmatrix}) = \begin{bmatrix} I & 0 \\ -R\exp(t\tilde{A}^T) & I \end{bmatrix} \begin{bmatrix} \exp(-t\tilde{A}^T) & 0 \\ 0 & \exp(t\tilde{A}) \end{bmatrix} \tag{5.24}$$

giving the following total factorisation for $S(t)=\exp(tM)$

$$S(t) = \begin{bmatrix} I & P \\ 0 & I \end{bmatrix} \begin{bmatrix} I & 0 \\ -R\exp(t\tilde{A}^T) & I \end{bmatrix} \begin{bmatrix} \exp(-t\tilde{A}^T) & 0 \\ 0 & \exp(t\tilde{A}) \end{bmatrix} \begin{bmatrix} I & -P \\ 0 & I \end{bmatrix} \tag{5.25}$$

Except for possibly the second factor on the right hand side of (5.25) applying the recipe of section 4 is a total triviality.

As to that second factor observe that

$$\frac{d}{dt}(\exp(t\begin{bmatrix} -\tilde{A}^T & 0 \\ -BB^T & \tilde{A} \end{bmatrix}) = \begin{bmatrix} -\exp(-t\tilde{A})A^T & 0 \\ -\frac{d}{dt}R & \exp(t\tilde{A})\tilde{A} \end{bmatrix}$$

$$= \begin{bmatrix} \exp(-t\tilde{A}^T) & 0 \\ -R & \exp(t\tilde{A}) \end{bmatrix} \begin{bmatrix} -\tilde{A}^T & 0 \\ -BB^T & \tilde{A} \end{bmatrix} \tag{5.26}$$

from which it follows that

$$\frac{dR}{dt} = -R\tilde{A}^T + \exp(\tilde{t}A)BB^T. \tag{5.27}$$

As a result

$$\frac{d}{dt}(R\exp(t\tilde{A}^T) = -R\tilde{A}^T\exp(t\tilde{A}^T) + \exp(t\tilde{A})BB^T\exp(t\tilde{A}^T) + R\tilde{A}^T\exp(t\tilde{A}^T) \tag{5.28}$$

$$= \exp(t\tilde{A})BB^T\exp(t\tilde{A}^T) \geqslant 0$$

and it follows that

$$R\exp(t\tilde{A}^T) \geqslant 0 \quad \text{all } t \tag{5.29}$$

which means that applying part (iii) of the recipe of section 4 (= part (iii) of the definition of the real form of the SSW representation) just involves solving a diffusion equation (no anti diffusion component); or, in other words that the inverse Fourier transformation involved will exist. Note also that if the initial condition ψ is Fourier transformable then, if P is nonnegative definite, the result of applying the parts of the recipe corresponding to the third and fourth factors on the RHS of 5.25 will still be a Fourier transformable function.

This concludes the description of the algorithm for propagating non-gaussian initial densities.

6. Identification

Given all that has been said above, this section can be mercifully short. The problem is the following. Given a linear system

$$dx = Axdt + Bdw, \ dy = Cxdt + dv \tag{6.1}$$

with *unknown* A,B,C, but constant A,B,C, we want to calculate the joint conditional density (given the observations $y(s)$, $0 \leq s \leq t$) for A,B,C,x. This can be approached as a nonlinear filtering problem by adding the equations

$$dA = 0, \ dB = 0, \ dC = 0 \tag{6.2}$$

or, more precisely, the equations stating (locally) that the free parameters remaining after specifying a local canonical form are constant but unknown. More generally one has the same setup and problem when, say, part of the parameters of (A,B,C) are known (or, generalizing a bit more, imperfectly known).

The approach, of course, will be the calculate the DMZ unnormalized version of the conditional density $\rho(x,A,B,t)$ given the observations $y(s)$, $0 \leq s \leq t$. Writing down the DMZ equation for the system (6.1)-(6.2) gives

$$d\rho = \pounds\rho dt + \sum_{j=1}^{P} (Cx)_j dy_j(t) \tag{6.3}$$

with $\pounds$ given by (5.3); i.e. exactly the same equation as occurred in section 5 for the case of known A,B,C. And, indeed the only difference is that in section 5 the A,B,C are known, while (6.3) should be seen as a family of equations parametrized by (the unknown parameters in) the A,B,C. Thus if $\rho(x,t|A,B,C)$ denotes the solution of (5.2) and $\rho(x,A,B,C,t)$ denotes the solution of (6.3) then

$$\rho(x,t|A,B,C) = \rho(x,A,B,C,t) \tag{6.4}$$

Now the bank of Kalman-Bucy filters for $\hat{x}$ parametrized by $(A,B,C) \in L_{m,n,p}^{co,cr}$ gives the probability density

$$\pi(x,t|A,B,C) = r(t,A,B,C)^{-1}\rho(x,t|A,B,C) \tag{6.5}$$

so that the normalization factor $r(t,A,B,C)$ can be calculated as $\int \rho(x,t,A,B,C)dx$.

By Bayes

$$\pi(x,A,B,C,t) = \pi(x,t|A,B,C)\pi(A,B,C,t)) \tag{6.6}$$

so that the normalization factor $r(t,A,B,C)$ is, so to speak, precisely equal to the difference between the solution of the DMZ equation (6.3) (or (5.2)) and the bank of Kalman filters producing $\pi(x,t|A,B,C)$. I.e. the marginal conditional density

$$\pi(A,B,C,t) = \int \pi(x,A,B,C,t)dx = \int \rho(x,A,B,C,t)dx \Big/ \int \rho(x,A,B,C,t)dxdAdBdC \tag{6.7}$$

is obtainable from the unnormalized version of the bank of Kalman-Bucy filters parametrized by (A,B,C). Given the relations between this bank of filters described in [13] and briefly recalled in section 7 below this may offer further opportunities.

Be that as it may the marginal density $\pi(A,B,C,t)$ which up to a normalization factor is equal to

$\int \rho(x,A,B,C,t)dx$ can be effectively calculated by the procedure of section 5 above with the only difference that $P=\mathrm{rc}(A,B,C)$ now has to be treated as a function. Once $\pi(A,B,C,t)$ (or in various cases some unnormalized version $\rho(A,B,C,t)$ is available a host of well known techniques such as maximum likelyhood become available.

If it is possible (as it will be in many cases) to work with a $\rho(A,B,C,t)=r(t)\pi(A,B,C,t)$ there is no (immediate) need to descend to the quotient manifold $L^{co,cr}_{m,n,p}(\mathbf{R}) / GL_n(\mathbf{R})$.

7. On the relation between the 'real form' of the SSW representation and the Kalman-Bucy filter

We have seen that the essential difficulty in obtaining the (unnormalized) conditional density $\rho(x,t)$ lies in 'solving' $\exp(t\mathcal{L})\psi$ where $\mathcal{L}$ is the second order differential operator (5.3). Now $\mathcal{L}$ corresponds in a fundamental way with the $2n \times 2n$ matrix

$$\begin{pmatrix} -A^T & -C^TC \\ -BB^T & A \end{pmatrix} \tag{7.1}$$

Not very surprisingly this matrix in turn is very much related to the matrix Riccati equation part of the Kalman-Bucy filter. Indeed, consider the matrix differential equation

$$\begin{pmatrix} \dot{X} \\ \dot{Y} \end{pmatrix} = \begin{pmatrix} -A^T & -C^TC \\ -BB^T & A \end{pmatrix} \begin{pmatrix} X \\ Y \end{pmatrix} \tag{7.2}$$

and, assuming that $X(t)$ is invertible, let

$$-P = YX^{-1}. \tag{7.3}$$

Then

$$\begin{aligned} \dot{P} &= -\dot{Y}X^{-1} + YX^{-1}\dot{X}X^{-1} = (+BB^TX - AY)X^{-1} + YX^{-1}(-A^TX - C^TCY)X^{-1} \\ &= +BB^T + AP + PA^T - PC^TCP \end{aligned}$$

which is the covariance equation of the Kalman-Bucy filter.

References

1. T.E. Duncan, Probability densities for diffusion processes with applications to nonlinear filtering theory, PhD thesis, Stanford, 1967.
2. M. Hazewinkel, (Fine) moduli (spaces) for linear systems: what are they and what are they good for. In: C.I. Byrnes, C.F. Martin (eds), Geometric methods for linear system theory (Harvard, June 1979), Reidel, 1980, 125-193.
3. M. Hazewinkel, The linear systems Lie algebra, the Segal-Shale-Weil representation and all Kalman-Bucy filters, J. Syst. Th. & Math. Sci. **5** (1985), 94-106.
4. M. Hazewinkel, Lectures on linear and nonlinear filtering. In: W. Schiehlen, W. Wedig (eds), Analysis and estimation of stochastic mechanical systems, CISM course June 1987, Springer (Wien), 1988, 103-135.
5. M. Hazewinkel, S.I. Marcus, On Lie algebras and finite dimensional filtering, Stochastics **7** (1982), 29-62.
6. M. Hazewinkel, J.C. Willems (eds), Stochastic systems: the mathematics of filtering and identification and applications, Reidel, 1981.
7. O. Hijab, Stabilization of control systems, Springer, 1987.
8. R.E. Howe, On the role of the Heisenberg group in harmonic analysis, Bull. Amer. Math. Soc. **3** (1980), 821-844.
9. S.I. Marcus, Algebraic and geometric methods in nonlinear filtering, SIAM J. Control and Opt. **22** (1984), 817-844.
10. R.E. Mortensen, Optimal control of continuous time stochastic systems, PhD thesis, Berkeley,

1966.
11. J. WEI, E. NORMAN, On the global representation of the solutions of linear differential equations as products of exponentials, Proc. Amer. Math. Soc. **15** (1964), 327-334.
12. M. ZAKAI, On the optimal filtering of diffusion processes, Z. Wahrsch. verw. Geb. **11** (1969), 230-243.

MISCELLANEOUS

A STOCHASTIC GROWTH MODEL ON RANDOM GRAPHS TO UNDERSTAND THE DYNAMICS OF AIDS-EPIDEMIC*

Ph. Blanchard

Theoretische Physik and BiBoS
Universität Bielefeld
D-4800 Bielefeld 1

* This work has been supported by Bundesgesundheitsamt und Bundesminister für Forschung und Technologie.

1. INTRODUCTION

Mathematical analysis of epidemics has a long history starting in 1662 with Graunt's numerical study of the Bills of Mortality for London. As early as 1760 Daniel Bernoulli [1] wrote the first mathematical paper describing the effects of an epidemic of smallpox.

The "mass-action" principle for the spread of infection in a population was first proposed 1906 by Hamer [2]. In its continuous time form this principle states that if $S(t)$ and $I(t)$ are the respective numbers of susceptibles and infectives in a given population then the rate of change of the infectives is proportional to the number of possible encounters $S(t)\,I(t)$ between susceptibles and infectives i.e.

$$\frac{dI(t)}{dt} = \beta\, S(t)\, I(t) \quad ,$$

β being the infection parameter.

For understanding the epidemiology of malaria Ross [3] proposed in the early 1900s a model of this type taking into account the coupling between the human and mosquito populations of a region. Since then deterministic and probabilistic epidemic models have relied on this mass-action effect in one form or another. For a recent survey of standard models we refer to [4] [5] [25].

AIDS, acquired immune deficiency syndrome, is the final stage of the disease caused by the human immunodeficiency virus (HIV), a retrovirus with a very long asymptomatic period [6]. Three modes of HIV transmissions have been observed: all forms of sexual intercourse (man-woman, man-man, anal-genital, oral-genital), sharing of HIV-infected blood and perinatal infection. Sexual transmission is the most common mode of transmission for HIV and will remain so in the future. Due to the chromosomal integration of the proviral DNA into the host cell an HIV-infected person remains infected and infectious for life.

For sexually transmitted diseases standard epidemiological models of the mass-action type have been used. See e.g. Weyer [7], Hethcoke and York [8], Hyman and Stanley [9] and Dietz's model [10] [15] which takes also into account the effects of the duration of partnerships. However for HIV epidemiology a mathematical model is needed which reflects the complex structure of sexual contacts between individuals and is able to take into account large variations in behaviour as well as in the disease progression, reflecting the population risk structure (age, sexual activity, drug use, ...). See [14] [20] [23,24] [26,27]. Therefore the model which has been proposed in [11] by Krüger, Bolz and myself is of completely different spirit. We consider the individuals of a given population as vertices of a graph, whose edges are supposed to represent their sexual contacts in the time interval $[0,T]$ during which the model is applied. Obviously from an epidemiological point of view only global statistical informations about the sexual activity of the population are relevant. This fact is reflected in our model by the use of a random graph as the fundamental underlying structure, on which the epidemic will develop as a stochastic process. This approach is novel in linking the characteristics of the random graph and its dynamics to social and medical data as well as in exploiting the richness of the structure of stochastic processes over random graphs to model a complicated epidemiological situation. Let us emphasize that the standard modelling of epidemics using ordinary differential equations appears as a special limiting case of this model. Indeed, if we consider only complete graphs or complete n-partite graphs, sexual contacts are now realized uniformly in the sense that everybody has contact with everybody.

To generate a graph which reflects the mesh of sexual contact of real life we need a function which tells us for all n, $n \in \mathbb{N}$, how many people have had n different sexual contacts during the time T. The set of all graphs satisfying a given contact distribution will be called a "random graph space". Furthermore for the investigation of the HIV epidemics we have to introduce more structure on the set of relevant graphs. Indeed we must take into account that the infection risk of an individual depends on both the behaviour of the individual and the prevalence of HIV infections in the risk-groups (homosexual males, prostitutes, bisexual males, intravenous drug users, heterosexual males and females, females having contact with bisexual males ...) with which the individual has sexual contacts or share needles. In the United States and Europe the most active groups play a very important role at least in the first stages of the epidemic [21] [22]. If mixing is small then the individuals belonging to high-risk groups are nearly all infected before the HIV-infection moves into lower-risk groups. If there is a large mixing many more lower activity individuals will be infected. Since many prostitutes are drug users and most drug users are heterosexual the spread of HIV-infection in the group of intravenous drug users exposes to a potential danger the heterosexual community.

Therefore the initial data we need are the contact rate functions for each subgroup as well as the mixing rates between two different groups. Of course by the nature of human finding sexual partner the population's age structure in each group is important. A model of the HIV-transmission in Africa would require also including at least blood transfusions and vertical transmission from infected mother to child.

To generate the dynamics of HIV infection we need the knowledge of the times at which edges are activated (realization of a sexual contact). Again we use a distribution expressing the duration of a partnership of an individual of a certain behavioral risk-group of a given age. In this way we obtain time valued edges for the time interval $[0,T]$. Finally we have to take into account some epidemiological and medical parameter. The most important are:

- infection probability per contact inside the different risk-groups
- distribution of the unusually long incubation (of the order of 8 - 9 years) period with a wide standard deviation.
- variable infectiousness (the infectivity is much more lower in the middle stage of HIV-infection)
- proportion of virus carriers that develops AIDS
- mortality
- influence of the different type of HIV-virus and genetic variability

of HIV DNA sequences
- influence of other venereal diseases (e.g. genital ulcers).

Since the dynamics of the AIDS-epidemic involves long time scales individuals will leave a given behavioral risk group and this flow between the different risk groups is an additional source of mixing. Let us emphasize that probability enters in the model as it enters into real life. First the underlying population risk structure is varied randomly according to sociological data expressing the sexual activity of individuals of different risk-groups. Secondly the dynamics are also generated at random using probability distributions of the sexual and epidemiological parameters.

The dynamics we consider are discrete but with a short time increment (about one day), and the computer simulation of the spread of the HIV infection requires little computation time. Most of the time in the computer simulation is spent on selecting a good random graph. For more details see [11] [12].

This stochastic model allows for a larger set of observables than standard epidemiological models and gives therefore a much better description of details of the infection spread (influence of the population risk structure, incubation time, variability of infectiousness ...) and qualitative insights even when many relevant data are lacking or badly known. Given the present state of knowledge of relevant parameters reliable long term predictions of AIDS are (as with any other model) not yet possible but different interventions strategies (safer sex, behavioral changes, testing and counseling programs ...) can be tested. For more details we refer again to [11] [12]. See also [13] where May and Anderson discuss how it is possible to estimate some important features of the epidemic indirectly from mathematical models.

As a side remark it is worth mentioning that the same mathematical framework can be used to describe a large class of propagation and growth phenomena: the edges can be viewed as communication lines in a network or as relationships in a social network.

2. RANDOM GRAPH EPIDEMICS

As heuristically explained in Section 1, the basic probability space we must introduce is a space of labeled random graphs. For the definition of the relevant graph theoretical concepts we refer to [17] [18]. Let $V = \{x_1,\dots,x_N\}$ to the set of vertices (individuals) which splits into several relevant groups V_i according to the population risk structure

$$V = \bigcup_{i \in I} V_i \ .$$

Since edges between vertives describe a sexual contact, the basic given data generating the underlying graphs are the degree distributions inside the V_i as well as the mixings between pairs (V_i,V_j) of subgroups of the vertices. Let us therefore introduce the following integer valued functions

$$n^o : \mathbb{N} \to \mathbb{N} \ , \quad n^o(z) = \# \{x_i \in V \mid d(x_i) = z\}$$

$$n^o_i : \mathbb{N} \to \mathbb{N} \ , \quad n^o_i \equiv n^o \restriction_{V_i}$$

$$n^o_{i,j} : \mathbb{N} \to \mathbb{N} \times \mathbb{N} \ , \quad n^o_{ij}(z) = (\# \{x_k \in V_i, d_j(x_k) = z \ , \ \# \{x_e \in V_j, d_i(x_e) = z\}) \ ,$$

where $\# \{\cdot\}$ denotes the cardinality of $\{\cdot\cdot\}$ and $d(x)$ the vertex degree of x. Introducing the adjacency matrix ϕ of the graph

$$\phi(x_n,x_m) \equiv \begin{cases} 1 & \text{if } (x_i,x_j) \text{ is an edge of } G \\ 0 & \text{otherwise} \end{cases}$$

we have

$$d_j(x_n) \equiv \sum_{x_m \in V_j} \phi(x_n,x_m) \ .$$

All graphs G which satisfy the above distribution of sexual contact rates n^o_i, n^o_{ij} with $i,j \in I$ are the elements of the random graph space $G^{n^o_i}_{n^o_{i,j}}$, which becomes a probability space by assigning to each element $G \in G^{n^o_i}_{n^o_{i,j}}$ the same probability

$$P(G) = \frac{1}{\#\{G \in G^{n_i^o}_{n^o_{i,j}}\}} \quad .$$

Conditions on n_i^o and n_{ij}^o can be given ensuring that $G^{n_i^o}_{n_{ij}^o}$ is not empty [11]. A more difficult taste is the estimation of the number of elements of $G^{n_i^o}_{n_{ij}^o}$ since already for regular graphs the size of the automorphism group is in most of the cases not explicitly known. For asymptotic estimates for $N \to +\infty$ we refer to [19] where a similar random graph space is considered.

For some applications it can be useful to introduce a larger class of random graphs for which only the number of edges connecting different subpopulations and the number of vertices having edges to different subgroups are given; see [11].

One way to describe the dynamics of the epidemic is to introduce two times on each graph V on the random graph space

- the system time indexed by $t \in [0,T]$
- the proper time of the elements of V (age of individuals).

$\tau(x_i) \in [C_i, C_i + T] \equiv A_{x_i}$, C_i being the age of x_i at the initial $t = 0$ according to the life time distribution of the population.

Here we will consider the situation where T is on the order of 15 years and therefore we do not need to consider births of individuals in our model because the age of the individuals at the first time of a sexual contact is expected to be not less than 15 years.

The vertex degree $d(x_i)$ describes now the number of sexual partners of x_i during $[0,T]$. We now define two functions which characterize the time dependence of the sexual activity of x_i:

$$\varphi : x_i \longmapsto \{\tau_k\} \cup \{0\} \quad \tau_k \in A_{x_i} \ , \quad k = 1,\dots n \geq d(x_i)$$

$$\psi_i : \tau_k \longmapsto k(x_i, 1) \ , \quad \psi(0) = x_i$$

In this basic model the fraction of HIV-infecteds who develop AIDS is assumed to be one.

with

$$k(x_i,1) \equiv \{y \in V | \phi(x_i,y) = 1\} \cup \{x_i\} .$$

The proper times τ_k can be viewed as the times at which x_i changes its sexual partner and $k(x_i,1)$ as the set of the potential sexual partners of x_i , x_i being allowed to be a sexual partner of x_i . If we introduce time intervals I_{ij} by

$$I_{ij} \equiv \bigcup_{\psi(\tau_k) = x_j} [\tau_k, \tau_{k+1}) ,$$

the length $|I_{ij}|$ of I_{ij} is to be understood as the time span during which x_i has x_j as a sexual partner. Clearly this interpretation requests a symmetry relation $|I_{ij}| = |I_{ij}|$ or at least we have to ensure that

$$I_{ij} - \min I_{ij} = I_{ji} - \min I_{ji} .$$

We remark on this point in [11].

The function φ is generated from data describing the sexual behaviour inside the group V_m to which x_i belongs. The function ψ_i is an "almost" random function expressing how the sexual partners of individual x_i are chosen from the entire population. Ideally the partner selection should be also based on sociological datas. For a discussion of the effect that partner selection (random or biased partner choice) has on the shape of the epidemic curve see [12].

To account for the spread of the HIV-infection and of the AIDS-epidemic at time t let us now introduce the following time dependent division of the population expressing the progression of the diseases

$$I_t^3(V) \subset I_t^2(V) \subset I_t^1(V) \subset I_t^o(V) \subset V$$

$$I_t^o(V) \equiv \{x \in V \mid x \text{ is HIV-infected at time } t\}$$

$$I_t^1(V) = \{x \in I_t^o(V) \mid x \text{ knows about the infection at time } t\}$$

$$I_t^2(V) = \{x \in I_t^1(V) \mid x \text{ develops AIDS at time } t\}$$

$$I_t^3(V) = \{x \in I_t^2(V) \mid x \text{ is dead at time } t\} .$$

At each time t the $I_t^j(V)$, $j = 0,1,2,3$ are subgraphs of V and e.g. $|I_t^o(V)| = \#\{x \in I_t^o(V)\}$ is to be understood as the total number of individuals which are HIV-infected at time t .

For simplicity we make now the following assumption for the probabilities of transmission per partnership (depending only of the state of the individuals) for all t

$$\begin{array}{llll} p_o & \text{for} \quad x \in I_t^o(V) & x \notin I_t^j(V) & j = 1,2,3 \\ p_1 & \text{for} \quad x \in I_t^1(V) & x \notin I_t^j(V) & j = 2,3 \\ p_2 & \text{for} \quad x \in I_t^2(V) & x \notin I_t^3(V) & \\ 0 & \text{for} \quad x \in I_t^3(V) \ . & & \end{array}$$

Let us emphasize that it is not a problem in our model to use probabilities of transmission depending on the population risk structure and to account for the fact that the infectiousness of individuals carrying HIV varies as the course of the diseases progresses.

Let us now describe a way to construct easily the dynamics of the epidemics of the $I_t^j(V)$. We take a discrete time step Δt (about 1 day) and start with a given family of sets $I_{t_m}^o$, $I_{t_m}^1$, $I_{t_m}^2$ and $I_{t_m}^3$. Let $t_{m+1} = t_m + \Delta t$. Given these initial sets we need next the infection times $\tau^k(x_i)$ for the initially infected individuals to define the algorithm generating the dynamics of the epidemic on random graphs.

$I_{t_{m+1}}^o$ is now generated in the following way:

Let $x_i \in I_{t_m}^o$, $\tau(x_i,t_{m+1}) \equiv \tau(x_i) + t_{m+1} \in I_{ij}$ and define

$$\chi_t^o(x_i) = \begin{cases} 1 & \text{if} \quad x_i \in I_t^o(V) \\ 0 & \text{if} \quad x_i \notin I_t^o(V) \end{cases}$$

then

$$\chi_{t_{m+1}}^o(x_j) = \begin{cases} 1 & \text{if} \quad x_j \in I_{t_m}^o(V) \\ \sigma_{t_{m+1}}(x_i) & \text{if} \quad x_j \notin I_{t_m}^o(V) \text{ and } \tau(x_j,t_{m+1}) \in I_{ji} \\ 0 & \text{if} \quad x_j \notin I_{t_m}^o(V) \text{ and } \tau(x_j,t_{m+1}) \notin I_{ij} \end{cases}$$

In the above formula $\{\sigma_{t_m}(x_i)\}$ is a random sequence belonging to $\Pi\{0,1\}$ with $P[\sigma_t(x_i) = 1] = \tilde{\sigma}_t(x_i)$ where $\tilde{\sigma}_t(x_i)$ is defined according to our choice of the transmission probabilities

$$\tilde{\sigma}_t(x_i) = \begin{cases} p_o & \text{if } x_i \in I_t^o(V) \quad x \notin I_t^j(V) \quad j = 1,2,3 \\ p_1 & \text{if } x_i \in I_t^1(V) \quad x \notin I_t^j(V) \quad j = 2,3 \\ p_2 & \text{if } x_i \in I_t^2(V) \quad x \notin I_t^3(V) \end{cases} .$$

To generate in a similar way the next steps of the dynamics $(x_i \in I^o_{t_m}(V) \longrightarrow x_i \in I^1_{t_{m+1}}(V) \longrightarrow x_i \in I^2_{t_{m+p}}(V))$ we must account for the latency period of seroconversion i.e. the time span between first HIV-infection and the earliest time for which an HIV-test can be positive and for the long asymptotic incubation period after infection with the HIV-virus that causes AIDS. See [11]. The unusually long and extremely variable incubation period induces smearing effects, called "transients" by M.G. Koch and J.J. Gonzalez [26] who reduce spuriousely the epidemic spread. It follows from this construction that the spread of the epidemics can be understood as an abstract discrete Markov process taking values in subgraphs $\{I_n^I\}_{\underline{\alpha}}$ representing the infected people at time t_m such that

$$I_n^I(w) \subset I_{n+1}^I(w)$$

where the $\underline{\alpha}$ denote the medical parameters responsible for the spread of the disease (transmission probabilities per partnership, seroconversion time, distribution of the incubation period, mean life expectation after an individual develops AIDS, ...).

Remark

The realization of the dynamics in actual computer simulation is for practical reasons slightly different; see [12]. Moreover in computer simulations we use probabilities of transmission per contact and not per partnership.

3. SIMPLIFIED MATHEMATICAL MODELS

In this last section we will rigorously study some simplified cases of special random graphs for which the dynamics of the epidemic can be studied analytically.

3.1 Complete Bipartite Graph

Let us first show how the well known Lotka-Volterra equation in its discretized version arises as a special case of our model if we restrict ourself to complete graphs. A graph of order N is called complete if it has all $\binom{N}{2} = \frac{N(N-1)}{2}$ possible edges. In a bipartite graphs there are two kinds of vertices, male and female, and every edge has one male vertex and one female vertex. If there are N male vertices and M female and all MN possible edges (sexual contacts) are present we have a complete bipartite graph detoned by $K_{M,N}$.

Let $K_{N_M N_F}$ be the complete bipartite graph with N_M males and N_F females. Let n count the discrete time steps after which on the average an edge contact between two individuals is distroyed and a new one is created. Let I_n^M and I_n^F be the number of infected males and females at time n . Suppose that $N_M \leq N_F$. At each time we assume a unique and complete pairing for N_M into a subset of N_F of size N_M . We conclude that

$$E[I_{n+1}^M] = E[I_n^M] + p_1 \quad E[{}^*I_n^F]$$

$$E[I_{n+1}^F] = E[I_m^F] + p_2 \quad E[{}^*I_m^M] \; .$$

p_1 and p_2 being the transmission probabilities per contact female → → male and male → female and $E[{}^*I_n^{F,M}]$ denoting the expected number of infected females or males having contact with non-infected males or females, respectively. Remember the asymmetry $N_F \geq N_M$. Since to a good approximation $E[I_n^M \; I_n^F] = E[I_n^F] \; E[I_n^M]$ we have

$$E[{}^*I^{F,M}] = N_M \frac{E[I_n^{F,M}]}{N_{F,M}} \; \frac{N_{M,F} - E[I_n^{M,F}]}{N_{M,F}} \; .$$

Introducing the variables $m_n = \frac{E[I_n^M]}{N_M}$, $f_n = \frac{E[I_n^F]}{N_F}$ we get

$$m_{n+1} = m_n + p_1 f_n(1 - m_n)$$
$$f_{n+1} = f_n + p_2 \frac{N_M}{N_F} m_n(1 - f_n) .$$

In the same way we get arbitrary generalized Lotka-Volterra systems by considering n-partite complete graphs for which the sexual activity in each subgroup of the population is uniform, taking into account also birth and death rates. The n-partite complete graph is to be understood in our model as a division of the population into n homogeneous subgroups.

3.2 Regular graphs of low degree

If every vertex of a graph V has exactly the same degree, the degree $d(x)$ being the number of vertices adjacent to a vertex x , say r , V is called regular of degree r or r-regular.
If $d(x_i) = 2$ for all $x_i \in V$ then the whole graph splits into disconnected components, each being a cycle. Therefore, irrespective of the kind of transmission dynamics, the velocity of the spread of the dynamics is slower than linear.

The case where, for all x_i , $d(x_i) = n \geq 3$, n being fixed, is more delicate and has been studied in [11] in the simplest situation for which there is only one infected individual at initial time $t = 0$ and all x_i change sexual partners simultaneously. Let I_n denote the set of individuals which are first infected at time n . Since the graph has in the initial stages of the epidemic the structure of a tree we conclude that for $d(x_i) = 3$

$$|I_n| = |I_{n-1}| + |I_{n-2}|.$$

After that cycles are present and multiple infections can develop along the epidemic cascade (edges between $I_{n-1} \cup I_{n-2}$). By these multiple infections we get a nonlinear correction to the (generalized) Fibonacci sequences, which make the study of the dynamics of the epidemic quite interesting even in this simplified approach. For regular graphs of degree 3 the dynamics can be described in the form of an infection of a 3-dimensional mapping; for more details see [11].

4. COMPUTER SIMULATION

In this last section let us take a look at our modelling of HIV-infection using random graphs from a simulational point of view. First of all, let us again emphasize the flexibility of our model, which can be suited to a variety of qualitatively different epidemiological assumptions. For more details see [11] [12]. The underlying configuration which represents the individuals of the model population, divided into subgroups according to the population risk structure, is a realization of the random graphs defined by the sociological parameters. The discrete dynamics is a realization of the stochastic process governed by the sociological and medical parameters (eventual realization of one of the potential sexual contacts for each individual, eventual transmission of HIV-infection during this contact). The natural time step for the dynamics is 1 day, while the time scale for re-configuring, i.e. selecting new partners, and measuring quantities characterizing the state of the model population is of the order of several months.

We have started to study a simple model which is modest in its demand of memory and computation time, yet complex enough to map main features of the real development. This simple model includes the following 7 groups: homosexual men, bisexual men, heterosexual men, heterosexual women, male intravenous drug users, female intravenous drug users, heterosexual women having contact to bisexual men. It comprises 20000 individuals having 5 partners at most. Number of partners, frequency of contacts and probability of infection per contact are specific for a group but homogeneous over this group. Changes in behaviour are not yet be included.

In [12] first results for this simplified situation are presented while work on the more elaborate model (as described in [11]) is in progress. These results show that scaling can be used and that the graphs structure does play a role in the dynamics of the epidemic. Indeed depending on the values of the transmission probabilities per contact we use, the graph structure representing the risk structure of the population has an influence or not. For low transmission probabilities we observe a saturation effect, which is entirely due to the underlying graph structure. At the contrary, for higher transmission probabilities the actual graph structure is no longer determining. Much more realistic models will be possible once data accumulates.

Acknowledgements

I am very grateful to G. Bolz, T. Krüger, C. Römer for the joy of collaboration and to Prof. J. Gonzalez, Prof. J. L'age-Stehr, Prof. M.A. Koch, Dr. M.G. Koch, Dr. B. Voigt and Prof. J. Weyer for very helpful discussions.

REFERENCES

1. D. Bernoulli, "Essai d'une nouvelle analyse de la mortalité causée par la petite vérole et des avantages de l'inoculation pour la prévenir", Mém. Math. Phys. Acad. Roy. Sci. Paris 1 - 45 (1760).
2. W.H. Hamer, "Epidemic disease in England - the evidence of variability and of persistence of type", The Lancet. II. 733 - 739 (1906)
3. R. Ross, "The prevention of malaria", Murray 1911
4. N. Bailey, "The mathematical theory of infections diseases and its applications", Charles Griffin 1975
5. J.C. Frauenthal, "Mathematical Modelling in Epidemiology", Springer 1980
6. M.G. Koch, "AIDS - Vom Molekül zur Pandemie - Spektrum der Wissenschaft Verlag, Heidelberg 1987

7. J. Weyer, private communication and "Den Toten können wir nicht helfen", Modellrechnung zur Prognose der Ausbreitung von AIDS - Kölner Universitäts-Journal 4, 58 - 61 (1987).

8. H.W. Hethcote, J.A. Yorke, Lect. Notes in Biomathematics 56, 1 - 105 Springer 1985.

9. J.M. Hyman, E.A. Stanley, "Using Mathematical Models to understand the AIDS-Epidemic", Los Alamos Preprint LA-UR-87-3078, 1988

10. K. Dietz, "The dynamics of spread of HIV infection in the hetero-sexual population"

11. Ph. Blanchard, G. Bolz, T. Krüger, "Simulation on Random Graphs of the epidemic dynamics of sexually transmitted diseases", BiBoS Preprint 291/87, Bielefeld.

12. G. Bolz, "Simulation on Random Graphs of the Epidemic Dynamics of Sexually Transmitted Diseases", Bielefeld 1988.

13. R.M. May, R.M. Anderson, "Transmission dynamics of HIV infection", Nature 326 137 - 142 (1987).

14. J.V. Niehoff, N. Sönnichsen, "AIDS - Eigenschaften einer echten Epidemie", 2. Klin. Med. 42 Heft 24, 2141 - 2145.

15. K. Dietz, "Schwellenwrte für die Persistenz der HIV-Infektion", Institut für Medizinische Biometrie, Preprint Tübingen 1987.

16. J.J. Gonzalez, M.G. Koch, "On the role of transients for the Prognostic Analysis of the AIDS Epidemic"

17. O. Ore, "Graphs and their Uses", New York - Toronto 1963

18. A. Bollobas, "Random graphs", Academic Press 1985

19. A. Beuder, E.R. Canfield, "The asymptotic number of labelled graphs with given degree sequences", J. Combinatorial Theory (A) 24 1978 296 - 307.

20. M.G. Koch, J. L'age-Stehr, J.J. Gonzalez, D. Dörner, "Die Epidemiologie von AIDS", Spektrum der Wissenschaft August 1987 38 - 51.

21. M.A. Koch, private communication and Bericht über die dem AIDS-Fallregister des BGA gemeldeten Fälle, Dez. 1987.

22. J. L'age-Stehr, "AIDS - Die epidemiologische Situation in Deutschland", Deutsches "Ärzteblatt - Ärztliche Mitteilungen", Heft 45 3027 - 3030 Nov. 1987.

23. D. Dörner, a) Ein Simulationsprogramm für die Ausbreitung von AIDS Projekt "Systemdenken" DFG 200/5 III/86 b) Addendeum zum a) Universität Bamberg.

24. M.G. Koch, U.v. Welck, ASSP, Aids-Spread Simulations and Projections, Projektstudie März 1987.

25. N.T.J. Bailey, "Introduction to the Modelling of Veneral Disease", J. Math. Biology 8, 301 - 322 (1979).

26. J. Weyer, B.C. Schmidt, B. Körner, Ein Mehrgruppenmodell zur Simulation der epidemischen Dynamik von AIDS, AIFO 3, 154 - 156, März 1988.

27. K. Dietz, K.P. Hadeler, Epidemiological models for sexually transmitted diseases, J. Math. Biol. (1988) 26, 1 - 25.

ON THE SCATTERING PROBLEM FOR MARKOV FIELDS

V.D. Koshmanenko
Institute of Mathematics, Ukrainian Acad.Sci.
Kiev, 252601, GSP, Repin str.3
USSR

ABSTRACT

We give a variant of the Haag-Ruelle scattering theory for Markov fields. Our method is Euclidean. We formulate four conditions for Schwinger functions which imply the existence of the wave and scattering operators. Our definition of the wave operators is not based on the unitary group of the perturbed field. Instead of it we use the semi-group.

1.

At first we explain our definition of the wave operators in an abstract situation.

Let H be a Hilbert space and h_o, h be a pair of positive selfadjoint operators in H. We suppose that h_o has only absolutely continuous spectrum, i.e. $\sigma(h_o) \equiv \sigma_{ac}(h_o)$. The wave operators for h_o, h are usually defined by means of the unitary group $\exp(-ith_o)$, $\exp(-ith)$ as follows

$$w^{\pm}(h,h_o) = \underset{t\to\pm\infty}{\text{s-lim}}\, e^{ith} e^{-ith_o}, \qquad (1.1)$$

supposing that these limits exist [1]. In physical applications the existence of the wave operators implies an important fact about spectrum of h. Namely $\sigma_{ac}(h_o) \subseteq \sigma(h)$.

We assert that the wave operators $w^{\pm}(h,h_o)$ can be defined without using the unitary group $\exp(-ith)$. Let us notice that the construction of exp(-ith) is rather a difficult problem. In our definition of wave operators [2,3,4], exp(-ith) is replaced by the following operator function

$$F_t(h) = \frac{1}{2\pi i}\int_0^\infty e^{-\tau h} F(t-i\tau)d\tau \quad , \tag{1.2}$$

where F(z) , z=t+iτ, is certain analytical function in the closed complex half-plane $\overline{\mathbb{C}^-}$. We write $F\in\mathbf{Z}^+$ if the following three conditions are fulfilled:

$$\text{(i)} \int_{\Gamma_t^+} e^{-iz\lambda} F(z)dz = 0, \qquad t\in\mathbb{R}^1 , \ \lambda \geqq 0 , \tag{1.3}$$

where Γ_t^+ is the boundary of the sector $G_t^+=\{z\in\overline{\mathbb{C}^-} \ |\mathrm{Re}\, z \leqq t, \mathrm{Im} z \geqq 0\}$.

$$\text{(ii)} \ \tilde{F}(\lambda) = \frac{1}{2\pi i}\int_{\mathbb{R}^1} e^{-i\lambda t} F(t+i\tau)_{|\tau=0} dt \neq 0$$

for some $\lambda \geqq 0$.

(iii) $\tilde{F}(\lambda)$ is continuous and $\tilde{F}(\lambda) \to 0$, as $\lambda \to \infty$. We write $F\in\mathbf{Z}^-$ if (ii), (iii) are the same as above and Γ_t^+ in (i) is replaced by Γ_t^- that is the boundary of the sector $G_t^- = \overline{\mathbb{C}^-} \setminus G_t^+$. In [3,4] we constructed examples of the functions F and showed that the sets $\mathbf{Z}^+$, $\mathbf{Z}^-$ are sufficiently rich for our aims.

The following property of $F\in\mathbf{Z}^+(\mathbf{Z}^-)$ is very important. From (1.3) it follows

$$\tilde{F}(\lambda) = \lim_{t\to\pm\infty} \frac{1}{2\pi i}\int_0^\infty e^{-i\lambda(t-i\tau)} F(t-i\tau)d\tau. \tag{1.4}$$

The last fact implies the asymptotic equality of two operator functions:

$$F_t(h) - \exp(ith)\tilde{F}(h) \to 0, \quad t \to \pm\infty \tag{1.5}$$

in the strong sense. Due to (1.4),(1.5) we have

<u>Theorem 1</u>. Let for all $F \in \mathbb{Z}^+(\mathbb{Z}^-)$ and all $\phi \in H$ there exist

$$s - \lim_{t \to \pm\infty} F_t(h) e^{-ith_o} \phi = \phi_F^{\pm} \tag{1.6}$$

such that

$$\| \phi_F^{\pm} \| = \| \tilde{F}(h_o)\phi \|. \tag{1.7}$$

Then the closures of the isometrical maps

$$w_F^{\pm}(h,h_o) : \tilde{F}(h_o)\phi \to \phi_F^{\pm}$$

coincide with usual wave operators

$$w_F^{\pm}(h,h_o) = w^{\pm}(h,h_o). \tag{1.8}$$

On the other side if the wave operators (1.1) exist, then (1.6),(1.7) are fulfilled and the equality (1.8) holds.

Note that in (1.6) we can also replace $\exp(-ith_o)$ by $F_{-t}(h_o)$. In this way one can obtain certain conditions for existence of the wave operators.

2.

Using the construction of $w^{\pm}(h,h_o)$ presented above we can give a sense to the wave operators for a pair of Markov fields in terms of corresponding Schwinger functions without continuing them to real times. We shall first formulate four conditions needed for our construction.

Let $\phi_o(f)$, $\phi(f)$ be a pair of Markov fields indexed by

$S(\mathbb{R}^d)$, $d \geq 1$, where $S(\mathbb{R}^d)$ is the Schwartz space of test functions. We suppose that $\phi_o(f)$ is the free Markov field on $\mathbb{R}^d$, i.e. it is a Gaussian stochastic field of mass $m > 0$ and zero mean with covariance $E(\phi_o(x)\phi_o(y)) = G_m(x-y)$ which is the kernel of the operator $(-\Delta+m^2)^{-1}$ in $L_2(\mathbb{R}^d, dx)$.

We denote the Schwinger functions corresponding to $\phi_o(f)$ by σ_n^o, $n=0,1,\ldots$. For $\phi(f)$ we assume existence of all Schwinger functions $\sigma_n(f_1,\ldots,f_n)=E(\phi(f_1)\ldots\phi(f_n))$, $n=1,2,\ldots$ which are translation invariant. We put $\sigma_o=1$ and introduce another family of Schwinger functions:

$$S_{-1} = 1,\ldots, \quad S_n(\underline{y}) = \sigma_{n+1}(\underline{x}) \;, \quad n=0,1,\ldots,$$

$$\underline{y} = (y_1,\ldots,y_n) \;, \quad y_l = x_{l+1} - x_l \in \mathbb{R}^d \;, \quad 1 \leq l \leq n \;,$$

$$\underline{x} = (x_1,\ldots,x_{n+1}).$$

Our first condition is

A. <u>Θ-positivity.</u>

$$\sum_{n,m=0}^{\infty} \int S_{n+m-1}(-\overleftarrow{\Theta\underline{y}}, -\Theta x + x', \underline{y}') u_n(x,\underline{y}) \times$$

$$\times \overline{u_m(x',\underline{y}')} d\underline{y}dxdx'd\underline{y}' \geq 0 \;, \tag{2.1}$$

where $x, x' \in \mathbb{R}^d$, $\underline{y} \in \mathbb{R}^{d(n-1)}$, $\underline{y}' \in \mathbb{R}^{d(m-1)}$,

$$\Theta x = (-x^o, \bar{x}) \;, \quad x^o \in \mathbb{R}^1 \;, \quad \bar{x} \in \mathbb{R}^{d-1} \;,$$

$$\overleftarrow{\Theta\underline{y}} = (\Theta y_{n-1},\ldots,\Theta y_1) \;, \quad u_l \in S(\mathbb{R}_+^{dl}), \; l=n \text{ or } m,$$

and a sequence $(c,u_1,u_2,\ldots)$ is finite.

Here $S(\mathbb{R}_+^{dl})$ denotes the subspace in $S(\mathbb{R}^{dl})$ where $\mathbb{R}_+^{dl} :=$ $:=\{\underline{x} \in \mathbb{R}^{dl} \,|\, x_1^o,\ldots,x_l^o > 0\}$.

Using (2.1), certain Hilbert space H_σ and a selfadjoint operator H on it may be constructed. H_σ and H correspond to the physical Hilbert space and Hamiltonian in the Osterwalder-Schrader reconstruction theorem [5]. For free Markov field the condition (2.1) is immediately satisfied. We denote H_{σ_o} and H_o the state space and Hamiltonian corresponding to $\phi_o(f)$. Our second condition is

B. Analyticity. The functions $S_n((y_1^o,\bar{y}_1),\ldots,(y_n^o,\bar{y}_n))$ have the analytical continuation in each variable $y_1^o,\ldots,y_n^o$ onto the complex half-plane $\mathbb{C}^-$ such that after smoothing in $\bar{y}_1,\ldots,\bar{y}_n$ the following estimates are fulfilled

$$|S_n((z_1,f_1),\ldots,(z_n,f_n))| \leq c_{f_1,\ldots,f_n} \times$$

$$\times\ (1+|\underline{z}|)^{\alpha_n}\ (1+\min_{1\leq l\leq n}|y_l|)^{-\beta_n} \qquad (2.2)$$

Here $z_l = y_l^o + i\eta_l$, $f_1,\ldots,f_n \in S(\mathbb{R}^{d-1})$, and $c_{f_1,\ldots,f_n}$, α_n and β_n are some constants. This condition is valid if $S_n(\underline{z,\bar{y}})$ is the Fourier-Laplace transform of some tempered distribution $W_n(\underline{q})$ with support in $\mathbb{R}_+^{dn}$.

Due to (2.2) we can construct the pre-scattering states in H_σ of the form

$$F_t(H)\exp(-itH_o)\underline{u}\ , \qquad \underline{u} = (c,u_1,u_2,\ldots),$$

$c \in \mathbb{C}$, $u_l \in S(\mathbb{R}^{dl})$. To prove the existence of their strong limits as $t \to \pm\infty$ we need yet another two conditions.

C. Convolution. Let S_n^T be the truncated Schwinger functions or in other words the semi-invariants of the Markov field $\phi(f)$. We suppose that $S_n^T((y_1^o,\bar{y}_1),\ldots,(y_n^o,\bar{y}_n))$, $n\geq 2$, as functions of the variables $\bar{y}_1,\ldots,\bar{y}_n \in \mathbb{R}^{d-1}$ belong to convolution algebra Θ_c^*, i.e.

$$\int S_n(\underline{y})u_n(y_1^o,\ldots,y_n^o)d\underline{y}^o \in \Theta_c^*(\mathbb{R}^{n(d-1)}) \qquad (2.3)$$

for all $u_n \in S(\mathbb{R}^n)$.

Our next condition is connected with the so-called one-particle problem. For its formulation we introduce subspace $H_{\sigma_2} \subset H_\sigma$ which is constructed only by means of two-point Schwinger function σ_2. This subspace is invariant under the action of H. We denote restriction of H to H_{σ_2} by h. We use $H_{\sigma_2^o}$ and h_o to denote the corresponding objects for $\phi_o(f)$.

D. One-particle problem. There exist the wave operators for h and h_o in the sense of the scattering theory in two Hilbert spaces:

$$s - \lim_{t\to\pm\infty} F_t(h)\, J_{\sigma_2}\, e^{-ith_o} = w^\pm(h,h_o;J_{\sigma_2}). \qquad (2.4)$$

Here $J_{\sigma_2} : S(\mathbb{R}^d) \to H_{\sigma_2}$ denotes the canonical inclusion operator.

Now we can state our main result.

Theorem 2. Let $\phi_o(f), \phi(f)$, $f \in S(\mathbb{R}^d)$ be a pair of Markov fields. $\phi_o(f)$ is the free Markov field of mass m>0 and $\phi(f)$ is such that all its moments are finite. Further we assume that the four conditions A-D formulated above are fulfilled. Then for Hamiltonians H_o,H there exist the wave operators $W^\pm(H,H_o;J_\sigma)$ in the sense of the scattering theory in two Hilbert spaces (J_σ is the identification operator).

We will conclude by outline of the construction of $W^\pm(H,H_o;J_\sigma)$. Roughly speaking, they are defined as closures of the maps $F(H_o)\underline{u} \to u_F^\pm$, where

$$\underline{u}_F^{\pm} = s - \mathcal{H} - \lim_{t\to\pm\infty} F_t(H)J_\sigma e^{-tiH_o} u. \qquad (2.5)$$

The precise structure of the pre-scattering states is more complicated (for details see [4]). We note only that the right hand side of (2.5) we construct directly in terms of Schwinger functions without explicit use of H. In addition we are using the functions $F \in \mathbb{Z}^+ \,(\mathbb{Z}^-)$ restricted to the $\mathbb{C}_\varepsilon^- = \mathbb{C}^- \setminus \{\, z\in\mathbb{C}^- \,|\mathrm{Im}|z|<\varepsilon\}$. Thus in (2.5) we are actually performing two limits $t\to\pm\infty$ and $\varepsilon\to 0$.

REFERENCES

1. Kato, T. "Perturbation Theory for Linear Operators, Springer-Verlag, Berlin,1966
2. Koshmanenko, V.D. "Wave operators for semi-group and transition functions",Doklady Acad.Sci.USSR,276, N2, 347-350(1984)
3. Koshmanenko, V.D. "Construction wave operators on perturbed semi-group", Ukrainian Math.J.,37,N5, 634-636 (1985)
4. Koshmanenko, V.D. Wave and Scattering Operators in Euclidean Approach", Preprint 85-42, Institute of Math.,Kiev (1985)
5. Osterwalder,K., Schrader, R., "Axioms for Euclidean Green's functions, I. Commun.Math.Phys.31,83-112(1973) II. Commun.Math.Phys.42,281-305(1975)

FEYNMAN INTEGRAL FOR A MODEL OF QED IN ONE SPACE-TIME DIMENSION*

J.Löffelholz

Karl Marx University
Department of Physics and NTZ
7010 Leipzig, GDR

ABSTRACT

We consider the imaginary time propagator for the system of a charged Schrödinger particle interacting with a quantum oscillator. Using reflection positivity of the associated complex measure $d\lambda$, we recover the Hamiltonian dynamics. The main point is to identify the relation $m\dot{x} = p - eA$.

*) Dedicated to the memory of R.Høegh-Krohn

1. INTRODUCTION

Quantum electrodynamics is plagued with infinities. In two space-time dimensions the model was studied 25 years ago by Schwinger[1]. The regularization of the Fermion determinant leads to chiral anomaly. After "integrating" out the Fermions one is left with a formulation in terms of a probability measure[2]. However, the need of complex measures to describe the interaction of gauge fields with matter seems to be fundamental.

So we looked for some caricature of electromagnetism to have a solvable problem within standard quantum mechanics. The model we propose to analyze is given by the equations of motion

$$\left.\begin{aligned} m\ddot{x} &= eE \\ \ddot{A}+\beta^2 A &= e\dot{x} \end{aligned}\right\}, \tag{1.1}$$

where $E=-\dot{A}$. It is related to the polaron theory[3]. The characteristic feature is the coupling term $e\dot{x}A$ in the classical Lagrangean which implies that, for $e\neq 0$, the imaginary time propagator

$$e^{-tH}(xA, x'A'), \quad t\geqslant 0, \tag{1.2}$$

is not positive. We can never associate a genuine random process $t\to (x(t),A(t))\in\mathbb{R}^2$, and the usual construction of the path integral breakes down[4]. Let

$$d\nu = e^{-u^2/2}du \otimes d\varphi_\beta(A) \tag{1.3}$$

be the probability measure corresponding to the free Hamiltonian H_0. More precisely, the particle is governed by the Wiener measure $d\omega(\xi/u) = d\xi \otimes e^{-u^2/2} du$, where $\xi = x(0)$ and $t \in (-\infty,\infty) \to u(t) = \dot{x}$ is white noise. For simplicity we put $\hbar$, c and also m equal to one. Below we prove that the cutoff interaction functional, for $g = e \cdot \chi_{[-\sigma,\sigma]}$,

$$\mathcal{J}(g) = e \cdot \int_{-\sigma}^{\sigma} ds\, u(s) A(s) \tag{1.4}$$

is measurable with respect to $d\nu$. The moments of the complex-valued measure $\exp i\mathcal{J}(g) \cdot d\nu$ satisfy the famous reflection positivity[5]. We find a trick for their calculation introducing a dummy variable $z = u - ig \cdot A$. Finally, we discuss the limit $g \nearrow e$, a renormalized Feynman-Kac formula and the structure of the physical Hilbert space.

2. HAMILTONIAN

On the classical level our model which we call shortly $(\mathrm{QED})_1$, is defined by

$$\mathcal{L} = \left(\frac{m}{2}\dot{x}^2 + \frac{\dot{A}^2 - \beta^2 A^2}{2}\right) + e\dot{x}A . \tag{2.1}$$

The above Lagrangean is PCT-invariant. One may add a term proportional to $(xA)^{\cdot}$. See ref.[6] The canonical momentum $p = m\dot{x} + eA$ and the total energy $H = \frac{m}{2}\dot{x}^2 + I_\beta$, where $I_\beta = \frac{1}{2}(\dot{A}^2 + \beta^2 A^2)$, are constants of motion. Let $m=1$. We substitute $\dot{x} = p - eA$ into "Ampere's law" and find that $A_p = A - ep/\gamma^2$ is a free oscillator of frequency $\gamma = +\sqrt{\beta^2 + e^2}$. Then for the particle trajectory we obtain the solution

$$m^*(x(t)-x(0)) = pt - \frac{e}{\beta^2}(E(t)-E(0)), \qquad (2.2)$$

where $m^* = \gamma^2/\beta^2$. To quantize the system we introduce operators $E = i\frac{\partial}{\partial A}$ and $V_p = \exp(iepE/\gamma^2), p \in (-\infty,\infty)$. Since A commutes with p the Hamiltonian of the model

$$H = V_p^*\left(\frac{p^2}{2m^*} + \frac{E^2+\gamma^2A^2}{2}\right)V_p \qquad (2.3)$$

can be realized in $\mathcal{H} = L^2(\mathbb{R}^2, dp\,dA)$. To get a pure discrete spectrum we put the particle for a moment in a box, say $x \in [-L,L]$, impose periodic boundary conditions, and define $H(L)$. Now $p = \frac{\pi k}{L}$, $k = 0, \pm 1, \ldots$, and hence $H(L)$ admits a mass gap $\varepsilon = \min\{\frac{1}{2m^*}(\frac{\pi}{L})^2, \gamma\}$. Compare Ito's constructive approach[7] to $(QED)_2$. From (2.3) we conclude that the imaginary time propagator reads

$$e^{-tH(L)}(xA, x'A') = \sum_p \frac{e^{ip(x-x')}}{2\pi}\left(e^{-\frac{tp^2}{2m^*}} \cdot K_\gamma^t(A_p, A_p')\right), \qquad (2.4)$$

where $t \geq 0$ and $K_\gamma^t(\cdot,\cdot)$ is the Mehler kernel for I_γ. In the limit $L \nearrow +\infty$ we recover the result of Cheng[8] for real time, modulo a term $ie\cdot(x'A'-xA)$. We convince ourselves that the above expression is complex-valued! Following Nelson[9] we define a multiplicative functional $W(0,t) = \exp ie\mathcal{J}(h)$, where $h = \chi_{[0,t]}$, to describe the interaction within the interval $0 \leq s \leq t$.

LEMMA

Let T_t, $t \in (-\infty,\infty)$, denote the operators of imaginary time translation and Ω_β the ground state for the free oscillator. Then, for $t \geq 0$ holds

$$0 \leq (\ell \otimes f \cdot \Omega_\beta, e^{-tH(L)} \ell \otimes f \cdot \Omega_\beta) \tag{2.5}$$
$$= \frac{e^{-t\beta/2}}{2\pi} \sum_p |\ell_p|^2 \iint e^{-ip.u(h)} f^* W(0,t) T_t f \, d\nu .$$

Proof: We observe that $T_t \ell = \sum_p \ell_p \cdot e^{-ip(\xi + u(h))}$. On the r.h.s. integration over white noise $u(s)$, $0 \leq s \leq t$, gives a factor $\exp\{-\frac{1}{2} \|h(p-eA)\|^2\}$. Finally, we apply the Trotter and Feynman-Kac formulae. As a simple exercise we derive the particle propagator. With $Z(eh) = \int W(0,t) \, d\nu$ we find for $t \geq 0$

$$Z(eh)^{-1} \cdot \iint e^{-ip.u(h)} W(0,t) \, d\nu = \exp\{-\tfrac{p^2}{2}(h, \frac{1}{1+e^2 \cdot G_\beta} h)\} \tag{2.6}$$

and

$$\frac{(\ell \otimes \Omega_\beta, e^{-tH(L)} \ell \otimes \Omega_\beta)}{(1 \otimes \Omega_\beta, e^{-tH(L)} 1 \otimes \Omega_\beta)} = \frac{1}{2\pi} \sum_p |\ell_p|^2 e^{-\frac{\tau(t) p^2}{2m^*}} , \tag{2.7}$$

where $\tau(t) = t + \frac{e^2}{\gamma \cdot \beta^2}(1 - e^{-t\gamma^2})$. Above G_β stands for the Green's function for the elliptic operator $\beta^2 - \frac{d^2}{dt^2}$. To calculate the partition function, i.e. the nominator on the l.h.s. we may use the convergent Taylor expansion for $\ln\{\det(1 + e^2 \cdot h G_\beta h)\}$. The result is

$$Z(eh) = \frac{e^{-t(\gamma - \beta)/2}}{(1 + \alpha . \, th \, t\gamma)^{\frac{1}{2}}} , \quad \alpha = \frac{\gamma^2 + \beta^2}{2\gamma\beta} . \tag{2.8}$$

By the spectral theorem we may separate $|(\Omega(L), 1 \otimes \Omega_\beta)|^2$ for $t \nearrow \infty$, where $\Omega(L) = 1 \otimes \Omega_{\gamma}$ is the true vacuum for the coupled system. Unfortunately, when we remove the box cutoff $\Omega(\infty) \notin \mathcal{H}$. The origin is the fact that the conditional Wiener process $\{t \to x(t) : x(0) = \xi\}$ is <u>not</u> stationary[10]. The expression $d\omega(\xi/u)$, $\xi \in (-\infty, \infty)$, albeit formally translation invariant is not a probability measure! To improve the situation one may introduce a bare particle potential.

3. POSITIVITY

We now separate $d\xi$ and consider only the translation invariant part $d\nu$ living on a σ-algebra $\Xi \times \mathcal{O}$. For $g = e \cdot \chi[-\sigma, \sigma]$ the operator $C = g G_\beta g$ is trace class and hence[11]

$$d\lambda_g = Z(g)^{-1} \cdot \{ \exp i J(g) \, d\nu \} \tag{3.1}$$

exists. By inspection $\| J(g) \|^2 = \operatorname{Tr} C < \infty$. Moreover, the diamagnetic inequality $0 \leq Z(g) \leq 1$ holds. We call $d\xi \otimes d\lambda$, which is the formal limit $g \nearrow e$, the "Feynman integral" for the model $(\mathrm{QED})_1$.

<u>COROLLARY</u>

$$d\mu_g = d\lambda_g |_{\Xi} \tag{3.2}$$

is positive and absolutely continuous with respect to white noise. As $g \nearrow e$ we recover the Gaussian measure $d\mu$ of mean zero and covariance $\langle u(s) u(t) \rangle = \delta(t-s) - e^2 G_\gamma(s,t)$.

Proof: Trivial. The measure $d\mu$ allows us to calculate all particle Green's functions w.r.t. ground state $\Omega(L)$ of the coupled system. We claim that $d\xi \otimes d\mu$ is given by the imaginary time action

$$S_{eff} = \frac{1}{2}\int dt\left(\dot{x}^2 + e^2 x^2\right) - \frac{e^2}{2}\left(x, \beta^2 \cdot G_\beta x\right), \tag{3.3}$$

studied by Khandekar et al.[12] We used $-\ddot{G}_\beta = 1 - \beta^2 \cdot G_\beta \geq 0$. After averaging over the oscillator configurations $t \to x(t)$ describes a random process with memory. For real time S_{eff} leads to a classical integro-differential equation.

Alternatively, the restriction $d\varphi = d\lambda / \alpha$ gives the dynamics when $p = 0$. If $e = 0$ we have a product measure and hence the correlation $\langle u(s) A(t) \rangle_0$ vanishes. To calculate mixed moments of $d\lambda_g(u, A)$ we expand the random functions u and $\alpha = gA$ in a complete orthonormal basis $\{e_n\}_{n=1}^{\infty}$ of $L^2(\mathbb{R}, dt)$, and use the identity

$$\int u_1 u_2 \dots u_n \, e^{iu\cdot\alpha} e^{-u^2/2} du = \frac{(-i)^n \partial^n}{\partial\alpha_1 \dots \partial\alpha_n} e^{-\alpha^2/2}. \tag{3.4}$$

From

$$\langle u(b) A(a) \rangle_g = \frac{i}{Z(g)} \int A(g\cdot b) A(a) \, e^{-\frac{1}{2}(A, g^2 A)} d\varphi_\beta(A), \tag{3.5}$$

taking the limit $g \nearrow e$, we get the complex-valued correlation function $+ie\cdot(b, G_\gamma a)$. Similarly we may proceed for higher order moments. We observe that $\alpha \to \langle u(b)^2 \rangle_\alpha$ is not positive. Schwartz' inequality gives the lower bound

$$\langle u(b)^2 \rangle_\alpha \geq \langle u(b)^2 \rangle_0 \cdot (1 - \alpha^2) e^{-\alpha^2/2}. \tag{3.6}$$

When we calculate $\langle \exp i(u(b) + A(a)) \rangle$ we find a little surprise: In terms of the variables A and $z = u - ieA$ the formal measure $d\lambda$ factorizes into $e^{-z^2/2}dz \otimes d\varphi(A)$. There is an analogy to the Schwinger model where the Green's functions for the Euclidean fields ϕ and $^*F = \epsilon^{ij}\partial_i A_j$ imply that

$$\eta = {}^*F + i\frac{e}{\sqrt{\pi}}\cdot\phi \tag{3.7}$$

is a white noise. Below we try to understand the meaning of $z = u - ie\cdot A$ not entering any "Hinterwelt"[13]. In the next step we will check reflection positivity of the measure $d\lambda_g$. Then the crucial property will hold for $d\xi\otimes d\lambda$ as well. In the unperturbed case $t \to u(t) = \dot{x}$ is ultralocal[14] so that for $d\nu$ it is obvious. For $d\xi\otimes d\mu_g$ the proof depends on a suitable decomposition of the memory term in S_{eff}. This nice trick[15] generalizes to the cutoff measure.

THEOREM

Let $0 \le g(t) \le e$ satisfy $\vartheta g = g$, where ϑ is the imaginary time reflection and let Θ denote the corresponding unitary operator in $\mathcal{H} = L^2(\Xi\times\mathcal{O}l, d\nu)$. Then

$$Z(g)\cdot d\lambda_g = (e^{iR})^*\cdot\Theta e^{iR}\, d\nu. \tag{3.8}$$

In particular, for $g = e\cdot\chi_{[-\sigma,\sigma]}$ holds $e^{-iR} = W(0,\sigma)$ which is a $(\Xi\times\mathcal{O}l)_+$-measurable function. Hence $d\lambda_g$ is reflection positive.

Proof: Since ϑ anticommutes with time derivative $\frac{d}{dt}$ we have $\Theta u(h) = -u(\vartheta h)$ and hence $J(g) = -R + \Theta R$. Finally, we use $Z(g) \geq 0$. We remark that $\Theta J(g) = J(-g)$ which implies P_x(CT)-invariance of the model. But $\|J(e)\| = \infty$, i.e. in the limit $g \to e$ the polar decomposition (3.1) breakes down[16]. Let $\mathcal{U}(eh)$ denote the imaginary time propagator for $(I_\mathcal{H} - epA\cdot h) + \mathcal{E}$, where again $h = \mathcal{H}_{[0,t]}$ and $\mathcal{E}$ is chosen such that $H(L) + \mathcal{E} \geq 0$. We claim that

$$\lim_{g \to e} \iint e^{-ip\cdot u(h)} f^* \cdot T_t f \, d\lambda_g = e^{-tp^2/2} (f \cdot \Omega_\mathcal{H}, \mathcal{U}(eh) f \cdot \Omega_\mathcal{H}), \quad (3.9)$$

i.e. for states $\Psi = \sum_p \sum_a l_p f_a \, e^{-ipx} \otimes e^{-iaA} : \Omega_\mathcal{H}$ in $\mathcal{H}$ we are able to calculate the time evolution by means of the functional integral $d\xi \otimes d\lambda$. Actually, we cannot give a definite answer to the question concerning the existence of $d\lambda$ as a measure. But we have a hierarchy of Green's functions which for $L = +\infty$ satisfy Osterwalder-Schrader like axioms. Of course the semigroup $\{e^{-tH}, t \geq 0\}$ will <u>not</u> be hypercontractive.

We briefly describe the reconstruction of the physical Hilbert space $\mathcal{H}$. For $j = u(b) + A(a)$ with $a, b \in S(\mathbb{R}_+)$ we obtain

$$\langle j^* \cdot \Theta j \rangle = \frac{1}{2\mathcal{H}} \| ie\cdot \check{b} + \check{a} \|^2_{\mathbb{C}}, \quad (3.10)$$

where $\check{b} = \int_0^\infty ds \, e^{-s\mathcal{H}} b(s)$ is the Fourier-Laplace transform. There is a canonical mapping W which takes polynomials of those j into

$$\mathcal{N} = \{ \Psi \in \mathcal{H} : (\Psi, p^2 \Psi) = 0 \}. \quad (3.11)$$

Indeed,

$$\langle z(h)^{*} \cdot \textcircled{H} \left(z(h) j(t_1) j(t_2) \ldots j(t_m)\right) \rangle \approx (\Omega(L), p^2 \Psi_m), \tag{3.12}$$

where $\Psi_n = Wj(t_1)j(t_2)\ldots j(t_m) \in \mathcal{N}$ is a complex linear combination of vectors $1 \otimes A^m \cdot \Omega_\gamma$, $0 \leq m \leq n$. In particular, from (3.10) we get $Wj = (ie \cdot \check{b} + \check{a}) 1 \otimes A \cdot \Omega_\gamma$. Roughly speaking we have the identities $WA(0) = 1 \otimes A \cdot \Omega_\gamma$ and $Wu(0) = -[H, x] \Omega(L)$, i.e. the variable $i \cdot z$ is precisely the imaginary time counterpart of $p = m\dot{x} + eA$. We remark that because of $[H, p] = 0$ the subspace $\mathcal{N}$ of $\mathcal{H}$ is invariant with respect to the full dynamics. Conversely, the linear span of

$$\begin{aligned} \Psi &= W T_t e^{-ip\xi} \\ &= e^{-\frac{tp^2}{2m^*}} e^{-ipx} \otimes e^{-tI_\gamma}(V_p \Omega_\gamma) \end{aligned} \tag{3.13}$$

for $t \geq 0$, is dense in $\mathcal{H}$. With other words the approximate vacuum $\Omega(L)$ may be cyclic only for the algebra generated by $\{x, \exp itH\}$ [17]. At the end we observe that renormalized particle propagator is defined by the family $\{\pi e^{-tH} \pi, t \geq 0\}$, where π stands for the projection in $\mathcal{H}$ onto the proper subspace $\mathcal{P}$ spanned by $\Psi = e^{-ipx} \otimes \Omega_\gamma$. Clearly, $\mathcal{N} \cap \mathcal{P} = \{c \cdot \Omega(L)\}$.

4. SUMMARY

In this report we explained the mechanism by which from a complex-valued measure one can reconstruct a quantum mechanical model. It is more general than prescibed by Klein[18]. It would be interesting to consider a spin ½[19].

ACKNOWLEDGEMENTS

I thank the organizers of the School for invitation, hospitality and support to present this material. I also would like to thank S.Albeverio and R.Gielerak for recalling my attention to the paper of Yngvason.

REFERENCES

1. Schwinger,J., Theor.Physics, p.89, IAEA Vienna (1963)
2. Nielsen,N.K.,Schroer,B.,Nucl.Phys. B120, 62 (1977)
3. Feynman,R.P.,Phys.Rev. 97, 660 (1955)
4. Guerra,F., Constr.QFT, p.243, LNPh 25, Springer(1973)
5. Osterwalder,K.,Schrader,R.,Comm.Math.Phys.31,83(1973)
6. Caldeira,A.O.,Leggett,A.J.,Ann.Phys.149,374 (1983)
7. Ito,K., RIMS-243 preprint Univ.Kyoto (1977)
8. Cheng,B., J.Phys. A17, 819 (1984)
9. Nelson,E., J.Func.Anal. 12, 97 (1973)
10. Reed,M.C., Constr.QFT, p.2, LNPh 25, Springer (1973)
11. Glimm,J.,Jaffe,A.,Quantum Physics, Springer (1981)
12. Khandekar,D.C.,Lawande,S.V.,Bhagwat,K.V., J.Phys.A16, 4209 (1983)
13. Seiler,E.,Gauge theories as a problem of constr.QFT and stat.mechanics, LNPh.159, Springer (1982)
14. Hida,T.,Appl.Mat.Opt.12, 115 (1984), and: Høegh-Krohn,R.,Comm.Math.Phys. 38, 195 (1974)
15. Fröhlich,J.,Israel,R.Lieb,E.,Simon,B.,Comm.Math. Phys.60,1 (1980)
16. Yngvason,J.,Reports in Math.Phys.13, 101 (1978)
17. Löffelholz,J.,Univ.Wroclaw Preprint ITP 653 (1985)
18. Klein,A., Bull.Am.Math.Soc.82, 762 (1976)
19. De Angelis,G.F.,Jona-Lasinio,G.,J.Phys. A15,2053(1982).

QUANTUM WAVEGUIDES MODELLED BY GRAPHS

Pavel Exner and Petr Šeba
Laboratory of Theoretical Physics, Joint Institute for Nuclear Research, 141980 Dubna, USSR

ABSTRACT

We introduce the notion of quantum waveguide and show how these objects can be modelled in the long-wave approximation by appropriate graphs. Some applications in microelectronics are briefly described.

1. INTRODUCTION

This lecture does not deal with stochastic methods, but its subject represents a large field of potential applications for them. Some ideas about how to implement random impurities here may be derived from the lecture presented by Prof.Pastur.

We shall study motion of a quantum particle on stripes, tubes, layers and similar "manifolds" in $\mathbb{R}^n$; for obvious reasons such a system will be called quantum waveguide (or briefly, Q-guide). Motivation for study of this problem appeared with the advent of modern microelectronics. Various technologies have been developed recently which allow to construct ultrathin layers or "wires" of metallic or semiconductor materials and to build more complicated structures of them[1).

For simplicity, we restrict our attention to the case of a planar structure of variously curved and branched stripes. It can model either real quantum wires of a fixed thickness or, in some cases, a thin film structure over a cylinder-type substrate.

2. THE LONG-WAVE APPROXIMATION

A spectral and/or scattering analysis of a Q-guide requires solution of the corresponding Schrödinger equation (with appropriate boundary conditions). This task is in general very difficult, and at the same time it might not be necessary in some cases. In this connection, recall the point interaction methods used for description of low-energy processes, where the particle "sees" a point instead of an interaction potential which can be of a complicated structure but localized to a region small comparing to the de Broglie wavelength[2]. We shall conjecture now that the same idea could be applied to the propagation of particles in Q-guides. One would expect that in the low-energy limit the influence of the detailed shape will be suppressed and the particle will feel only the global "topological" structure. Our aim here is to demonstrate that this is really the case.

We restrict ourselves to the two simplest cases of a curved and bifurcated Q-guide. In the first of them, we show that the curvature of the Q-guide yields the potential by which the scattring in the bends is modelled. A bifurcated Q-guide can be in the long-wave approximation replaced by the corresponding graph. We shall show how an appropriate Hamiltonian can be constructed.

3. THE CURVED Q-GUIDES

We shall consider a free spinless particle living on a curved planar strip Ω of a width d. We denote by s,u the natural longitudinal and transversal coordinates, respectively. The boundary $u = 0$ will serve as the reference curve Γ. The strip Ω is fully characterized by its width d and the (signed) curvature $\gamma(s)$ of Γ. To avoid non-essential technicalities, we assume that

(a) the boundary of Ω is infinitely smooth and $\gamma(s) = O(|s|^{-3/2-})$ as $|s| \to \infty$.

The Hamiltonian of our problem is

$$H_\Omega = -\frac{\hbar^2}{2m}\Delta_D \,, \tag{3.1}$$

where Δ_D is the Laplacian with Dirichlet boundary conditions on $\partial\Omega$. Its spectral properties are the following[3]

Theorem 1 : Assume (a), then

(i) $\sigma_{ess}(H_\Omega) = [E_\infty, \infty)$, where $E_\infty = \hbar^2\pi^2/2md^2$,

(ii) there is $d_0 > 0$ such that for all $d < d_0$, H_Ω has at least one bound state in $[0, E_\infty)$,

(iii) the Birman-Schwinger alternative holds : H_Ω has at least one bound state below the bottom of $\sigma_{ess}(H_\Omega)$ iff

$$\int_{\mathbb{R}\times[0,d]} \left\{ -\frac{\gamma^2}{(1+u\gamma)^2} + \frac{u^2\gamma'^2}{(1+u\gamma)^4} \right\} \sin^2\!\left(\frac{\pi u}{d}\right) ds\, du < 0 \tag{3.2}$$

Replacement of Ω by the appropriate graph, i.e., by the curve Γ corresponds to the limit $d \to 0$. It can be performed formally passing to the coordinates s and $u_\varepsilon = \varepsilon u$ with $\varepsilon \to 0$; one gets the one-dimensional Schrödinger operator

$$H : \; H\psi = -\frac{\hbar^2}{2m}\psi'' - \frac{\hbar^2}{8m}\gamma(s)^2\psi \,. \tag{3.3}$$

If we denote by E its ground state, and by $E(d), E_\infty(d)$ the ground state and the first transversal-mode energy of H_Ω, respectively, then a straightforward estimate gives

Theorem 2 : $\lim_{d\to 0} (E(d) - E_\infty(d)) = E$.

This is the rigorous version on a known heuristic result[4].

Let us remark that the physical consequences of the results, in particular, possible existence of edge-confined-currents in sharply edged thin films[5)] are of a great interest.

4. THE BIFURCATED Q-GUIDES

It is even more difficult to get an exact solution for the bifurcated Q-guide (or Y-junction). On the other hand, the proposed description by a graph (Fig.1) is quite simple. Let us present some considerations supporting our conjecture.

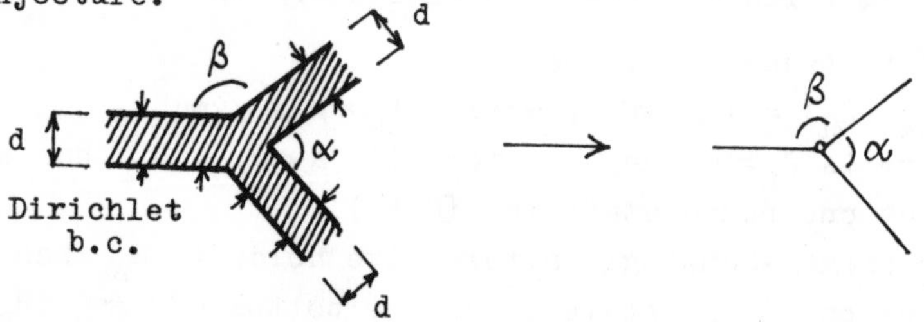

Fig.1 The long-wave approximation for a Y-junction

Quantum motion on a branching graph can be constructed using the theory of self-adjoint extensions. The state Hilbert space is $\mathcal{H} = L^2(\mathbb{R}^+) \oplus L^2(\mathbb{R}^+) \oplus L^2(\mathbb{R}^+)$ and all possible Hamiltonians are obtained as self-adjoint extensions of the symmetric operator

$$H_0 = H_{0,1} \oplus H_{0,2} \oplus H_{0,3} \quad , \tag{4.1}$$

where $H_{0,j} : H_{0,j}\psi_j = -(\hbar^2/2m)\psi_j''$ is defined on $D(H_{0,j}) = C_0^\infty(\mathbb{R}^+)$. The construction has been performed in Ref.6 ; here we review the results.

We restrict our attention to the Hamiltonians which commute with the operators P_{jk} permuting the halflines. Then there is a two-parameter family of them whose elements are specified by the following boundary conditions

$$f_1'(0) = Af_1(0) + Bf_2(0) + Bf_3(0) \ ,$$
$$f_2'(0) = Bf_1(0) + Af_2(0) + Bf_3(0) \ , \qquad (4.2)$$
$$f_3'(0) = Bf_1(0) + Bf_2(0) + Af_3(0)$$

for some real parameters A,B . There are also two one-dimensional families corresponding to the cases when the conditions (4.2) become singular, namely

$$f_1'(0) = f_2'(0) = f_3'(0) \equiv f'(0) \ , \qquad (4.3)$$
$$f_1(0) + f_2(0) + f_3(0) = Df'(0)$$

for $D \in \mathbb{R}$ and

$$f_1'(0) + f_2'(0) + f_3'(0) = 0 \ ,$$
$$f_2(0) - f_3(0) = C(f_3'(0) - f_2'(0)) \ , \qquad (4.4)$$
$$f_1(0) - f_3(0) = C(f_3'(0) - f_1'(0))$$

for $C \in \mathbb{R}$. The scattering matrix for each of these Hamiltonians can be easily calculated ; it equals

$$S = \begin{pmatrix} a & b & b \\ b & a & b \\ b & b & a \end{pmatrix} \qquad (4.5)$$

with

$$a = \frac{-1 + ikB - k^2(A^2 + AB - 2B^2)}{[1 - ik(A+2B)][1 - ik(A-B)]} \ ,$$

$$b = \frac{-2ikB}{[1 - ik(A+2B)][1 - ik(A-B)]} \ ,$$

where A,B refer to the boundary conditions (4.2), and

$$a = \frac{1 - ikD}{3 - ikD} \ , \quad b = \frac{-2}{3 - ikD} \ ,$$

$$a = \frac{3ikC - 1}{3(1+ikC)} \ , \quad b = \frac{2}{3(1+ikC)}$$

for the conditions (4.3) and (4.4), respectively.

What is the physical meaning of the parameters A,B, C,D ? A natural conjecture is that they are related to the angles α, β which characterize the junction. In particular, we conjecture that one of the "exceptional" Hamiltonians corresponds to each symmetric junction (with $\alpha = 2(\pi - \beta)$) , while the non-symmetric ones are described by elements of the two-parameter class (4.2). This guess is justified by comparison of the low-energy behaviour of the coefficients a,b to the reflection and transmission coefficients, respectively, calculated for symmetric and non-symmetric junctions numerically in the classical waveguide theory[7].

These considerations cannot substitute, of course, a rigorous formulation and proof of the limit $d \to 0$. It can be performed easily in the symmetric case and will be published in a subsequent paper ; the non-symmetric case remains an open problem.

5. APPLICATIONS

The long-wave approximation described above represents a useful way in which one can replace the the original problem by a mathematically much more simple one. Among its potential applications, those related to Aharonov-Bohm-type effects in microstructures are particularly interesting[8-10]. We limit ourselves here to one of them,

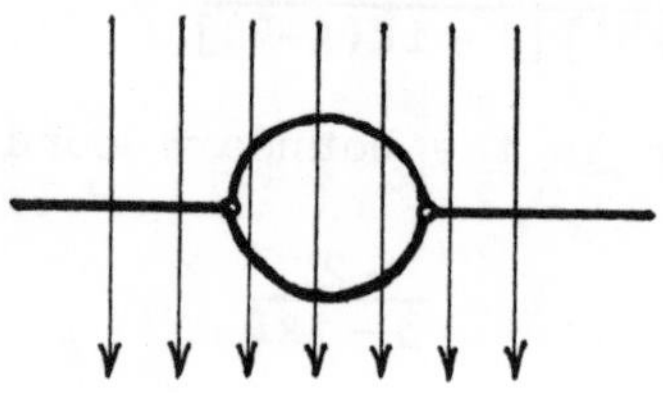

Fig.2 The semiconductor loop in an electric field

namely to the description of a semiconductor loop with two external leads in a homogeneous electric field (Fig.2). To calculate the transmission coefficient through the loop, one has to "sew" the wavefunctions

$$u_1(x) = e^{-ikx} + a\,e^{ikx} \quad , \quad u_4(x) = b\,e^{-ikx}$$

describing the electron behaviour on the leads, and the functions u_2, u_3 which solve the equations

$$-\frac{\hbar^2}{2m}\,u_j''(x) + V_j(x)u_j(x) = Eu_j(x)$$

on the arms of the loop. Here $E = \hbar^2k^2/2m$ and the potentials V_j are given by the electric field (they should contain also a curvature-dependent term, but it can be neglected if the loop has not too sharp edges). It is sufficient to consider the boundary conditions (4.2) when "sewing" the above mentioned wavefunctions, since the "exceptional" cases correspond to the following situations

$$A = \tfrac{1}{3}D - 2B \quad , \quad B \to \infty \quad , \tag{5.1}$$

and

$$A = B - C \quad , \quad B \to \infty \quad , \tag{5.2}$$

respectively and in numerical calculations one can replace these cases by (4.2) with A, B large enough.

The problem leads to a system of linear equations which yields the following expressions for the transmission coefficient[11)]

$$T(E) = |b|^2 = \frac{B_2^2}{1 + k^2A_2^2}\,|c_2 + d_2|^2 \quad , \tag{5.3}$$

where

$$\begin{pmatrix} c_1 \\ c_2 \end{pmatrix} = C_2(-k) \begin{pmatrix} d_1 \\ d_2 \end{pmatrix} \quad ,$$

$$\begin{pmatrix} d_1 \\ d_2 \end{pmatrix} = [\Pi_2^{-1} C_2(-k) - C_1(-k)\Pi_3^{-1}]^{-1} \begin{pmatrix} z_1 \\ z_2 \end{pmatrix}$$

with

$$z_1 = (A_1 - B_1) z_2 \quad , \quad z_2 = \frac{2ik}{1 + ik(B_1 - A_1)} \quad ,$$

$$C_1(k) = [B_1(1+ik(B_1-A_1))]^{-1} \times$$
$$\times \begin{pmatrix} A_1 + ik(B_1^2 - A_1^2) & (B_1 - A_1)\ (A_1+B_1)(1-ikA_1) + 2ikB_1^2 \\ 1 - ikA_1 & -A_1 - ik(B_1^2 - A_1^2) \end{pmatrix}$$

and $C_2(k)$ expresses similarly through the parameters A_2, B_2 specifying the second junction. Furthermore, Π_j are the transfer matrices

$$\begin{pmatrix} u_j(l_j) \\ u_j'(l_j) \end{pmatrix} = \Pi_j \begin{pmatrix} u_j(0) \\ u_j'(0) \end{pmatrix} \quad , \quad j = 2,3 \quad .$$

Transport properties of the loop are then given by Landauer formula : the conductance equals

$$G = \frac{e^2}{\pi\hbar} \frac{T(E)}{1 - T(E)} \qquad (5.4)$$

(recall that $\pi\hbar/e \approx 12906\,\Omega$). Using the WKB-approximation for Π_j , one can calculate G from (5.3) and (5.4) numerically for various shapes of the loop and parameters specifying the junctions. As an example, we present here (Fig.3) the results for a loop of 200 Å GaAs wire of the sketched shape (the corresponding energy $E \approx 0.05$ eV). With the above considerations in mind, we choose $A_1 = B_1 = -A_2 = -B_2 = 10^3$ Å to describe the junctions. We see that the conductance plot exhibits sharp minima at reasonably low field intensities. Changing the loop shape and the parameters, we obtain other pictures, however, generally they are of the same form. This quantum interference effect is particularly promising from the viewpoint of possible switching-device applications.

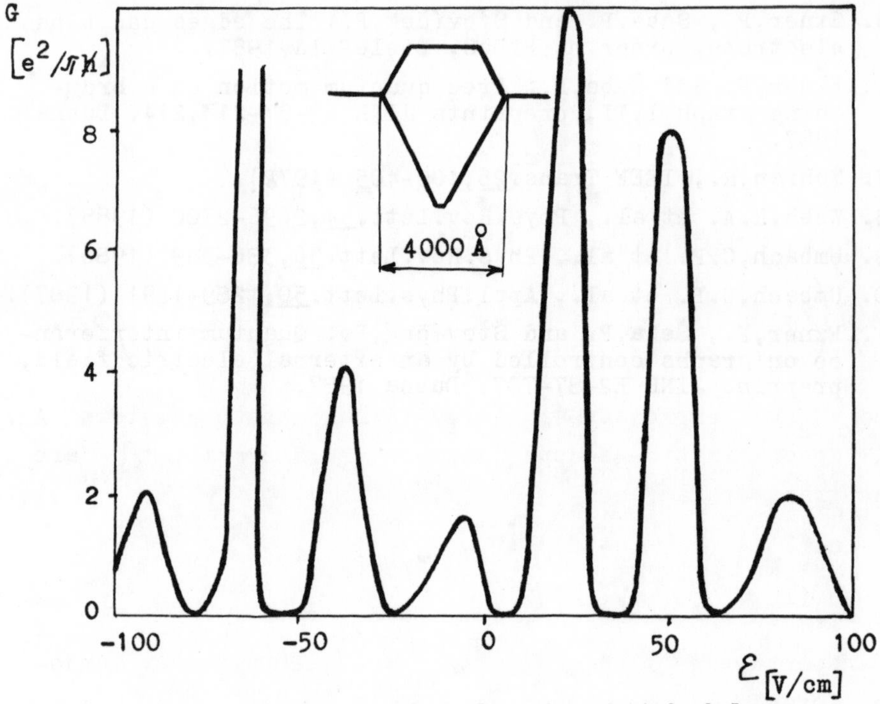

Fig.3   The conductance plot for the sketched loop

In conclusion, let us mention that the long-wave approximation described here can be used effectively in other fields too, e.g., in the electromagnetic waveguide theory, for killing the humming noise in air-conditioning systems etc.

REFERENCES

1. Sakaki,H., in Proc.Int.Symp.Foundations of Quantum Mechanics, Physical Soc.of Japan, Tokyo 1983 ; pp.94-110.
2. Albeverio,S., Gesztesy,F., Høegh-Krohn,R. and Holden,H.: Solvable Models in Quantum Mechanics, Springer, Berlin 1988.
3. Exner,P. and Šeba,P.: Bound states in curved quantum waveguides, preprint, Bochum 1987.
4. da Costa,R.C.T., Phys.Rev.A23,1982-1987 (1981).

5. Exner,P., Šeba,P. and Šťovíček,P.: The edges can bind electrons, preprint BiBoS, Bielefeld 1987.

6. Exner,P. and Šeba,P.: Free quantum motion on a branching graph I,II, preprints JINR E2-87-213,214, Dubna 1987.

7. Mehran,R., IEEE Trans.26,400-405 (1978).

8. Webb,R.A. et al., Phys.Rev.Lett.54,2696-2700 (1985).

9. Umbach,C.P. et al., Phys.Rev.Lett.56,386-389 (1986).

10. Umbach,C.P. et al., Appl.Phys.Lett.50,1289-1291 (1987).

11. Exner,P., Šeba,P. and Šťovíček,P.: Quantum interference on graphs controlled by an external electric field, preprint JINR E2-87-707, Dubna 1987.

SOME REMARKS ABOUT THE SPECTRAL DIMENSION OF FRACTALS

Ph. BLANCHARD
Universität Bielefeld and BiBos, R.F.A.
Ph. COMBE
Université d'Aix-Marseille II and Centre de Physique Théorique, Marseille, France
H. NENCKA
IFN Polskiej Akademii Nauk Poznan, Poland

To describe or modelize many irregular structures which appear in the nature B. Mandelbrot [1] [2] [3] introduced the notion of "fractals". From that time most of the attention has been focused on the geometrical properties of such irregular structures. In particular partial geometrical characterization of fractal structures can be done by introducing various notions of "dimensions". These dimensions can give only informations about the expensions of the fractals or include (partial) informations about the number of multi-points (see eg [4], [5]). For compact irregular structures a natural characterization of their expensions is provided by the Hausdorff dimension d_H. This is associated to a family of (non necessarily finite) Borel measures, the k-dimensional Hausdorff measures H_k ($k \in \mathbb{R}_+$). The Hausdorff measure H_{d_H} define a non identically vanishing finite Borel measure on the fractals [6].

In this way a fractal can be defined as a subset of $\mathbb{R}^d$, such that the Hausdorff dimension is different from the topological one [1] [2] [7].

However, in many physical problems as the study of low-energy excitations of ill-condensed matter [8] [9], percolation clusters [10] [11], diffusion on disordered lattices (see eg [12]), quantum properties of fractal lattices [13], linear and branched polymers [14], the important physical properties are not fully described in terms of Hausdorff dimension (or other "geometrical dimensions"). We need an other quantity connected with the density of state which is called the spectral dimension d_s.

In regular media the problem of determining for example spectra of low-energy excitations, diffusion, some quantum properties are associated with the existence of a Laplace operator. But how to define a Laplacian on a fractal structure which is non differentiable?

A natural way to answer this question is using stochastic methods. Indeed the Laplacian can be seen as the generator of the Brownian motion and the construction of the Brownian motion does not need a differentiable structure.

In the case of regular structure the Hausdorff dimension d_H is equal to the topological dimension d. That is d_H is an integer number. If we solve the eigenvalue problem

$$\Delta u = -\lambda u \quad \text{in } \Omega$$
$$u = 0 \quad \text{on } \partial\Omega \tag{1}$$

in a bounded region Ω of $\mathbb{R}^d$, with a smooth boundary $\partial\Omega$, by the famous Weyl's theorem [15] [16] the number of eigenvalues $N(\lambda)$ less or equal to λ behaves as

$$N(\lambda) \sim C_d \, |\Omega| \, \lambda^{d/2} \tag{2}$$

where $|\Omega|$ is the Lebesgue measure of Ω and C_d is a constant which depends only on the dimension d of the space.

In the case of a fractal structure $\mathcal{F}$, we have (in general) no longer the previous formula with d replaced by d_H but rather

$$N(\lambda) \sim C(\mathcal{F}) \, \lambda^{\frac{d_s}{2}} \tag{3}$$

where d_s is called the spectral dimension.

In solid state physics litterature this formula is very often written in terms of the density of states (see eg [17]). If we assume that $N(\lambda)$ has a density

$$N(d\lambda) = \nu(\lambda) \, d\lambda \tag{4}$$

introducing

$$\omega = \lambda^{1/2}$$

and ρ such that

$$\rho(\omega) \, d\omega = \nu(\lambda) \, d\lambda$$

the density of states ρ behaves as

$$\rho(\omega) \sim \omega^{d_s - 1} \tag{5}$$

For a large class of fractal curves we have the following inequality

$$1 \leq d_s \leq d_H \leq d \tag{6}$$

(a rigourous proof can be done for graph structures [18]).

If we think in terms of Brownian motion then

i - in a regular case

$$<x^2> \sim t \tag{7}$$

or physically speaking the diffusion length $\sigma_t = <x^2>^{1/2}$ behaves as

$$\sigma_t \sim t^{1/2} \tag{8}$$

ii - in the case of an irregular structures with Hausdorff dimension d_H and spectral dimension d_s we have

$$<x^2> \sim t^{d_s/d_H} \tag{9}$$

the diffusion length behaves as

$$\sigma_t \sim t^{1/d_w} \tag{10}$$

where $d_w = 2\,\frac{d_H}{d_s}$ is the (Hausdorff) dimension of the Brownian motion (on the fractal structures).

The notion of fractal introduced by B. Mandelbrot includes a very large class of structures and has the advantage not to confine this notion is a strict mathematical frame. To develop an harmonic analysis, we need a much more precise structure. In particular the additional property of self-similarly has been fruitful in the physical applications and provide a mathematical frame which is reminicent of the classical one, except that the translation invariance of Euclidean space is replaced by dilation symmetry. Self-similar fractals are obtained from some initial shape rescaled up to infinity by a family of similitude $\Sigma = (\sigma_1, \sigma_2, ..., \sigma_N)$ where

$$\sigma_i(x) = \nu_i^{-1}\, U_i(x) + \alpha_i \tag{11}$$

where $\nu_i > 1$, U_i is a unitary transformation of $\mathbb{R}^d$, and α_i is a translation [19].

In the case of a self-similar structure one can define a new dimension, the so called similarity dimension d_Σ by $\sum_{i=1}^{N} \nu^{-d_\Sigma} = 1$. That is

$$d_\Sigma = \frac{\text{Log } N}{\text{Log } \nu} \tag{12}$$

supposing that all $\nu_i = \nu$, $(i = 1,...N)$.

In many cases, where the rescaled shape are "not too much overlopped" [19] $d_H = d_\Sigma$ and d_Σ provides an easy way to determine the Hausdorff dimension d_H.

Planar curves can be classified by introducing to class according to the fact whether they have or not branching points. In the case where the curves are nonbranching the spectral dimension is $d_s = 1$. This is the consequence of the fact that one has in this case a one-to-one parametrization by [0,1]. However, in the case of parametrizable curves howing branching points we have to make a choice in order to construct a process. In this last case it seems natural to divide the families into two subclasses in respect to the fact whether the structure is finitely ramified or not. A finitely ramified fractal can be disconnected by removing finitely many points. A typical example of such similar fractal is the Sierpinski gasket [18, 20, 21].

Any continuous process X on the gasket G must pass-through the vertices of the triangles generated at each stage of iteration. The discrete time process $X^{(n)}$ obtained by looking at x at successive hits of the vertices on the n th stage $G^{(n)}$ of the iteration can be seen as a random walk on the graphe $G^{(n)}$. Then in order to consider a "Brownian motion" leaving on such a fractal structure it is natural to look at a sequence of Markovian random walks $X^{(n)}$ on graph $G^{(n)}(n=1,2...)$ such that the induced walk $X^{(n)}$ on $G^{(n)}$ by the walk $X^{(m)}$ on $G^{(m)}$, $m > n$ does not depend on $m > n$. For the Sierpinski gasket the symmetric random walk

$$P_x^{(m)}\ [X^{(m)}(0) = x] = 1 \tag{13}$$

$$P_x^{(m)}\ [X^{(m)}(n+1) = y \mid X^{(m)}(n) = z] = \begin{cases} N(z)^{-1} & , \text{ if } |y-z| = 2^{-n} \\ 0 & \text{otherwise} \end{cases}$$

for any x, y, z $\in G^{(n)}$

$$N(z) = \#\{ y \in G^{(m)} , |y-z| = 2^{-n} \}$$

is a good candidate and is the (unique) fixed point of the natural family of similitude

$\Sigma = \left\{ \sigma_i(\vec{x}) = \frac{1}{2}\vec{x} + \vec{\alpha}_i \right\}_{i=1,2,3}$ where

$\vec{\alpha}_1 = 0 , \quad \vec{\alpha}_2 = \frac{1}{2}\vec{\rho}_1 , \quad \alpha_3 = \frac{1}{2}\vec{\rho}_2$

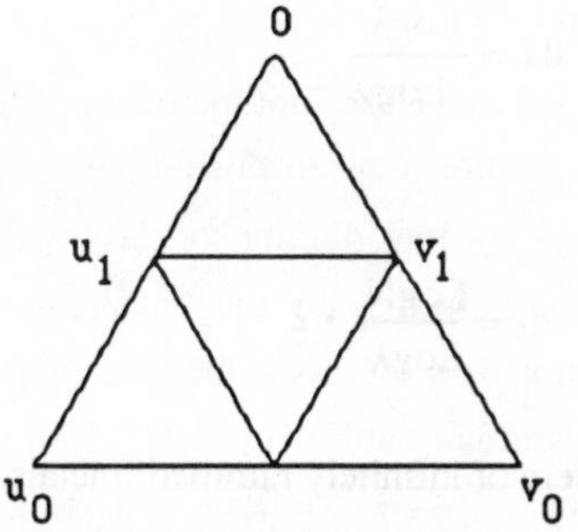

$\vec{e}_1 = \vec{ou_0}$
$\vec{e}_2 = \vec{ov_0}$

(Let us notice that there exist other walks which have the same property. In fact,as the gasket is invariant under the group of reflexions through the biscetor of the initial triangle, there exist many similarities generating the same fractal but, other fixed point probability [20]).

Now, it is clear that the "transition time" for the process $X^{(n)}$ cannot be the same as for one $X^{(m)}$, m > n. Hence, we have to do some renormalization procedure (as for the usual Brownian motion. For a large class of finitely ramefied fractals, the "nested fractals" [22] which are a generalization of the Sierpinski gasket, transition time $\tau_i^{(m)}$ of $X^{(m)}$ and $\tau_i^{(m+1)}$ of $X^{(m+1)}$ can be chosen in such a way that $\tau_i^{(n)}$ and $\lambda\tau_i^{(n+1)}$ with λ independant of i have the same distribution. In the case of the Sierpinski gasket, for the symmetric random walk, there is only one transition time, and the determination of λ is quite easy.If we consider a process starting from o, and denote by τ_n the first hitting time with vertices ($u_n = 2^{-n} u_0$, $v_n = 2^{-n} v_0$) of $G^{(n)}$ it can be shown that

$$\lambda = \frac{\mathbb{E}[\tau_n]}{\mathbb{E}[\tau_{n+1}]} = 5 \tag{14}$$

Using renormalization procedure we can determine the Brownian dimension

$$d_W = \frac{\text{Log}5}{\text{Log}2} \cdot 2 \tag{15}$$

and by Tauberian theorem the spectral dimension

$$d_S = \frac{\text{Log}3}{\text{Log}5} \tag{16}$$

More generaly for nested fractals we obtain

$$d_W = \frac{\text{Log}\lambda}{\text{Log}\nu} \tag{17}$$

and

$$d_S = \frac{\text{Log}N}{\text{Log}\lambda} \cdot 2 \tag{18}$$

Let us now just mention that the problem of infinitely ramified fractals, as in the cases of Kirpatric model [10] or of the Sierpinski carpet, is much more delicate to investigate and it is difficult to evaluate the spectral dimension.

REFERENCES

1. Mandelbrot, B., "Fractals, Form, Chance and Dimension" Freeman San Francisco (1977).

2. Mandelbrot, B., "The Fractals Geometry of Nature" Freeman San Francisco (1982).

3. Mandelbrot, B., "Les Objets Fractals" Flamarion Paris 1° édition 1975 (2° édition 1984).

4. Dubuc, S., "Modèle de Courbes Irrégulières" in "Fractals, Dimension non Entières et Applications" Ed by Cherbit, G. Masson, Paris 1987.

5. Tricot, C., C.R.A.S. Paris 293 série I (1981), 549-552. See also "Les Avatars Successifs de la Notion de Dimension Fractale", La Gazette des Sciences Mathématiques du Québec X N° 2 (1986) 3-22.

6. Billingsley, P., "Probability and Measure", J. Wiley and Son (1979).

7. Gefer, Y., Aharony, A., Mandelbrot, B., J. Phys. A Math. Gene. 16 1267-1278 (1983).

8. Dhar, D., J. Math. Phys. 18 (1977) 5.

9. Alexander, S., Orbach, R., Phys. Rev. Lett. 43, 625-631 (1982).

10. Kirpatrick, S., "Models of Disordered Materials" in "La Matière Mal Condensée", Les Houches 1978, Ed. by Ballian, R., Maynard, R., Toulouse, G., North-Holland (1979).

11. Ramal, R., Toulouse, G., J. Physique Lett. 44, L18-L22 (1983).

12. Haus, J.W., Kehr, K.W., Phys. Rep. 150 (1987).

13. Domany, E., Alexender, S., Bensimon, D., Kadanoff, L.P., Phys. Rev. B 28, 3110-3123 (1983).

14. Helman, J.S., Coniglio, A., Tsallis, C., Phys. Rev. Lett. 53, 1195-1197 (1984).

15. Weyl, H., Nachrichten der Königlichen Gesellsdraf der Wissenschaften zu Göttingen, Mathematisch Physikalische Klasse 110-117 (1911).

16. Kac, M., Am. Math. Mon. 73, 1-23 (1966).

17. Rammal, R., J. Physique 45, 191-206 (1984).

18. Kusuoka, S., "A Diffusion Process on a Fractal" in "Probabilistic Methods in Mathematical Physics", Taniguchi symp. Katutu 1985, 251-274, Kino Kuniya, North Holland (1987).

19. Hutchison, J.E., Ind. Uni. Math. Jornal 30, 713-747 (1981).

20. Goldstein, S., Random Walks and Diffusion on Fractals in Percolation Theory and Ergodic Theory of Infinite Particles systems IMA Volume in Mathematics and the Applications, Springer.

21. Barlow, M.T., Perkins, E.A., Brownian Motion on the Sierpinski Gasket, Preprint 1987.

22. Lindstrøm, T., Brownian Motion on a Class of Self Similar Fractals, to be published.

LIST OF PARTICIPANTS

S. ALBEVERIO	Bochum, Fed.Rep.of Germany
T. ARAK	Tartu, USSR
U. BEHN	Leipzig, German Dem.Rep.
V. BELAVKIN	Moscow, USSR
A. de BIVAR-WEINHOLTZ	Lisboa, Portugal
P. BLANCHARD	Bielefeld, Fed.Rep.of Germany
P. BLEHER	Moscow, USSR
M. BOŻEJKO	Wrocław, Poland
B. BRODA	Łódź, Poland
L. BUNIMOVICH	Moscow, USSR
E. CARLEN	Princeton, USA
A. CHEBOTARIOV	Moscow, USSR
W. CHOJNACKI	Warszawa, Poland
P. CLIFFORD	Oxford, Great Britain
P. COMBE	Marseille, France
A. CRUZEIRO	Lisboa, Portugal
R. DOBRUSHIN	Moscow, USSR
M. DUDYŃSKI	Warszawa, Poland
C. DE CARVALHO	Lisboa, Portugal
M. EKIEL-JEŻEWSKA	Warszawa, Poland
P. EXNER	Moscow, USSR
Z. GALASIEWICZ	Wrocław, Poland
P. GARBACZEWSKI	Wrocław, Poland
R. GIELERAK	Wrocław, Poland
A. GÓRSKI	Kraków, Poland
M. GORZELAŃCZYK	Wrocław, Poland
Z. HABA	Wrocław, Poland
J. HAŃĆKOWIAK	Wrocław, Poland
M. HAZEWINKEL	Amsterdam, Netherlands
J. HERCZYŃSKI	Warszawa, Poland

M. HERRMANN	Leipzig, German Dem. Rep.
T. HIDA	Nagoya, Japan
A. HILBERT	Bochum, Fed.Rep.of Germany
H. HOLDEN	Trondheim, Norway
J. IWANISZEWSKI	Toruń, Poland
K. IWATA	Bochum, Fed.Rep.of Germany
A. JADCZYK	Wrocław, Poland
L. JAKOBCZYK	Wrocław, Poland
B. JAKUBCZYK	Warszawa, Poland
B. JANCEWICZ	Wrocław, Poland
Z. JASKOLSKI	Wrocław, Poland
J. JEDRZEJEWSKI	Wrocław, Poland
Z. JUREK	Wrocław, Poland
A. KALICINSKA	Białystok, Poland
A. KARGOL	Kielce,Poland
W. KARWOWSKI	Wrocław, Poland
K. KHANIN	Moscow, USSR
Y. KONDRATIEV	Kiev, USSR
V. KOSHMANIENKO	Kiev, USSR
R. KOTECKY	Praha, Czechoslovakia
P. KREE	Paris, France
J.T. LEWIS	Dublin, Ireland
J. LÖFFELHOLZ	Leipzig, German Dem.Rep.
J. ŁUCZKA	Katowice, Poland
J. LUKIERSKI	Wrocław, Poland
J. MADAJCZYK	Warszawa, Poland
A. MAJEWSKI	Gdańsk, Poland
A. MAKOWSKI	Toruń, Poland
E. MALEC	Kraków, Poland
D. MERLINI	Milano, Italy
B. MINCER	Wrocław, Poland
P. MORAWIEC	Wrocław, Poland
M. MOZRZYMAS	Wrocław, Poland

R. MÜLLER	Leipzig, German Dem.Rep.
B. NACHTERGAELE	Leuven, Belgium
T. NADZIEJA	Wrocław, Poland
G. NAPPO	Heidelberg, Fed.Rep.of Germany
H. NENCKA	Poznań, Poland
L. PASTUR	Kharkov, USSR
A. PASZKIEWICZ	Łódź, Poland
S. PIROGOV	Moscow, USSR
Z. POPOWICZ	Wrocław, Poland
D. PROROK	Wrocław, Poland
R. QUEZADA	Kraków, Poland
S. RABSZTYN	Gliwice, Poland
K. SCHIELE	Leipzig, German Dem.Rep.
E. SEILER	München, Fed.Rep.of Germany
D. SHIRKOV	Moscow, USSR
S. SHLOSMAN	Moscow, USSR
J. SOBCZYK	Wrocław, Poland
K. SOBCZYK	Warszawa, Poland
P. STASZEWSKI	Toruń. Poland
K. STEFAŃSKI	Toruń, Poland
A. THIELMANN	Warszawa, Poland
A. TRUMAN	Swansea, Great Britain
A. VERBEURE	Leuven, Belgium
A. WERON	Wrocław, Poland
M. WOLF	Wrocław, Poland
J. ZAMBRINI	Coventry, Great Britain
T. ZASTAWNIAK	Kraków, Poland